Data Science in Theory and Practice

Data Science in Theory and Practice

Techniques for Big Data Analytics and Complex Data Sets

Maria Cristina Mariani
University of Texas, El Paso
El Paso, United States

Osei Kofi Tweneboah
Ramapo College of New Jersey
Mahwah, United States

Maria Pia Beccar-Varela
University of Texas, El Paso
El Paso, United States

This first edition first published 2022

Registered Office
John Wiley & Sons, Inc., 111 River Street, Hoboken, NJ 07030, USA

Editorial Office
111 River Street, Hoboken, NJ 07030, USA

For details of our global editorial offices, customer services, and more information about Wiley products visit us at www.wiley.com

Wiley also publishes its books in a variety of electronic formats and by print-on-demand. Some content that appears in standard print versions of this book may not be available in other formats.

Library of Congress Cataloging-in-Publication Data applied for
ISBN: 9781119674689

Cover Design: Wiley
Cover Image: © nobeastsofierce/Shutterstock

Set in 9.5/12.5pt STIXTwoText by Straive, Chennai, India

10 9 8 7 6 5 4 3 2 1

Contents

List of Figures

List of Tables

Preface

This textbook is dedicated to practitioners, graduate, and advanced undergraduate students who have interest in Data Science, Business analytics, and Statistical and Mathematical Modeling in different disciplines such as Finance, Geophysics, and Engineering. This book is designed to serve as a textbook for several courses in the aforementioned areas and a reference guide for practitioners in the industry.

The book has a strong theoretical background and several applications to specific practical problems. It contains numerous techniques applicable to modern data science and other disciplines. In today's world, many fields are confronted with increasingly large amounts of complex data. Financial, healthcare, and geophysical data sampled with high frequency is no exception. These staggering amounts of data pose special challenges to the world of finance and other disciplines such as healthcare and geophysics, as traditional models and information technology tools can be poorly suited to grapple with their size and complexity. Probabilistic modeling, mathematical modeling, and statistical data analysis attempt to discover order from apparent disorder; this textbook may serve as a guide to various new systematic approaches on how to implement these quantitative activities with complex data sets.

The textbook is split into five distinct parts. In the first part of this book, foundations of Data Science, we will discuss some fundamental mathematical and statistical concepts which form the basis for the study of data science. In the second part of the book, Data Science in Practice, we will present a brief introduction to R and Python programming and how to write algorithms. In addition, various techniques for data preprocessing, validations, and visualizations will be discussed. In the third part, Data Mining and Machine Learning techniques for Complex Data Sets and fourth part of the book, Advanced Models for Big Data Analytics and Complex Data Sets, we will provide exhaustive techniques for analyzing and predicting different types of complex data sets.

We conclude this book with a discussion of ethics in data science: With great power comes great responsibility.

The authors express their deepest gratitude to Wiley for making the publication a reality.

El Paso, TX and Mahwah, NJ, USA
September 2021

Maria Cristina Mariani
Osei Kofi Tweneboah
Maria Pia Beccar-Varela

1

Background of Data Science

1.1 Introduction

Data science is one of the most promising and high-demand career paths for skilled professionals in the 21st century. Currently, successful data professionals understand that they must advance past the traditional skills of analyzing large amounts of data, statistical learning, and programming skills. In order to explore and discover useful information for their companies or organizations, data scientists must have a good grip of the full spectrum of the data science life cycle and have a level of flexibility and understanding to maximize returns at each phase of the process.

Data science is a "concept to unify statistics, mathematics, computer science, data analysis, machine learning and their related methods" in order to find trends, understand, and analyze actual phenomena with data. Due to the Coronavirus disease (COVID-19) many colleges, institutions, and large organizations asked their nonessential employees to work virtually. The virtual meetings have provided colleges and companies with plenty of data. Some aspect of the data suggest that virtual fatigue is on the rise. Virtual fatigue is defined as the burnout associated with the over dependence on virtual platforms for communication. Data science provides tools to explore and reveal the best and worst aspects of virtual work.

In the past decade, data scientists have become necessary assets and are present in almost all institutions and organizations. These professionals are data-driven individuals with high-level technical skills who are capable of building complex quantitative algorithms to organize and synthesize large amounts of information used to answer questions and drive strategy in their organization. This is coupled with the experience in communication and leadership needed to deliver tangible results to various stakeholders across an organization or business.

Data scientists need to be curious and result-oriented, with good knowledge (domain specific) and communication skills that allow them to explain very technical results to their nontechnical counterparts. They possess a strong quantitative background in statistics and mathematics as well as programming knowledge with

Data Science in Theory and Practice: Techniques for Big Data Analytics and Complex Data Sets, First Edition. Maria Cristina Mariani, Osei Kofi Tweneboah, and Maria Pia Beccar-Varela.

focuses in data warehousing, mining, and modeling to build and analyze algorithms. In fact, data scientists are a group of analytical data expert who have the technical skills to solve complex problems and the curiosity to explore how problems need to be solved.

1.2 Origin of Data Science

Data scientists are part mathematicians, statisticians and computer scientists. And because they span both the business and information technology (IT) worlds, they're in high demand and well-paid. Data scientists were not very popular some decades ago; however, their sudden popularity reflects how businesses now think about "Big data." Big data is defined as a field that treats ways to analyze, systematically extract information from, or otherwise deal with data sets that are too large or complex to be dealt with by traditional data-processing application software. That bulky mass of unstructured information can no longer be ignored and forgotten. It is a virtual gold mine that helps boost revenue as long as there is someone who explores and discovers business insights that no one thought to look for before. Many data scientists began their careers as statisticians or business analyst or data analysts. However, as big data began to grow and evolve, those roles evolved as well. Data is no longer just an add on for IT to handle. It is vital information that requires analysis, creative curiosity, and the ability to interpret high-tech ideas into innovative ways to make profit and to help practitioners make informed decisions.

1.3 Who is a Data Scientist?

The term "data scientist" was invented as recently as 2008 when companies realized the need for data professionals who are skilled in organizing and analyzing massive amounts of data. Data scientists are quantitative and analytical data experts who utilize their skills in both technology and social science to find trends and manage the data around them. With the growth of big data integration in business, they have evolved at the forefront of the data revolution. They are part mathematicians, statisticians, computer programmers, and analysts who are equipped with a diverse and wide-ranging skill set, balancing knowledge in several computer programming languages with advanced experience in statistical learning and data visualization.

There is not a definitive job description when it comes to a data scientist role. However, we outline here some stuffs they do:

- Collecting and recording large amounts of unruly data and transforming it into a more usable format.

- Solving business-related problems using data-driven techniques.
- Working with a variety of programming languages, including SAS, Minitab, R, and Python.
- Having a strong background of mathematics and statistics including statistical tests and distributions.
- Staying on top of quantitative and analytical techniques such as machine learning, deep learning, and text analytics.
- Communicating and collaborating with both IT and business.
- Looking for order and patterns in data, as well as spotting trends that enables businesses to make informed decisions.

Some of the useful tools that every data scientist or practitioner needs are outlined below:

- *Data preparation:* The process of cleaning and transforming raw data into suitable formats prior to processing and analysis.
- *Data visualization:* The presentation of data in a pictorial or graphical format so it can be easily analyzed.
- *Statistical learning or Machine learning:* A branch of artificial intelligence based on mathematical algorithms and automation. Artificial intelligence (AI) refers to the process of building smart machines capable of performing tasks that typically require human intelligence. They are designed to make decisions, often using real-time data. Real-time data are information that is passed along to the end user immediately it is gathered.
- *Deep learning:* An area of statistical learning research that uses data to model complex abstractions.
- *Pattern recognition:* Technology that recognizes patterns in data (often used interchangeably with machine learning).
- *Text analytics:* The process of examining unstructured data and drawing meaning out of written communication.

We will discuss all the above tools in details in this book. There are several scientific and programming skills that every data scientist should have. They must be able to utilize key technical tools and skills, including R, Python, SAS, SQL, Tableau, and several others. Due to the ever growing technology, data scientist must always learn new and emerging techniques to stay on top of their game. We will discuss the R and Python programming in Chapters 5 and 6.

1.4 Big Data

Big data is a term applied to ways to analyze, systematically extract information from, or otherwise deal with data sets that are too large or complex to be dealt with by classical data-processing tools. In particular, it refers to data sets whose size or type is beyond the ability of traditional relational databases to capture, manage, and process the data with low latency. Sources of big data includes data from sensors, stock market, devices, video/audio, networks, log files, transactional applications, web, and social media and much of it generated in real time and at a very large scale.

In recent times, the use of the term "big data" (both stored and real-time) tend to refer to the use of user behavior analytics (UBA), predictive analytics, or certain other advanced data analytics methods that extract value from data. UBA solutions look at patterns of human behavior, and then apply algorithms and statistical analysis to detect meaningful anomalies from those patterns' anomalies that indicate potential threats. For example detection of hackers, detection of insider threats, targeted attacks, financial fraud, and several others.

Predictive analytics deals with the process of extracting information from existing data sets in order to determine patterns and predict future outcomes and trends. Generally, predictive analytics does not tell you what will happen in the future. However, it forecasts what might happen in the future with some degree of certainty. Predictive analytics goes hand in hand with big data: Businesses and organizations collect large amounts of real-time customer data and predictive analytics and uses this historical data, combined with customer insight, to forecast future events. Predictive analytics helps organizations to use big data to move from a historical view to a forward-looking perspective of the customer. In this book, we will discuss several methods for analyzing big data.

1.4.1 Characteristics of Big Data

Big data has one or more of the following characteristics: high volume, high velocity, high variety, and high veracity. That is, the data sets are characterized by huge amounts (volume) of frequently updated data (velocity) in various types, such as numeric, textual, audio, images and videos (variety), with high quality (veracity). We briefly discuss each in detail. **Volume**: Volume describes the quantity of generated and stored data. The size of the data determines the value and potential insight, and whether it can be considered big data or not. **Velocity**: Velocity describes the speed at which the data is generated and processed to meet the demands and challenges that lie in the path of growth and development. Big data is often available in both stored and real-time. Compared to small data, big data are produced more continually (it could be nanosecond, second, minute, hours, etc.). Two types of velocity related to big data are the frequency of generation and the frequency of handling, recording, and reporting. **Variety**: Variety describes the type and formats of the data. This helps people who analyze it to effectively use the resulting insight. Big data draws from different formats and completes missing pieces through data fusion. Data fusion is a term used to describe the technique of integrating multiple data sources to produce more consistent, accurate, and useful information than that provided by any individual data source. **Veracity**: Veracity describes the quality of data and the data value. The quality of data obtained can greatly affect the accuracy of the analyzed results. In the next subsection we will discuss some big data architectures. A comprehensive study of this topic can be found in the application architecture guide of the Microsoft technical documentation.

1.4.2 Big Data Architectures

Big data architectures are designed to handle the ingestion, processing, and analysis of data that is too large or complex for classical data-processing application tools. Some popular big data architectures are the Lambda architecture, Kappa architecture and the Internet of Things (IoT). We refer the reader to the Microsoft technical documentation on Big data architectures for a detailed discussion on the different architectures. Almost all big data architectures include all or some of the following components:

- *Data sources*: All big data solutions begin with one or more data sources. Some common data sources includes the following: Application data stores such as relational databases, static files produced by applications such as web server log files, and real-time data sources such as the Internet of Things (IoT) devices.
- *Data storage*: Data for batch processing operations is typically stored in a distributed file store that can hold high volumes of large files in various formats. This kind of store is often called a data lake. A data lake is a storage repository that allows one to store structured and unstructured data at any scale until it is needed.
- *Batch processing*: Since data sets are enormous, often a big data solution must process data files using long-running batch jobs to filter, aggregate, and otherwise prepare the data for analysis. Normally, these jobs involve reading source files, processing them, and writing the output to new files. Options include running U-SQL jobs or using Java, Scala, R, or Python programs. U-SQL is a data processing language that merges the benefits of SQL with the expressive power of ones own code.
- *Real-time message ingestion*: If the solution includes real-time sources, the architecture must include a way to capture and store real-time messages for stream processing. This might be a simple data store, where incoming messages are stored into a folder for processing. However, many solutions need a message ingestion store to act as a buffer for messages and to support scale-out processing, reliable delivery, and other message queuing semantics.
- *Stream processing*: After obtaining real-time messages, the solution must process them by filtering, aggregating, and preparing the data for analysis. The processed stream data is then written to an output sink.

- *Analytical data store*: Several big data solutions prepare data for analysis and then serve the processed data in a structured format that can be queried using analytical tools. The analytical data store used to serve these queries can be a Kimball-style relational data warehouse, as observed in most classical business intelligence (BI) solutions. Alternatively, the data could be presented through a low-latency NoSQL technology, such as HBase, or an interactive Hive database that provides a metadata abstraction over data files in the distributed data store.
- *Analysis and reporting*: The goal of most big data solutions is to provide insights into the data through analysis and reporting. Users can analyze the data using mathematical and statistical models as well using data visualization techniques. Analysis and reporting can also take the form of interactive data exploration by data scientists or data analysts.
- *Orchestration*: Several big data solutions consist of repeated data processing operations, encapsulated in workflows, that transform source data, move data between multiple sources and sinks, load the processed data into an analytical data store, or move the results to a report or dashboard.

2

Matrix Algebra and Random Vectors

2.1 Introduction

The matrix algebra and random vectors presented in this chapter will enable us to precisely state statistical models. We will begin by discussing some basic concepts that will be essential throughout this chapter. For more details on matrix algebra please consult (Axler 2015).

2.2 Some Basics of Matrix Algebra

2.2.1 Vectors

Definition 2.1 (Vector) A vector $\mathbf{x}$ is an array of real numbers $x_1, x_2, \ldots, x_N$, and it is written as:

$$\mathbf{x} = \begin{bmatrix} x_1 \\ x_2 \\ \vdots \\ x_n \end{bmatrix}.$$

Definition 2.2 (Scaler multiplication of vectors) The product of a scalar c, and a vector is the vector obtained by multiplying each entry in the vector by the scalar:

$$c\mathbf{x} = \begin{bmatrix} cx_1 \\ cx_2 \\ \vdots \\ cx_n \end{bmatrix}.$$

Data Science in Theory and Practice: Techniques for Big Data Analytics and Complex Data Sets, First Edition. Maria Cristina Mariani, Osei Kofi Tweneboah, and Maria Pia Beccar-Varela.

Definition 2.3 (Vector addition) The sum of two vectors of the same size is the vector obtained by adding corresponding entries in the vectors:

$$\mathbf{x}+\mathbf{y}=\begin{bmatrix} x_1 \\ x_2 \\ \vdots \\ x_n \end{bmatrix}+\begin{bmatrix} y_1 \\ y_2 \\ \vdots \\ y_n \end{bmatrix}=\begin{bmatrix} x_1+y_1 \\ x_2+y_2 \\ \vdots \\ x_n+y_n \end{bmatrix}$$

so that $\mathbf{x}+\mathbf{y}$ is the vector with the ith element x_i+y_i.

2.2.2 Matrices

Definition 2.4 (Matrix) Let m and n denote positive integers. An m-by-n matrix is a rectangular array of real numbers with m rows and n columns:

$$\mathbf{A}=\begin{bmatrix} A_{1,1} & \cdots & A_{1,n} \\ \vdots & & \vdots \\ A_{m,1} & \cdots & A_{m,n} \end{bmatrix}.$$

The notation $A_{i,j}$ denotes the entry in row i, column j of A. In other words, the first index refers to the row number and the second index refers to the column number.

Example 2.1

$$\text{If } A=\begin{pmatrix} 1 & 4 & 8 \\ 0 & 4 & 9 \\ 7 & -1 & 7 \end{pmatrix},$$

then $A_{3,1}=7$.

Definition 2.5 (Transpose of a matrix) The transpose operation A^T of a matrix changes the columns into rows, i.e. in matrix notation $(A^T)_{i,j}=A_{j,i}$, where "T" denotes transpose.

Example 2.2

$$\text{If } A_{2\times 3}=\begin{pmatrix} 1 & 4 & 8 \\ 0 & 4 & 9 \end{pmatrix}, \text{ then } A^T_{3\times 2}=\begin{pmatrix} 1 & 0 \\ 4 & 4 \\ 8 & 9 \end{pmatrix}.$$

Definition 2.6 (Scaler multiplication of a matrix) The product of a scalar c, and a matrix is the matrix obtained by multiplying each entry in the matrix by the scalar:

$$\mathbf{cA} = \begin{bmatrix} cA_{1,1} & \cdots & cA_{1,n} \\ \vdots & & \vdots \\ cA_{m,1} & \cdots & cA_{m,n} \end{bmatrix}.$$

In other words, $(cA)_{i,j} = cA_{i,j}$.

Definition 2.7 (Matrix addition) The sum of two vectors of the same size is the vector obtained by adding corresponding entries in the vectors:

$$\begin{aligned}\mathbf{A} + \mathbf{B} &= \begin{bmatrix} A_{1,1} & \cdots & A_{1,n} \\ \vdots & & \vdots \\ A_{m,1} & \cdots & A_{m,n} \end{bmatrix} + \begin{bmatrix} B_{1,1} & \cdots & B_{1,n} \\ \vdots & & \vdots \\ B_{m,1} & \cdots & B_{m,n} \end{bmatrix} \\ &= \begin{bmatrix} A_{1,1} + B_{1,1} & \cdots & A_{1,n} + B_{1,n} \\ \vdots & & \vdots \\ A_{m,1} + B_{m,1} & \cdots & A_{m,n} + B_{m,n} \end{bmatrix}.\end{aligned}$$

In other words, $(A + B)_{i,j} = A_{i,j} + B_{i,j}$.

Definition 2.8 (Matrix multiplication) Suppose A is an m-by-n matrix and B is an n-by-p matrix. Then AB is defined to be the m-by-p matrix whose entry in row i, column j, is given by the following equation:

$$(AB)_{i,j} = \sum_{k=1}^{n} A_{i,k} B_{k,j}.$$

In other words, the entry in row i, column j, of AB is computed by taking row i of A and column j of B, multiplying together corresponding entries, and then summing. The number of columns of A must be equal to the number of rows of B.

Example 2.3

$$\text{If } A = \begin{bmatrix} 1 & 4 \\ 0 & 4 \\ 7 & -1 \end{bmatrix} \quad \text{and} \quad B = \begin{bmatrix} 1 & 1 \\ 2 & 1 \end{bmatrix}$$

then

$$AB = \begin{bmatrix} 1 & 4 \\ 0 & 4 \\ 7 & -1 \end{bmatrix} \begin{bmatrix} 1 & 1 \\ 2 & 1 \end{bmatrix} = \begin{bmatrix} 1(1) + 4(2) & 1(1) + 4(1) \\ 0(1) + 4(2) & 0(1) + 4(1) \\ 7(1) + -1(2) & 7(1) + -1(1) \end{bmatrix} = \begin{bmatrix} 9 & 5 \\ 8 & 4 \\ 5 & 6 \end{bmatrix}.$$

Definition 2.9 (Square matrix) A matrix A is said to be a square matrix if the number of rows is the same as the number of columns.

Definition 2.10 (Symmetric matrix) A square matrix A is said to be symmetric if $\mathbf{A}=\mathbf{A}^{\mathrm{T}}$ or in matrix notation $(A^T)_{i,j} = A_{i,j} = A_{j,i}$ all i and j.

Example 2.4 The matrix $A = \begin{bmatrix} 1 & 4 \\ 4 & 4 \end{bmatrix}$ is symmetric; the matrix $B = \begin{bmatrix} 1 & 6 \\ 4 & -4 \end{bmatrix}$ is not symmetric.

Definition 2.11 (Trace) For any square matrix A, the trace of A denoted by $\mathrm{tr}(A)$ is defined as the sum of the diagonal elements, i.e.

$$\mathrm{tr}(A) = \sum_{i=1}^{n} a_{ii} = a_{11} + a_{22} + \cdots + a_{nn}.$$

Example 2.5 Let A be a matrix with

$$A = \begin{bmatrix} 1 & 4 & 9 \\ 1 & 0 & 0 \\ 1 & 4 & -9 \end{bmatrix}.$$

Then

$$\mathrm{tr}(A) = \sum_{i=1}^{2} a_{ii} = a_{11} + a_{22} + a_{33} = 1 + 0 + (-9) = -8.$$

We remark that trace are only defined for square matrices.

Definition 2.12 (Determinant of a matrix) Suppose A is an n-by-n matrix,

$$A = \begin{bmatrix} a_{1,1} & \cdots & a_{1,n} \\ \vdots & & \vdots \\ a_{n,1} & \cdots & a_{n,n} \end{bmatrix}.$$

The determinant of A, denoted $\det A$ or $|A|$, is defined by

$$\det\ A = a_{i1}C_{i1} + a_{i2}C_{i2} + \cdots + a_{in}C_{in},$$

where C_{ij} are referred to as the "cofactors" and are computed from

$$C_{ij} = (-1)^{i+j} \det\ M_{i,j}.$$

The term M_{ij} is known as the "minor matrix" and is the matrix you get if you eliminate row i and column j from matrix A.

Finding the determinant depends on the dimension of the matrix A; determinants only exist for square matrices.

Example 2.6 For a 2 by 2 matrix

$$A = \begin{bmatrix} a & b \\ c & d \end{bmatrix}$$

we have

$$\det\ A = |A| = \begin{vmatrix} a & b \\ c & d \end{vmatrix} = ad - bc.$$

Example 2.7 For a 3 by 3 matrix

$$A = \begin{bmatrix} a_{11} & a_{12} & a_{13} \\ a_{21} & a_{22} & a_{23} \\ a_{31} & a_{32} & a_{33} \end{bmatrix}$$

we have

$$\begin{aligned} \det\ A = |A| &= \begin{vmatrix} a_{11} & a_{12} & a_{13} \\ a_{21} & a_{22} & a_{23} \\ a_{31} & a_{32} & a_{33} \end{vmatrix} \\ &= a_{11}(a_{22}a_{33} - a_{23}a_{33}) - a_{12}(a_{21}a_{33} - a_{23}a_{31}) \\ &\quad + a_{13}(a_{21}a_{32} - a_{22}a_{31}). \end{aligned}$$

Definition 2.13 **(Positive definite matrix)** A square $n \times n$ matrix A is called positive definite if, for any vector $u \in \mathbb{R}^n$ nonidentically zero, we have

$$u^T A u > 0.$$

Example 2.8 Let A be a 2 by 2 matrix

$$A = \begin{bmatrix} 9 & -2 \\ -2 & 6 \end{bmatrix}.$$

To show that A is positive definite, by definition

$$\begin{aligned} u^T A u &= [u_1, u_2] \begin{bmatrix} 9 & -2 \\ -2 & 6 \end{bmatrix} \begin{bmatrix} u_1 \\ u_2 \end{bmatrix} \\ &= 9u_1^2 - 4u_1u_2 + 6u_2^2 \\ &= (2u_1 - u_2)^2 + 5(u_1^2 + u_2^2) > 0 \text{ for } [u_1, u_2] \neq [0, 0]. \end{aligned}$$

Therefore, A is positive definite.

Definition 2.14 (Positive semidefinite matrix) A matrix A is called positive semidefinite (or nonnegative definite) if, for any vector $u \in \mathbb{R}^n$, we have

$$u^T A u \geq 0.$$

Definition 2.15 (Negative definite matrix) A square $n \times n$ matrix A is called negative definite if, for any vector $u \in \mathbb{R}^n$ nonidentically zero, we have

$$u^T A u < 0.$$

Example 2.9 Let A be a 2 by 2 matrix

$$A = \begin{bmatrix} -2 & 1 \\ 1 & -2 \end{bmatrix}.$$

To show that A is negative definite, by definition

$$\begin{aligned} u^T A u &= \begin{bmatrix} u_1, u_2 \end{bmatrix} \begin{bmatrix} -2 & 1 \\ 1 & -2 \end{bmatrix} \begin{bmatrix} u_1 \\ u_2 \end{bmatrix} \\ &= -2u_1^2 + 2u_1u_2 - 2u_2^2 \\ &= -(u_1 - u_2)^2 < 0 \text{ for } [u_1, u_2] \neq [0, 0]. \end{aligned}$$

Therefore, A is negative definite.

Definition 2.16 (Negative semidefinite matrix) A matrix A is called negative semidefinite if, for any vector $u \in \mathbb{R}^n$, we have

$$u^T A u \leq 0.$$

We state the following theorem without proof.

Theorem 2.1 A 2 by 2 symmetric matrix

$$A = \begin{bmatrix} a & b \\ c & d \end{bmatrix}$$

is:

1. positive definite if and only if $a > 0$ and $\det A > 0$
2. negative definite if and only if $a < 0$ and $\det A > 0$
3. indefinite if and only if $\det A < 0$.

2.3 Random Variables and Distribution Functions

We begin this section with the definition of σ-algebra.

Definition 2.17 (σ-algebra) A σ-algebra $\mathcal{F}$ is a collection of sets $\mathcal{F}$ of Ω satisfying the following condition:

1. $\emptyset \in \mathcal{F}$.
2. If $F \in \mathcal{F}$ then its complement $F^c \in \mathcal{F}$.
3. If $F_1, F_2, \ldots$ is a countable collection of sets in $\mathcal{F}$ then their union $\cup_{n=1}^{\infty} F_n \in \mathcal{F}$.

Definition 2.18 (Measurable functions) A real-valued function f defined on Ω is called measurable with respect to a sigma algebra $\mathcal{F}$ in that space if the inverse image of the set B, defined as $f^{-1}(B) \equiv \{\omega \in E : f(\omega) \in B\}$ is a set in σ-algebra $\mathcal{F}$, for all Borel sets B of $\mathbb{R}$. Borel sets are sets that are constructed from open or closed sets by repeatedly taking countable unions, countable intersections and relative complement.

Definition 2.19 (Random vector) A random vector $\mathbf{X}$ is any measurable function defined on the probability space $(\Omega, \mathcal{F}, \mathbf{p})$ with values in $\mathbb{R}^n$ (Table 2.1).

Measurable functions will be discussed in detail in Section 20.5.

Suppose we have a random vector $\mathbf{X}$ defined on a space $(\Omega, \mathcal{F}, \mathbf{p})$. The sigma algebra generated by $\mathbf{X}$ is the smallest sigma algebra in $(\Omega, \mathcal{F}, \mathbf{p})$ that contains all the pre images of sets in $\mathbb{R}$ through $\mathbf{X}$. That is

$$\sigma(\mathbf{X}) = \sigma(\{\mathbf{X}^{-1}(B) \mid \text{ for all } B \text{ Borel sets in } \mathbb{R}\}).$$

This abstract concept is necessary to make sure that we may calculate any probability related to the random variable $\mathbf{X}$.

Any random vector has a distribution function, defined similarly with the one-dimensional case. Specifically, if the random vector $\mathbf{X}$ has components $\mathbf{X} = (X_1, \ldots, X_n)$, its cumulative distribution function or cdf is defined as:

$$F_{\mathbf{X}}(\mathbf{x}) = \mathbf{P}(\mathbf{X} \leq \mathbf{x}) = \mathbf{P}(X_1 \leq x_1, \ldots, X_n \leq x_n) \text{ for all } \mathbf{x}.$$

Associated with a random variable $\mathbf{X}$ and its cdf $F_{\mathbf{X}}$ is another function, called the probability density function (pdf) or probability mass function (pmf). The terms pdf and pmf refer to the continuous and discrete cases of random variables, respectively.

Table 2.1 Examples of random vectors.

Experiment	**Random variable**
Toss two dice	$\mathbf{X}$ = sum of the numbers
Toss a coin 10 times	$\mathbf{X}$ = sum of tails in 10 tosses

Definition 2.20 (Probability mass function) The pmf of a discrete random variable $\mathbf{X}$ is given by

$$f_{\mathbf{X}}(\mathbf{x}) = \mathbf{P}(\mathbf{X} = \mathbf{x}) \text{ for all } \mathbf{x}.$$

Definition 2.21 (Probability density function) The pdf, $f_{\mathbf{X}}(\mathbf{x})$ of a continuous random variable $\mathbf{X}$ is the function that satisfies

$$F(\mathbf{x}) = F(x_1, \ldots, x_n) = \int_{-\infty}^{x_1} \cdots \int_{-\infty}^{x_n} f_{\mathbf{X}}(t_1, \ldots, t_n) dt_n \ldots dt_1.$$

We will discuss these notations in details in Chapter 20.

Using these concepts, we can define the moments of the distribution. In fact, suppose that $g : \mathbb{R}^n \to \mathbb{R}$ is any function, then we can calculate the expected value of the random variable $g(X_1, \ldots, X_n)$ when the joint density exists as:

$$E[g(X_1, \ldots, X_n)] = \int_{-\infty}^{\infty} \cdots \int_{-\infty}^{\infty} g(x_1, \ldots, x_n) f(x_1, \ldots, x_n) dx_1 \ldots dx_n.$$

Now we can define the moments of the random vector. The first moment is a vector

$$E[\mathbf{X}] = \mu_{\mathbf{X}} = \begin{pmatrix} E[X_1] \\ \vdots \\ E[X_n] \end{pmatrix}.$$

The expectation applies to each component in the random vector. Expectations of functions of random vectors are computed just as with univariate random variables. We recall that expectation of a random variable is its average value.

The second moment requires calculating all the combination of the components. The result can be presented in a matrix form. The second central moment can be presented as the covariance matrix.

$$\begin{aligned} \mathrm{Cov}(\mathbf{X}) &= E[(\mathbf{X} - \mu_{\mathbf{X}})(\mathbf{X} - \mu_{\mathbf{X}})^t] \\ &= \begin{pmatrix} \mathrm{Var}(X_1) & \mathrm{Cov}(X_1, X_2) & \ldots & \mathrm{Cov}(X_1, X_n) \\ \mathrm{Cov}(X_2, X_1) & \mathrm{Var}(X_2) & \ldots & \mathrm{Cov}(X_2, X_n) \\ \vdots & \vdots & \ddots & \vdots \\ \mathrm{Cov}(X_n, X_1) & \mathrm{Cov}(X_n, X_2) & \ldots & \mathrm{Var}(X_n) \end{pmatrix}, \end{aligned} \tag{2.1}$$

where we used the transpose matrix notation and since the $\mathrm{Cov}(X_i, X_j) = \mathrm{Cov}(X_j, X_i)$, the matrix is symmetric.

We note that the covariance matrix is positive semidefinite (nonnegative definite), i.e. for any vector $u \in \mathbb{R}^n$, we have $u^T \mathbf{X} u \leq 0$.

Now we explain why the covariance matrix has to be semidefinite. Take any vector $u \in \mathbb{R}^n$. Then the product

$$u^T \mathbf{X} = \sum u_i X_i \tag{2.2}$$

is a random variable (one dimensional) and its variance must be nonnegative. This is because in the one-dimensional case, the variance of a random variable is defined as $\mathrm{Var}(X) = E(X - E[X])^2$. We see that the variance is nonnegative for every random variable, and it is equal to zero if and only if the random variable is constant. The expectation of (2.2) is $E[u^T\mathbf{X}] = u^T\mu_{\mathbf{X}}$. Then we can write (since for any number a, $a^2 = aa^T$)

$$\begin{aligned}\mathrm{Var}(u^t\mathbf{X}) &= E\ [(u^T\mathbf{X} - u^T\mu_{\mathbf{X}})^2] \\ &= E\ [(u^T\mathbf{X} - u^T\mu_{\mathbf{X}})(u^t\mathbf{X} - u^T\mu_{\mathbf{X}})^t] \\ &= E\ [u^T(\mathbf{X} - \mu_{\mathbf{X}})(\mathbf{X} - \mu_{\mathbf{X}})^t(u^T)^t] \\ &= u^T\mathrm{Cov}(\mathbf{X})\ u.\end{aligned}$$

Since the variance is always nonnegative, the covariance matrix must be nonnegative definite (or positive semidefinite). We recall that a square symmetric matrix $A \in \mathbb{R}^{n\times n}$ is positive semidefinite if $u^tAu \geq 0$, $\forall u \in \mathbb{R}^n$. This difference is in fact important in the context of random variables since you may be able to construct a linear combination $u^T\mathbf{X}$ which is not always constant but whose variance is equal to zero.

The covariance matrix is discussed in detail in Chapter 3.

We now present examples of multivariate distributions.

2.3.1 The Dirichlet Distribution

Before we discuss the Dirichlet distribution, we define the Beta distribution.

Definition 2.22 (Beta distribution) A random variable X is said to have a Beta distribution with parameters α and β if it has a pdf $f(x)$ defined as:

$$f(x) = \begin{cases} \frac{\Gamma(\alpha+\beta)}{\Gamma(\alpha)\Gamma(\beta)}x^{\alpha-1}(1-x)^{\beta-1}, & \text{if } 0 < x < 1, \\ 0, & \text{if otherwise,} \end{cases}$$

where $\alpha > 0$ and $\beta > 0$.

The Dirichlet distribution $\mathrm{Dir}(\boldsymbol{\alpha})$, named after Johann Peter Gustav Lejeune Dirichlet (1805–1859), is a multivariate distribution parameterized by a vector $\boldsymbol{\alpha}$ of positive parameters $(\alpha_1, \ldots, \alpha_n)$.

Specifically, the joint density of an n-dimensional random vector $\mathbf{X} \sim \mathrm{Dir}(\boldsymbol{\alpha})$ is defined as:

$$f(x_1, \ldots, x_n) = \frac{1}{\mathbf{B}(\boldsymbol{\alpha})}\left(\prod_{i=1}^{n} x_i^{\alpha_i - 1}\mathbf{1}_{\{x_i>0\}}\right)\mathbf{1}_{\{x_1+\cdots+x_n=1\}},$$

where $1_{\{x_1+\cdots+x_n=1\}}$ is an indicator function.

Definition 2.23 (Indicator function) The indicator function of a subset A of a set X is a function

$$1_A : X \to \{0, 1\}$$

defined as

$$1_A(x) = \begin{cases} 1, & \text{if } x \in A, \\ 0, & \text{if } x \notin A. \end{cases}$$

The components of the random vector $\mathbf{X}$ thus are always positive and have the property $X_1 + \cdots + X_n = 1$. The normalizing constant $\mathbf{B}(\boldsymbol{\alpha})$ is the multinomial beta function, that is defined as:

$$\mathbf{B}(\boldsymbol{\alpha}) = \frac{\prod_{i=1}^{n} \Gamma(\alpha_i)}{\Gamma\left(\sum_{i=1}^{n} \alpha_i\right)} = \frac{\prod_{i=1}^{n} \Gamma(\alpha_i)}{\Gamma(\alpha_0)},$$

where we used the notation $\alpha_0 = \sum_{i=1}^{n} \alpha_i$ and $\Gamma(x) = \int_0^\infty t^{x-1} e^{-t}\, dt$ for the Gamma function.

Because the Dirichlet distribution creates n positive numbers that always sum to 1, it is extremely useful to create candidates for probabilities of n possible outcomes. This distribution is very popular and related to the multinomial distribution which needs n numbers summing to 1 to model the probabilities in the distribution. The multinomial distribution is defined in Section 2.3.2.

With the notation mentioned above and α_0 as the sum of all parameters, we can calculate the moments of the distribution. The first moment vector has coordinates:

$$E[X_i] = \frac{\alpha_i}{\alpha_0}.$$

The covariance matrix has elements:

$$\operatorname{Var}(X_i) = \frac{\alpha_i(\alpha_0 - \alpha_i)}{\alpha_0^2(\alpha_0 + 1)},$$

and when $i \neq j$

$$\operatorname{Cov}(X_i, X_j) = \frac{-\alpha_i \alpha_j}{\alpha_0^2(\alpha_0 + 1)}.$$

The covariance matrix is singular (its determinant is zero).

Finally, the univariate marginal distributions are all beta with parameters $X_i \sim \text{Beta}(\alpha_i, \alpha_0 - \alpha_i)$. All these are in the reference (see Balakrishnan and Nevzorov 2004).

Please refer to Lin (2016) for the proof of the properties of the Dirichlet distribution.

2.3.2 Multinomial Distribution

We begin with a definition of the binomial distribution.

Definition 2.24 (Binomial distribution) A random variable X is said to have a binomial distribution with parameters n and p if it has a pmf shown below

$$P(x; p, n) = \binom{n}{k} (p)^x (1-p)^{(n-x)} \text{ for } x = 0, 1, \ldots, n,$$

where p is the probability of success on an individual trial and n is number of trials in the binomial experiment.

The multinomial distribution is a generalization of the binomial distribution. Specifically, assume that n independent distributions may result in one of the k outcomes generically labeled $S = \{1, 2, \ldots, k\}$, each with corresponding probabilities $(p_1, \ldots, p_k)$. Now define a vector $\mathbf{X} = (X_1, \ldots, X_k)$, where each of the X_i counts the number of outcomes i in the resulting sample of size n. The joint distribution of the vector $\mathbf{X}$ is

$$f(x_1, \ldots, x_k) = \frac{n!}{x_1! \ldots x_k!} p_1^{x_1} \ldots p_k^{x_k} \mathbf{1}_{\{\mathbf{x_1} + \cdots + \mathbf{x_k} = \mathbf{n}\}}.$$

In the same way as the binomial probabilities appear as coefficients in the binomial expansion of $(p + (1-p))^n$, the multinomial probabilities are the coefficients in the multinomial expansion $(p_1 + \cdots + p_k)^n$, so they sum to 1. This expansion in fact gives the name of the distribution.

If we label the outcome i as a success and everything else a failure, then X_i simply counts successes in n independent trials and thus $X_i \sim \text{Binom}(n, p_i)$. Thus, the first moment of the random vector and the diagonal elements in the covariance matrix are easy to calculate as np_i and $np_i(1-p_i)$, respectively. The off-diagonal elements (covariances) are not that complicated to calculate either. However, for multinomial random vectors, the first two moments are difficult to compute. The one-dimensional marginal distributions are binomial; however, the joint distribution of $(X_1, \ldots, X_r)$, the first r components, is not multinomial. Instead, suppose we group the first r categories into 1 and we let $Y = X_1 + \cdots + X_r$. Because the categories are linked, that is, $X_1 + \cdots + X_k = n$, we also have that $Y = n - X_{r+1} - \cdots - X_k$. We can easily verify that the vector $(Y, X_{r+1}, \ldots, X_k)$, or equivalently $(n - X_{r+1} - \cdots - X_k, X_{r+1}, \ldots, X_k)$, will have a multinomial distribution with associated probabilities $(p_Y, p_{r+1}, \ldots, p_k) = (p_1 + \cdots + p_r, p_{r+1}, \ldots, p_k)$.

Next consider the conditional distribution of the first r components given the last $k - r$ components. That is, the distribution of

$$(X_1, \ldots, X_r) \mid X_{r+1} = n_{r+1}, \ldots, X_k = n_k.$$

This distribution is also multinomial with the number of elements $n - n_{r+1} - \cdots - n_k$ and probabilities $(p'_1, \ldots, p'_r)$, where $p'_i = \frac{p_i}{p_1 + \cdots + p_r}$.

2.3.3 Multivariate Normal Distribution

A vector $\mathbf{X}$ is said to have a k-dimensional multivariate normal distribution (denoted $MVN_k(\mu, \sum)$, where N_k is k-dimensional multivariate normal distribution) with mean vector $\mu = (\mu_1, \ldots, \mu_k)$ and covariance matrix $\sum = (\sigma_{ij})_{ij \in \{1,\ldots,k\}}$ if its density can be written as

$$f(\mathbf{x}) = \frac{1}{(2\pi)^{k/2} \det(\sum)^{1/2}} e^{-\frac{1}{2}(\mathbf{x}-\mu)^T \sum^{-1}(\mathbf{x}-\mu)},$$

where we used the usual notations for the determinant, transpose, and inverse of a matrix. The vector of means μ may have any elements in $\mathbb{R}$, but, just as in the one-dimensional case, the standard deviation has to be positive. In the multivariate case, the covariance matrix $\sum$ has to be symmetric and positive definite.

The multivariate normal defined thus has many nice properties. The basic one is that the one-dimensional distributions are all normal, that is, $X_i \sim N(\mu_i, \sigma_{ii})$ and $\text{Cov}(X_i, X_j) = \sigma_{ij}$. This is also true for any marginal. For example, if $(X_r, \ldots, X_k)$ are the last coordinates, then

$$\begin{pmatrix} X_r \\ X_{r+1} \\ \vdots \\ X_k \end{pmatrix} \sim MVN_{k-r+1}\left(\begin{pmatrix} \mu_r \\ \mu_{r+1} \\ \vdots \\ \mu_k \end{pmatrix}, \begin{pmatrix} \sigma_{r,r} & \sigma_{r,r+1} & \cdots & \sigma_{r,k} \\ \sigma_{r+1,r} & \sigma_{r+1,r+1} & \cdots & \sigma_{r+1,k} \\ \vdots & \vdots & \ddots & \vdots \\ \sigma_{k,r} & \sigma_{k,r+1} & \cdots & \sigma_{k,k} \end{pmatrix}\right).$$

So any particular vector of components is normal.

Conditional distribution of a multivariate normal is also a multivariate normal. Given that $\mathbf{X}$ is a $MVN_k(\mu, \sum)$ and using the vector notations above assuming that $\mathbf{X}_1 = (X_1, \ldots, X_r)$ and $\mathbf{X}_2 = (X_{r+1}, \ldots, X_k)$, then we can write the vector μ and matrix $\sum$ as

$$\mu = \begin{pmatrix} \mu_1 \\ \mu_2 \end{pmatrix} \quad \text{and} \quad \sum = \begin{pmatrix} \sum_{11} & \sum_{12} \\ \sum_{21} & \sum_{22} \end{pmatrix},$$

where the dimensions are accordingly chosen to match the two vectors (r and $k - r$). Thus, the conditional distribution of $\mathbf{X}_1$ given $\mathbf{X}_2 = \mathbf{a}$, for some vector $\mathbf{a}$ is

$$\mathbf{X}_1 | \mathbf{X}_2 = \mathbf{a} \sim MVN_r\left(\mu_1 - \sum_{12} \sum_{22}^{-1} (\mu_2 - \mathbf{a}),\ \sum_{11} - \sum_{12} \sum_{22}^{-1} \sum_{21}\right).$$

Furthermore, the vectors $\mathbf{X}_2$ and $\mathbf{X}_1 - \sum_{21} \sum_{22}^{-1} \mathbf{X}_2$ are independent. Finally, any affine transformation $AX + b$, where A is a $k \times k$ matrix and b is a k-dimensional constant vector, is also a multivariate normal with mean vector $A\mu + b$ and covariance matrix $A \sum A^T$. Please refer to the text by Axler (2015) and Johnson and Wichern (2014) for more details on the Multinomial distribution and Multivariate normal distributions.

2.4 Problems

1 If A and B are two matrices, prove the following properties of the trace of a matrix.

(a) $\text{tr}(AB) = \text{tr}(BA)$.

(b) $\text{tr}(A + B) = \text{tr}(A) + \text{tr}(B)$.

(c) $\text{tr}(cA) = c\text{tr}(A)$, for a any constant c.

2 If A and B are two matrices, prove the following properties of the determinant of a matrix.

(a) $\det A = \det A^T$.

(b) $\det(AB) = \det A \cdot \det A = \det(BA)$.

3 Let

$$A = \begin{pmatrix} 1 & 4 & 8 \\ 0 & 4 & 9 \end{pmatrix}, \quad B = \begin{pmatrix} 2 & 4 & -3 \\ 1 & 8 & 9 \end{pmatrix}.$$

(a) Find $A + B$.

(b) Find $A - B$.

(c) Find $A'A$.

(d) Find AA'.

4 Let

$$A = \begin{pmatrix} 1 & 4 & 8 \\ 0 & 4 & 9 \end{pmatrix}, \quad B = \begin{pmatrix} 2 & 4 \\ 1 & 8 \\ -3 & 9 \end{pmatrix}.$$

(a) Find AB.

(b) Find BA.

(c) Compare $\text{tr}(AB)$ and $\text{tr}(BA)$.

5 Let

$$A = \begin{pmatrix} 5 & 4 & 8 \\ 0 & 4 & 3 \end{pmatrix}, \quad B = \begin{pmatrix} 0 & 4 \\ 1 & 3 \\ -3 & 2 \end{pmatrix}.$$

(a) Find A^{-1}.

(b) Find $(BA)^{-1}$.

6 Show that the real symmetric matrix

$$A = \begin{pmatrix} 3 & -1 & 0 \\ -1 & 3 & -1 \\ 0 & -1 & 3 \end{pmatrix}$$

is positive definite for any non-zero column vector.

7 Prove that if A and B are positive definite matrices then so is $A + B$.

8 For what values of a is the following matrix positive semidefinite?

$$A = \begin{pmatrix} 4 & -1 & a \\ -1 & 4 & -1 \\ a & -1 & 4 \end{pmatrix}.$$

9 Decide whether the following matrices are positive definite, negative definite, or neither. Please explain your reasoning.

$$A = \begin{pmatrix} 3 & -1 & -1 \\ -1 & 3 & 1 \\ -1 & 1 & 3 \end{pmatrix}, \quad B = \begin{pmatrix} 1 & 2 & 4 \\ 2 & 6 & 1 \\ 4 & 1 & 9 \end{pmatrix},$$

$$C = \begin{pmatrix} 1 & 3 & 0 & 0 \\ 3 & 4 & -1 & 0 \\ 0 & -1 & 5 & -1 \\ 0 & 0 & -1 & 2 \end{pmatrix}.$$

10 For random variables X and Y, show that

$$V(Y) = E[V(Y \mid X)] + V(E[Y \mid X]).$$

The variance $V(Y \mid X)$ is the variance of the random variable $Y \mid X$, while the same holds for the random variable $E[Y \mid X]$.

3

Multivariate Analysis

3.1 Introduction

Multivariate analysis is the statistical analysis of several variables at once. This is when multiple measurements are made on each experimental unit, and for which the relationship among multivariate measurements and their structure are important to the experiment's understanding. Experimental units are what you apply the treatments to. Many problems in the analysis of life science are multivariate in nature. However the analysis of large multivariable data sets is a major challenge for many research fields. Applications of multivariate techniques are vast. Some includes behavioral and biological sciences, finance, geophysics, medicine, ecology, and many other fields. The materials in this chapter will form the basis of discussion for what will be discussed later in this text.

3.2 Multivariate Analysis: Overview

We begin with the formal definition of multivariate analysis.

Definition 3.1 (Multivariate analysis) Multivariate analysis consists of a collection of techniques that can be used when several measurements are made on each experimental unit.

These measurements (i.e. data) must frequently be arranged and displayed in various ways. We now discuss the concepts underlying the first steps of data organization.

Multivariate data arise whenever an investigator, practitioner, or researcher seeks to study some physical phenomenon and selects a number $p \geq 1$ of variables to record. We will use the notation x_{jk} to indicate the particular value of the kth variable that is observed on the jth unit (i.e. subject). Hence, n measurements on

Data Science in Theory and Practice: Techniques for Big Data Analytics and Complex Data Sets, First Edition. Maria Cristina Mariani, Osei Kofi Tweneboah, and Maria Pia Beccar-Varela.

p variables can be displayed as a rectangular array called data matrix $\mathbf{X}$, of n rows and p columns:

$$\mathbf{X} = \begin{bmatrix} x_{1,1} & x_{1,2} & \cdots & x_{1,k} & \cdots & x_{1,p} \\ x_{2,1} & x_{2,2} & \cdots & x_{2,k} & \cdots & x_{2,p} \\ \vdots & \vdots & & \vdots & \vdots & \vdots \\ x_{j,1} & x_{j,2} & \cdots & x_{j,k} & \cdots & x_{j,p} \\ \vdots & \vdots & & \vdots & \vdots & \vdots \\ x_{n,1} & x_{n,2} & \cdots & x_{n,k} & \cdots & x_{n,p} \end{bmatrix}.$$

The rectangular array $\mathbf{X}$ contains the data consisting of all of the observations on all of the variables.

Example 3.1 (A data array) A selection of three receipts from Bestbuy was obtained in order to investigate the nature of movie sales. Each receipt provided, among other things, the number of movies sold and the total amount of each sale. Let the first variable be total dollar sales and the second variable be number of movies sold. Then we can take the corresponding numbers on the receipts as three measurements on two variables. From the above description, we obtain the tabular form of the data as follows:

$$\begin{array}{lccc} \text{Variable 1 (dollar sales)}: & 48 & 22 & 50 \\ \text{Variable 2 (number of movies)}: & 3 & 1 & 2 \end{array}$$

Then the data matrix $\mathbf{X}$ is

$$\mathbf{X} = \begin{bmatrix} 48 & 3 \\ 22 & 1 \\ 50 & 2 \end{bmatrix},$$

with three rows and two columns.

We now present some descriptive statistics. We will begin with the mean vectors.

3.3 Mean Vectors

Let $\mathbf{X}^T = [x_1, x_2, \ldots, x_p]$ be p random variables, where T denotes the transpose of a matrix. The expected value of the random $p \times 1$ vector is defined as the vector of expectations i.e.

$$E(\mathbf{X}) = E\begin{bmatrix} x_1 \\ x_2 \\ \vdots \\ x_p \end{bmatrix} = \begin{bmatrix} E(x_1) \\ E(x_2) \\ \vdots \\ E(x_p) \end{bmatrix}.$$

More generally, if $\mathbf{Z}_{n \times p} = [z_{jk}]$ is a matrix of random variables, then the $E(\mathbf{Z})$ is the matrix of expectations with elements $[E(z_{jk})]$, i.e.:

$$\mathbf{Z} = E\begin{bmatrix} z_{1,1} & z_{1,2} & \cdots & z_{1,k} & \cdots & z_{1,p} \\ z_{2,1} & z_{2,2} & \cdots & z_{2,k} & \cdots & z_{2,p} \\ \vdots & \vdots & & \vdots & \vdots & \vdots \\ z_{j,1} & z_{j,2} & \cdots & z_{j,k} & \cdots & z_{j,p} \\ \vdots & \vdots & & \vdots & \vdots & \vdots \\ z_{n,1} & z_{n,2} & \cdots & z_{n,k} & \cdots & z_{n,p} \end{bmatrix}$$

$$= \begin{bmatrix} E(z_{1,1}) & E(z_{1,2}) & \cdots & E(z_{1,k}) & \cdots & E(z_{1,p}) \\ E(z_{2,1}) & E(z_{2,2}) & \cdots & E(z_{2,k}) & \cdots & E(z_{2,p}) \\ \vdots & \vdots & & \vdots & \vdots & \vdots \\ E(z_{j,1}) & E(z_{j,2}) & \cdots & E(z_{j,k}) & \cdots & E(z_{j,p}) \\ \vdots & \vdots & & \vdots & \vdots & \vdots \\ E(z_{n,1}) & E(z_{n,2}) & \cdots & E(z_{n,k}) & \cdots & E(z_{n,p}) \end{bmatrix}.$$

For a random vector $\mathbf{X}^T = [x_1, x_2, \ldots, x_p]$, the mean vector consists of the means of each variable:

$$E(\mathbf{X}) = E\begin{bmatrix} x_1 \\ x_2 \\ \vdots \\ x_p \end{bmatrix} = \begin{bmatrix} E(x_1) \\ E(x_2) \\ \vdots \\ E(x_p) \end{bmatrix} = \begin{bmatrix} \mu_1 \\ \mu_2 \\ \vdots \\ \mu_p \end{bmatrix} = \mu,$$

where $E(X_i) = \mu_i$. For example let $x_{11}, x_{21}, \dots, x_{n1}$ be n measurements on the first variable. Then the **sample mean** for the first variable $\overline{x}_1$ is defined as follows:

$$\overline{x}_1 = \frac{1}{n}\sum_{j=1}^{n} x_{j1}.$$

The sample mean can be computed from the n measurements on each of the p variables. Therefore, in general for p sample means, we have:

$$\overline{x}_k = \frac{1}{n}\sum_{j=1}^{n} x_{jk}, \quad k = 1, 2, \dots, p.$$

Example 3.2 Consider the following data matrix introduced in Example 3.1:

$$\mathbf{X} = \begin{bmatrix} 48 & 3 \\ 22 & 1 \\ 50 & 2 \end{bmatrix}.$$

Each receipt yields a pair of measurements, total dollar sales, and number of movies sold. To find the sample mean $\overline{x}$, we calculate the average of each column as follows:

$$\overline{x}_1 = \frac{1}{3}\sum_{j=1}^{3} x_{j1} = \frac{1}{3}(48 + 22 + 50) = 40,$$

$$\overline{x}_2 = \frac{1}{3}\sum_{j=1}^{3} x_{j2} = \frac{1}{3}(3 + 1 + 2) = 2.$$

Therefore,

$$\overline{\mathbf{X}} = \begin{bmatrix} \overline{x}_1 \\ \overline{x}_2 \end{bmatrix} = \begin{bmatrix} 40 \\ 2 \end{bmatrix}.$$

This implies that the average dollar sales for two movies is \$40.00. Therefore, the average amount of dollars that cost a movie is 20 dollars.

3.4 Variance–Covariance Matrices

The variance–covariance matrix (or simply the covariance matrix) of a random vector $\mathbf{X}$ is given by

$$\text{Cov}(\mathbf{X}) = E[(\mathbf{X} - E(\mathbf{X}))(\mathbf{X} - E(\mathbf{X}))^T],$$

where $E(\mathbf{X})$ is the mean vector.

The sample covariance matrix $S = (s_{jk})$ is the matrix of sample variances and covariances of the p variables.

$$\mathbf{S} = (s_{ik}) = \begin{bmatrix} s_{1,1} & s_{1,2} & \cdots & s_{1,p} \\ s_{2,1} & s_{2,2} & \cdots & s_{2,p} \\ \vdots & \vdots & & \vdots \\ s_{p,1} & s_{p,2} & \cdots & s_{p,p} \end{bmatrix}. \tag{3.1}$$

In (3.1), the covariance matrix consists of the variances of the variables along the main diagonal and the covariances between each pair of variables in the other matrix positions. The sample covariance of the ith and kth variables, s_{ik}, is calculated using the ith and kth columns of $\mathbf{X}$:

$$s_{ik} = \frac{1}{n-1}\sum_{j=1}^{n}(x_{ji} - \overline{x}_i)(x_{jk} - \overline{x}_k), \quad i = 1, 2, \dots, p, \ k = 1, 2, \dots, p, \tag{3.2}$$

where n is the number of measurements.

For example if $i = 1, k = 2$:

$$s_{12} = \frac{1}{n-1}\sum_{j=1}^{n}(x_{j1} - \overline{x}_1)(x_{j2} - \overline{x}_2) \tag{3.3}$$

and if $i = 1, k = 1$:

$$s_{11} = \frac{1}{n-1}\sum_{j=1}^{n}(x_{j1} - \overline{x}_1)^2 \tag{3.4}$$

we have the sample variance.

The sample covariance measures the association between the ith and kth variables. The sample covariance reduces to the sample variance when $i = k$ as observed in (3.4). We note that the sample covariance matrix (3.1) is symmetric, i.e. $s_{ik} = s_{ki}$ for all i and k because of its definition. Other names used for the covariance matrix are variance matrix, variance–covariance matrix, and dispersion matrix. In finance the concept of covariance is applied in portfolio theory, in the diversification method, that reduces the risk by choosing assets that do not present a high positive covariance with each other.

If $\mathbf{X}$ is a random vector taking on any possible value in a multivariate population, the **population covariance matrix** is defined as

$$\sum = \text{cov}(\mathbf{X}) = \begin{bmatrix} \sigma_{1,1} & \sigma_{1,2} & \cdots & \sigma_{1,p} \\ \sigma_{2,1} & \sigma_{2,2} & \cdots & \sigma_{2,p} \\ \vdots & \vdots & & \vdots \\ \sigma_{p,1} & \sigma_{p,2} & \cdots & \sigma_{p,p} \end{bmatrix}. \tag{3.5}$$

Just like the sample covariance case defined in (3.1), the diagonal elements $\sigma_{jj} = \sigma_j^2$ are the population variances of the **X**'s, and the off-diagonal elements σ_{ik} are the population covariances of all possible pairs of **X**s, i.e. X_{ik} for $i \neq k$.

The notation $\sum$ for the covariance matrix is widely used and seems natural because $\sum$ is the uppercase version of σ.

Example 3.3 Consider the following data matrix introduced in Example 3.1:

$$\mathbf{X} = \begin{bmatrix} 48 & 3 \\ 22 & 1 \\ 50 & 2 \end{bmatrix}.$$

Each receipt yields a pair of measurements, total dollar sales, and number of movies sold. Since there are three receipts, we have a total of three observations on each variable. We find the sample variances and covariance $\mathbf{S}_n$ as follows:

$$\begin{aligned}
s_{11} &= \frac{1}{2}\sum_{j=1}^{3}(x_{j1} - \overline{x}_1)^2 \\
&\frac{1}{2}((48 - 40)^2 + (22 - 40)^2 + (50 - 40)^2) = 244, \\
s_{22} &= \frac{1}{2}\sum_{j=1}^{3}(x_{j2} - \overline{x}_2)^2 \\
&\frac{1}{2}((3 - 2)^2 + (1 - 2)^2 + (2 - 2)^2) = 1, \\
s_{12} &= \frac{1}{2}\sum_{j=1}^{3}(x_{j1} - \overline{x}_1)(x_{j2} - \overline{x}_2) \\
&\frac{1}{2}((48 - 40)(3 - 2) + (22 - 40)(1 - 2) + (50 - 40)(2 - 2)) = 13, \\
s_{21} &= s_{12}.
\end{aligned}$$

Therefore,

$$\mathbf{S}_n = \begin{bmatrix} 244 & 13 \\ 13 & 1 \end{bmatrix}.$$

3.5 Correlation Matrices

A correlation matrix is a table showing correlation coefficients between variables. Correlation is a statistical technique that can show whether and how strongly pairs of variables are related. The sample correlation between the ith and kth variables is defined as

$$r_{ik} = \frac{s_{ik}}{\sqrt{s_{ii}}\sqrt{s_{kk}}}, \tag{3.6}$$

where

$$s_{ik} = \frac{1}{n-1}\sum_{j=1}^{n}(x_{ji} - \bar{x}_i)(x_{jk} - \bar{x}_k), \quad i = 1, 2, \ldots, p \text{ and } k = 1, 2, \ldots, p,$$

$$s_{ii} = \frac{1}{n-1}\sum_{j=1}^{n}(x_{ji} - \bar{x}_i)^2, \quad i = 1, 2, \ldots, p,$$

$$s_{kk} = \frac{1}{n-1}\sum_{j=1}^{n}(x_{jk} - \bar{x}_k)^2, \quad k = 1, 2, \ldots, p.$$

Substituting s_{ik}, s_{ii} and s_{kk} into (3.6) and canceling terms, we obtain

$$r_{ik} = \frac{\sum_{j=1}^{n}(x_{ji} - \bar{x}_i)(x_{jk} - \bar{x}_k)}{\sqrt{\sum_{j=1}^{n}(x_{ji} - \bar{x}_i)^2}\sqrt{\sum_{j=1}^{n}(x_{jk} - \bar{x}_k)^2}} \tag{3.7}$$

for $i = 1, 2, \ldots, p$ and $k = 1, 2, \ldots, p$. We note that the sample correlation is symmetric since $r_{ik} = r_{ki}$ for all i and k.

The sample correlation coefficient is a measure of the linear association between two variables and does not depend on the units of measurement, i.e. when you construct the sample correlation coefficient, the units of measurement that are used cancel out. The **sample correlation matrix** is analogous to the covariance matrix with correlations in place of covariances:

$$\mathbf{R} = (r_{ik}) = \begin{bmatrix} 1 & r_{1,2} & \cdots & r_{1,p} \\ r_{2,1} & 1 & \cdots & r_{2,p} \\ \vdots & \vdots & & \vdots \\ r_{p,1} & r_{p,2} & \cdots & 1 \end{bmatrix}. \tag{3.8}$$

The **population correlation matrix** similar to (3.8) is defined as follows:

$$\mathbf{P} = (\rho_{ik}) = \begin{bmatrix} 1 & \rho_{1,2} & \cdots & \rho_{1,p} \\ \rho_{2,1} & 1 & \cdots & \rho_{2,p} \\ \vdots & \vdots & & \vdots \\ \rho_{p,1} & \rho_{p,2} & \cdots & 1 \end{bmatrix}, \tag{3.9}$$

where

$$\rho = \frac{\sigma_{ik}}{\sqrt{\sigma_{ii}}\sqrt{\sigma_{kk}}}.$$

We note that even though the signs of the sample correlation and the sample covariance are the same, the correlation is easier to interpret because its magnitude is bounded. It is bounded within the closed interval $-1 \leq r \leq 1$. To summarize, the sample correlation r has the following properties:

1. The value of the sample correlation r must lie between -1 and $+1$ inclusive. $+1$ indicates perfect linear relationship and -1 indicates perfect inverse relationship.
2. The sample correlation r measures the strength of the linear association between two variables. If r equals to zero, it implies no linear association between the components. Otherwise, the sign of r indicates the direction of the association. If r is positive, it means that as one variable gets larger the other gets larger. If r is negative, it means that as one gets larger, the other gets smaller (often called an "inverse" correlation). A larger value of r implies greater linear strength. This is an indication that both variables move in the opposite direction if one variable increases, the other variable decreases with the same magnitude (and vice versa).

Example 3.4 Consider the following data matrix introduced in Example 3.1:

$$\mathbf{X} = \begin{bmatrix} 48 & 3 \\ 22 & 1 \\ 50 & 2 \end{bmatrix}.$$

Each receipt yields a pair of measurements, total dollar sales, and number of movies sold. We find the sample correlation $\mathbf{R}$ as follows:

$$\begin{aligned} r_{12} &= \frac{s_{12}}{\sqrt{s_{11}}\sqrt{s_{22}}} \\ &= \frac{13}{\sqrt{244}\sqrt{1}} = 0.8321, \\ r_{21} &= r_{12}. \end{aligned}$$

Therefore,

$$\mathbf{R} = \begin{bmatrix} 1 & 0.832 \\ 0.832 & 1 \end{bmatrix}.$$

In this example, we observe the variables x_1 and x_2 are highly positively correlated since $r = 0.832$. This implies that if dollar sales (x_1) increases, the number of movies sold (x_2) also increases.

3.6 Linear Combinations of Variables

Most often, we are interested in linear combinations of the variables $x_1, x_2, \dots, x_p$. In this section, we investigate the means, variances, and covariances of linear combinations.

Let $a_1, a_2, \ldots, a_p$ be constants and consider the linear combination of the elements of the vector $\mathbf{X}$,

$$z = a_1x_1 + a_2x_2 + \cdots + a_px_p = \mathbf{a}^T\mathbf{X}, \tag{3.10}$$

where $\mathbf{a}^T = (a_1, a_2, \ldots, a_p)$. If the same coefficient vector $\mathbf{a}$ is applied to each $\mathbf{x}_i$ in a sample, we have

$$z_i = a_1x_{i1} + a_2x_{i2} + \cdots + a_px_{ip} = \mathbf{a}^T\mathbf{x}_i, i = 1, 2, \ldots, n. \tag{3.11}$$

For example, if $i = 1$, we have

$$\begin{aligned} z_1 &= \mathbf{a}^T\mathbf{x}_1 \\ &= (a_1, a_2, \ldots, a_p)\begin{pmatrix} x_{11} \\ x_{12} \\ \vdots \\ x_{12} \end{pmatrix}. \end{aligned}$$

3.6.1 Linear Combinations of Sample Means

The sample mean of z can be found either by averaging the n values $z_1 = \mathbf{a}^T\mathbf{x}_1$, $z_2 = \mathbf{a}^T\mathbf{x}_2, \ldots, z_n = \mathbf{a}^T\mathbf{x}_n$ or as a linear combination of $\overline{\mathbf{X}}$, the sample mean vector of $\mathbf{x}_1, \mathbf{x}_2, \ldots, \mathbf{x}_n$.

$$\begin{aligned} \bar{z} &= \frac{1}{n}\sum_{i=1}^{n} z_i \\ &= \bar{z}_1 + \bar{z}_2 + \cdots + \bar{z}_n \\ &= \mathbf{a}^T\overline{\mathbf{X}} \\ &= (a_1, a_2, \ldots, a_p)\begin{pmatrix} \bar{x}_1 \\ \bar{x}_2 \\ \vdots \\ \bar{x}_p \end{pmatrix} \\ &= a_1\bar{x}_1 + a_2\bar{x}_2 + \cdots + a_p\bar{x}_p. \end{aligned} \tag{3.12}$$

3.6.2 Linear Combinations of Sample Variance and Covariance

The sample variance of $z_i = \mathbf{a}^T\mathbf{x}_i, i = 1, 2, \ldots, n$ can be found as the sample variance of $z_1, z_2, \ldots, z_n$ or directly from $\mathbf{a}$ and $\mathbf{S}$, where $\mathbf{S}$ is the sample covariance matrix of $x_1, x_2, \ldots, x_n$:

$$s_z^2 = \frac{\sum_{i=1}^{n}(z_i - z)^2}{n-1} = \mathbf{a}^T\mathbf{S}\mathbf{a}. \tag{3.13}$$

We recall from Section 2.3 that variance is always nonnegative. Thus, we have $s_z^2 \geq 0$, and therefore, $\mathbf{a}^T\mathbf{S}\mathbf{a} \geq 0$, for every $\mathbf{a}$.

If we define another linear combination $u = \mathbf{b}^T\mathbf{X} = b_1x_1 + b_2x_2 + \cdots + b_px_p$, where $b^T = (b_1, b_2, \ldots, b_p)$ is a vector of constants different from $\mathbf{a}^T$, then the sample covariance of $z = \mathbf{a}^\mathbf{T}\mathbf{X}$ and $u = \mathbf{b}^\mathbf{T}\mathbf{X}$ is given by

$$s_{zu} = \frac{\sum_{i=1}^{n}(z_i - \bar{z})(u_i - \bar{u})}{n-1} = \mathbf{a}^T\mathbf{S}\mathbf{b}, \tag{3.14}$$

where n is the number of measurements.

Please refer to Johnson and Wichern (2014) for the proof of (3.14).

3.6.3 Linear Combinations of Sample Correlation

The sample correlation between $z = \mathbf{a}^T\mathbf{X}$ and $u = \mathbf{b}^T\mathbf{X}$ is obtained as follows:

$$\begin{aligned} r_{zu} &= \frac{s_{zu}}{\sqrt{s_z^2}\sqrt{s_u^2}} \\ &= \frac{\mathbf{a}^T\mathbf{S}\mathbf{b}}{\sqrt{(\mathbf{a}^T\mathbf{S}\mathbf{a})}\sqrt{(\mathbf{b}^T\mathbf{S}\mathbf{b})}}. \end{aligned} \tag{3.15}$$

We note that the sample results in Section 3.6 have population counterparts. We briefly state them below:

The population mean of $z = \mathbf{a}^T\mathbf{X}$ is defined as follows:

$$E(z) = E(\mathbf{a}^T\mathbf{X}) = \mathbf{a}^TE(\mathbf{X}) = \mathbf{a}^T\mu,$$

where μ denotes the population mean vector. The population variance of z is defined as follows:

$$\sigma_z^2 = \text{var}(\mathbf{a}^T\mathbf{X}) = \mathbf{a}^T\text{cov}(\mathbf{X})\mathbf{a} = \mathbf{a}^T\sum\mathbf{a},$$

where $\sum$ denotes the population covariance matrix which is defined in (3.5) as

$$\sum = \text{cov}(\mathbf{X}) = \begin{bmatrix} \sigma_{1,1} & \sigma_{1,2} & \cdots & \sigma_{1,p} \\ \sigma_{2,1} & \sigma_{2,2} & \cdots & \sigma_{2,p} \\ \vdots & \vdots & & \vdots \\ \sigma_{p,1} & \sigma_{p,2} & \cdots & \sigma_{p,p} \end{bmatrix}.$$

Let $u = \mathbf{b}^T\mathbf{X}$, where $\mathbf{b}$ is a vector of constants different from $\mathbf{a}$. The population covariance of $z = \mathbf{a}^T\mathbf{X}$ and $u = \mathbf{b}^T\mathbf{X}$ is defined as

$$\text{cov}(z, u) = \text{cov}(\mathbf{a}^T\mathbf{X}, \mathbf{b}^T\mathbf{X}) = \sigma_{zu} = \mathbf{a}^T\sum\mathbf{b},$$

where $\sum$ denotes the population covariance matrix which is defined in (3.5).

Finally, the population correlation of $z = \mathbf{a}^T\mathbf{X}$ and $u = \mathbf{b}^T\mathbf{X}$ is defined as

$$\begin{aligned}\rho_{zu} &= \text{corr}(\mathbf{a}^T\mathbf{X}, \mathbf{b}^T\mathbf{X}) \\ &= \frac{\sigma_{zu}}{\sqrt{\sigma_z^2}\sqrt{\sigma_u^2}} \\ &= \frac{\mathbf{a}^T \sum \mathbf{b}}{\sqrt{\mathbf{a}^T \sum \mathbf{a}}\sqrt{\mathbf{b}^T \sum \mathbf{b}}},\end{aligned}$$

where $\sum$ denotes the population covariance matrix which is defined in (3.5).

Remarks 3.1 If **A** is a scalar matrix and **X** are random vectors, then **AX** represents several linear combinations. The population mean vector and covariance matrix are given by

$$E(\mathbf{AX}) = \mathbf{A}E(\mathbf{X}) = \mathbf{A}\mu,$$

$$\text{cov}(\mathbf{AX}) = \mathbf{A}\sum \mathbf{A}^T,$$

where μ denotes the population mean vector and $\sum$ denotes the population covariance matrix which is defined in (3.5).

The proof of Remark 3.1 is left as an exercise. Please see Problem 2 .

Please refer to Johnson and Wichern (2014), Rencher (2002), and Axler (2002) and references therein for more details of multivariate analysis.

3.7 Problems

1 The following are four measurements on the variables x_1, x_2, and x_3:

x_1	x_2	x_3
9	12	5
1	6	0
0	14	3
5	8	7

Use the above information to answer the following questions:

(a) Find the sample mean vector, $\bar{x}$.

(b) Find the sample covariance matrix, $\mathbf{S}_n$.

(c) Find the sample correlation matrix, **R**.

2 For random vectors **X** and **Y**, scalar matrices **A** and **B**, and scalar vectors **a** and **b**, prove the following:

(a) $E(\mathbf{AX}) = \mathbf{A}\mu$.

(b) $\text{cov}(\mathbf{AX}) = \mathbf{A}\sum\mathbf{A}^T$.

(c) $E(\mathbf{AX} + \mathbf{b}) = \mathbf{A}\mu + \mathbf{b}$.

(d) $\text{cov}(\mathbf{AX} + \mathbf{b}) = \mathbf{A}\sum\mathbf{A}^T$, where $\sum$ denotes the population covariance matrix which is defined in (3.5).

(e) $\text{cov}(\mathbf{X}, \mathbf{Y}) = \text{cov}(\mathbf{Y}, \mathbf{X})$.

(f) $\text{cov}(\mathbf{a} + \mathbf{A}\,\mathbf{X}, \mathbf{b} + \mathbf{BY}) = \mathbf{A}\text{cov}(\mathbf{X},\mathbf{Y})\mathbf{B}^T$.

3 Consider the five pairs of measurements (x_1, x_2):

x_1	3	4	2	6	8
x_2	5	5.5	4	4	10

Calculate

(a) The sample means $\bar{x}_1$ and $\bar{x}_2$.

(b) The sample variances s_{11} and s_{22}.

(c) The sample covariances s_{12}.

4 Consider the data matrix:

$$\mathbf{X} = \begin{bmatrix} -1 & 4 & 3 \\ 0 & 4 & 2 \\ 3 & 2 & 5 \end{bmatrix}.$$

Calculate the matrix of deviations (residuals), $\mathbf{X} - 1\overline{\mathbf{X}}'$. Is this matrix of full rank?

Note: A matrix is of full rank if all rows and columns are linearly independent. A square matrix is full rank if and only if its determinant is nonzero.

5 Calculate the sample covariance matrix S using the data matrix in Problem 4.

6 Consider the data matrix:

$$\mathbf{X} = \begin{bmatrix} -1 & 4 & 3 \\ 0 & 4 & 2 \\ 3 & 2 & 5 \\ 20 & 8 & 3 \\ 13 & 2 & 5 \end{bmatrix}.$$

Obtain the mean corrected data matrix, sample covariance matrix and verify that the columns are linearly dependent.

7 If $z_i = ax_i$ for $i = 1, 2, \ldots, n$, show that $\overline{z} = a\overline{x}$, where a is a constant.

8 If $z_i = ax_i$ for $i = 1, 2, \ldots, n$, show that $s_z^2 = a^2 s^2$, where a is a constant.

9 The data in Table 3.1 (Elston and Grizzle 1962) consist of measurements y_1, y_2, y_3, and y_4 of the ramus bone at four different ages on each of 20 boys.
(a) Find $\overline{\mathbf{y}}$.
(b) Find $\mathbf{S}_n$.
(c) Find $\mathbf{R}$.

Table 3.1 Ramus Bone Length at Four Ages for 20 Boys.

	Age			
Individual	8 yr (y_1)	$8\frac{1}{2}$ yr (y_2)	9 yr (y_3)	$9\frac{1}{2}$ yr (y_4)
1	47.8	48.8	49.0	49.7
2	46.4	47.3	47.7	48.4
3	46.3	46.8	47.8	48.5
4	45.1	45.3	46.1	47.2
5	47.6	48.5	48.9	49.3
6	52.5	53.2	53.3	53.7
7	51.2	53.0	54.3	54.5
8	49.8	50.0	50.3	52.7
9	48.1	50.8	52.3	54.4
10	45.0	47.0	47.3	48.3
11	51.2	51.4	51.6	51.9
12	48.5	49.2	53.0	55.5
13	52.1	52.8	53.7	55.0
14	48.2	48.9	49.3	49.8
15	49.6	50.4	51.2	51.8
16	50.7	51.7	52.7	53.3
17	47.2	47.7	48.4	49.5
18	53.3	54.6	55.1	55.3
19	46.2	47.5	48.1	48.4
20	46.3	47.6	51.3	51.8

10 For the data in Table 3.1, define $z = y_1 + y_2 + 2y_3 - 4y_4$ and $u = -2y_1 + 4y_2 + y_3 - 2y_4$.

(a) Find $\bar{z}$, $\bar{u}$, s_z^2, and s_u^2 using (3.12) and (3.13).

(b) Find s_{zu} and r_{zu} using (3.14) and (3.15).

4

Time Series Forecasting

4.1 Introduction

In the traditional statisitical handling of time series data, making predictions about the future is called "extrapolation". However, modern fields focus on the topic and refer to it as time series forecasting. Time series forecasting involves fitting models to historical data and using them to predict future observations. One important distinction in forecasting is that the future is completely unknown and must only be estimated from prior information. In time series forecasting, we analyze the past behavior of a variable in order to predict its future behavior.

Definition 4.1 (Time series) If a random variable X is indexed in time, usually denoted by t, the observations $\{X_t, t \in \mathbb{T}\}$ is called a time series, where $\mathbb{T}$ is a time index set (for example, $\mathbb{T} = \mathbb{Z}$, the integer set).

In other words, a time series is a series of data points indexed in time order.

Most commonly, a time series is a collection of observations made sequentially through time. Thus, it is a sequence of discrete-time data. Examples of time series includes the Dow Jones Industrial Averages, historical data on sales, inventory, customer counts, interest rates, costs, etc. Time series are usually plotted via line charts. Time series are often used in statistics, signal processing, pattern recognition, mathematical finance, weather forecasting, earthquake prediction, and largely in several domain of applied sciences and engineering which involves temporal measurements. Methods for analyzing time series constitute an important area of research. Figure 4.1 is an example of an earthquake time series corresponding to a set of magnitude 3.0 – 3.3 aftershocks of a recent magnitude 5.2 intraplate earthquake which occurred in Clifton, Arizona, on 26 June 2014,

Data Science in Theory and Practice: Techniques for Big Data Analytics and Complex Data Sets, First Edition. Maria Cristina Mariani, Osei Kofi Tweneboah, and Maria Pia Beccar-Varela.

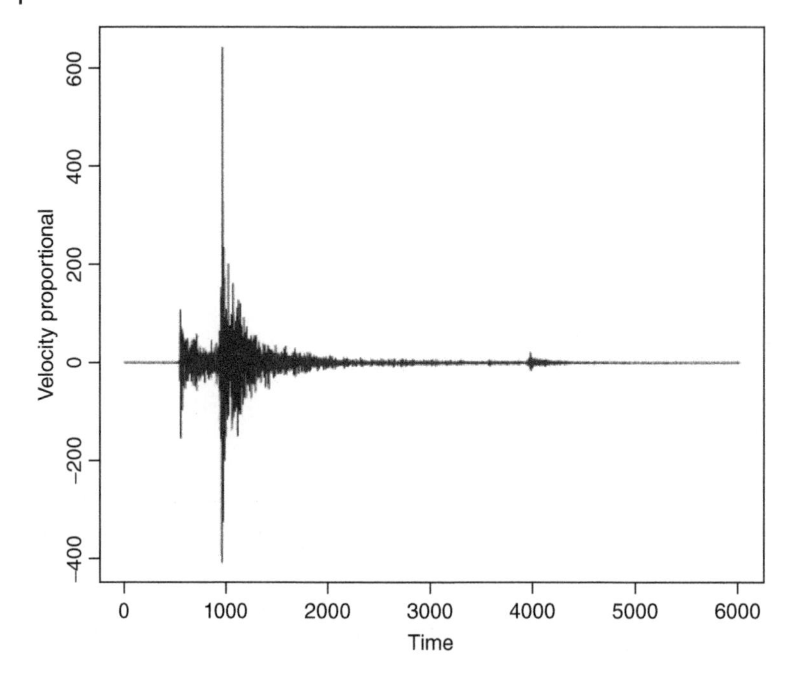

Figure 4.1 Time series data of phase arrival times of an earthquake.

and Figure 4.2 is an example of a high-frequency financial returns time series (per minute) corresponding to the Bank of America Corporation (BAC) stock index.

Please see Mikosch (1998), Chartfiled (2003), and Yaffee and McGee (1999) for more details of time series forecasting.

We begin the next section by presenting some definitions that will be useful throughout this chapter.

4.2 Terminologies

We begin with the definition of a discrete time series.

Definition 4.2 (Discrete time series) A time series $\{X_t\}$ is said to be discrete when observations are taken only at specific times, usually equally spaced.

An example is the realization of a binary process. The binary process is a special type of time series that arise when observations can take one of only two values, usually denoted by 0 and 1. They occur in many fields including communication theory. The monthly maximum and minimum temperatures in NYC is another example of two discrete time series.

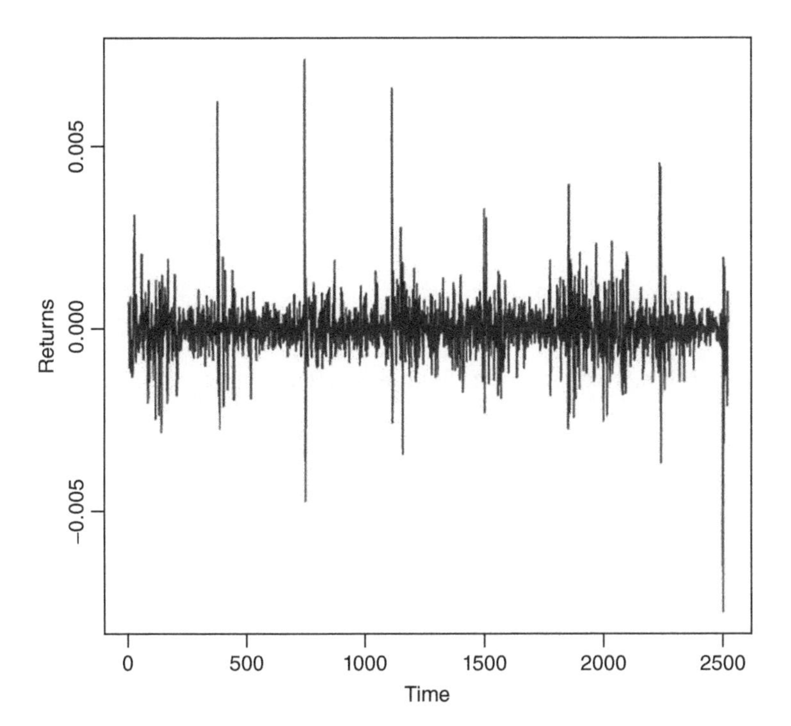

Figure 4.2 Time series data of financial returns corresponding to Bank of America (BAC) stock index.

Definition 4.3 (Continuous time series) A time series $\{X_t\}$ is said to be continuous when observations are made continuously through time.

The term "continuous" is used for series of this type even when the measured observation is a continuous variable. In this chapter, we will be discussing discrete time series, where the observations are taken at equal time intervals.

When modeling finite number of random variables, the covariance matrix is usually computed to summarize the dependence between these variables. For a time series $\{X_t\}_{t=-\infty}^{\infty}$, we need to model the dependence over infinite number of random variables. The concepts of autocovariance and autocorrelation functions provide us a tool for this purpose.

Definition 4.4 (Autocovariance function) The autocovariance function of a time series $\{X_t\}$ with $\mathrm{Var}(X_t) < \infty$ is defined by

$$\gamma_X(s,t) = \mathrm{Cov}(X_s, X_t) = E[(X_s - E[X_s])(X_t - E[X_t])].$$

With autocovariance functions, we can define the covariance stationarity, or weak stationarity.

Definition 4.5 (Stationarity) The time series $\{X_t, t \in \mathbb{Z}\}$ (where $\mathbb{Z}$ is the set on integers) is stationary if

1. $E[X_t^2] < \infty$ for all $t \in \mathbb{Z}$.
2. $E[X_t] = \mu$ for all $t \in \mathbb{Z}$.
3. $\gamma_X(s,t) = \gamma_X(s+h, t+h)$ for all $s, t, h \in \mathbb{Z}$.

Based on Definition 4.5, we can rewrite the autocovariance function of a stationary process as

$$\gamma_X(h) = \mathrm{Cov}(X_t, X_{t+h}) \text{ for } t, h \in \mathbb{Z}.$$

Definition 4.6 (Autocorrelation function) The autocorrelation function of a stationary time series $\{X_t\}$ is defined by

$$\rho_X(h) = \frac{\gamma_X(h)}{\gamma_X(0)},$$

where $\gamma_X(h) = \mathrm{Cov}(X_t, X_{t+h})$ and $\gamma_X(0) = \mathrm{Cov}(X_t, X_t)$.

Remarks 4.1 When the time series X_t is stationary, we must have

$$\rho_X(h) = \rho_X(-h).$$

Definition 4.7 (Strict stationary) The time series $\{X_t, t \in \mathbb{Z}\}$ is said to be strict stationary if the joint distribution of $(X_{t_1}, X_{t_2}, \ldots, X_{t_k})$ is the same as $(X_{t_1+h}, X_{t_2+h}, \ldots, X_{t_k+h})$.

A stationary data is one whose statistical properties such as mean, variance, autocorrelation, etc., are all constant over time. A nonstationary data on the other hand is a time series variable exhibiting a significant upward or downward trend over time.

In general most statistical forecasting methods are based on the assumption that the time series can be rendered approximately stationary through the use of mathematical transformations. This is because with stationary data series one can simply forecast that its statistical properties will be the same in the future as they have been in the past. The predictions for the stationarized series can then be "untransformed," by reversing whatever mathematical transformations were previously performed to obtain predictions for the original series. Nonstationary data on the other hand are unpredictable and cannot be modeled or forecasted.

The results obtained by using nonstationary time series may be false in that they may indicate a relationship between two variables where actually one does not exist. In order to receive consistent, reliable results, the nonstationary data needs to be transformed into stationary data. Examples of nonstationary data include the population of United States, income, price changes, and several others.

Definition 4.8 (Seasonal data) A seasonal data is a time series variable exhibiting a repeating patterns at regular intervals over time.

Seasonality is always of a fixed and known frequency. For example, retail sales tend to peak for the Christmas season and then decline after the holidays. So the time series of retail sales will typically show increasing sales from September through December and declining sales in January and February. Seasonality is very common in economic time series. It is less common in scientific and engineering data. Other examples of seasonal data include heating cost during cold winters and hot summers and sun-spots cycles. They are classified as seasonal data because its frequency is unchanging and associated with some aspect of the calendar.

4.3 Components of Time Series

The factors that are responsible for bringing changes in a time series are called the components of the time series. Time series consist of four components namely: seasonal, trend, cyclical, and random. The seasonal, trend, and cyclical components are responsible to form the pattern underlying the time series data, though not all components characterize every data set. An important purpose of time series analysis is to isolate each of these three components and demonstrate how each affects the value of the time series Y_t over time, including prediction for the future. The identification of this pattern, the predictable aspect of a time series, however, is complicated by the presence of random error. This is because all-time series may include random error, trend, cyclical, and seasonal components or combinations of these components are observed together with the error component.

We now discuss each component in detail. We begin with the seasonal variations.

4.3.1 Seasonal

We recall from Section 4.2 that seasonality occurs when the time series exhibits fixed and known frequency during the same month (or months) every year, or during the same quarter every year. When the series is characterized by a substantial regular annual variation, one must control for the seasonality in order

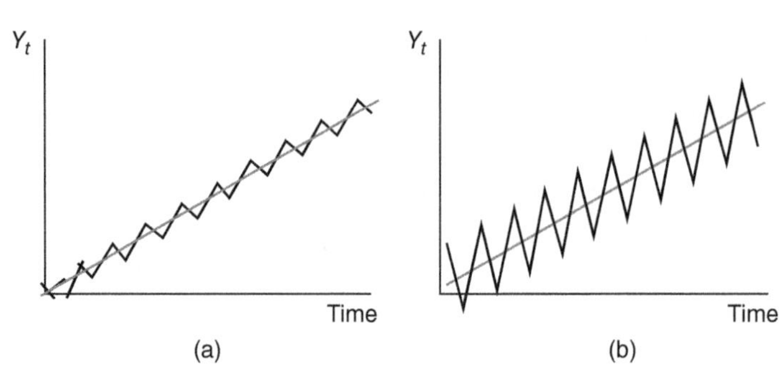

Figure 4.3 Seasonal trend component.

to forecast. Seasonality, the periodic annual changes in the series, may follow from yearly changes in weather such as temperature, humidity or precipitation, and the increase in retail sales during the month of December. Forecasting with such series requires seasonal adjustment (deseasonalization), which is discussed in more detail in Section 4.5. Two seasonal components in the absence of cycles are illustrated in Figure 4.3. Cycles usually occurs when the times series exhibits rises and falls that are not defined on a fixed frequency.

4.3.2 Trend

Trend is defined as the long-term pattern of a time series. Trends are very important when extracting, fitting and forecasting time series. They can be grouped into two types namely, deterministic trend and stochastic trend. A deterministic trend may derive from a definition that prescribes a well-defined formula for increment or decrement as a function of time, such as contractual interest. Another example is the gross domestic product (GDP) per person in a country increasing year after year.

A stochastic trend is due to random shift of level, perhaps the cumulative effect of some force that endows the series with a long-run change in level. In the GDP example presented in the deterministic case, when a "shock" occurs to the process generating GDP, due to a recession, for example, and GDP gets knocked off its long-run growth path, if GDP starts a new trend after a recession, its trend is said to be stochastic, driven by random shocks. Another example of a stochastic, nonlinear historical trend is the growth after 1977 in the number of international terrorist incidents until a peak was reached in 1987, after which this number declined see Yaffee and McGee (1999).

Regression may be used to test and model a trend. First, one plots the series versus time. If the trend appears to be linear (see Figure 4.4), one can regress it against a measure of time. If one finds a significant and or substantial relationship with time, the magnitude of the coefficient of time is evidence of a linear trend.

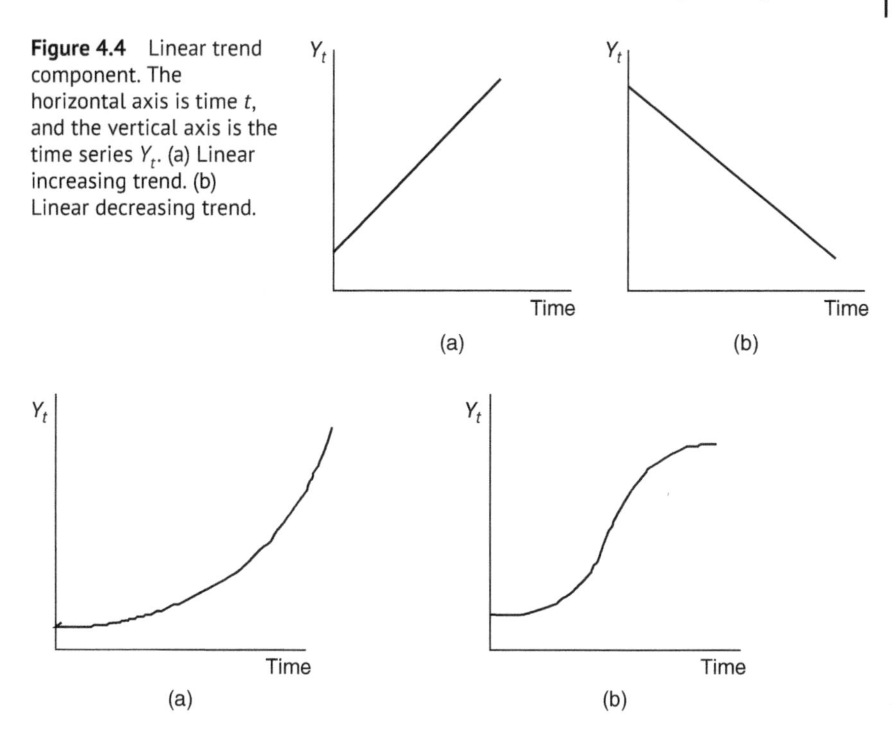

Figure 4.4 Linear trend component. The horizontal axis is time t, and the vertical axis is the time series Y_t. (a) Linear increasing trend. (b) Linear decreasing trend.

Figure 4.5 Nonlinear trend component. The horizontal axis is time t and the vertical axis is the time series Y_t. (a) Nonlinear increasing trend. (b) Nonlinear decreasing trend.

Some trends may appear to be nonlinear (see Figure 4.5). For example, one can construct a plot of the number of international terrorist incidents between 1977 and 1996 against time as a quadratic trend, depending on how time is parameterized. When nonlinear relationships exist, one can transform them into linear ones prior to modeling by a natural log transformation of the dependent variable. If a time series does not show an increasing or decreasing pattern, then the series is stationary in the mean. Techniques for achieving stationarity are discussed in Section 4.4.

4.3.3 Cyclical

Any pattern showing an up and down movement around a given trend is identified as a cyclical pattern. The duration of a cycle depends on the situation or type of business or industry being analyzed. For most business and economic data, the cyclical component is measured in periods of many years, usually decades, and so is usually not present in the typical time series analysis. This component reflects broad swings about either side of the trend line as seen in Figure 4.6.

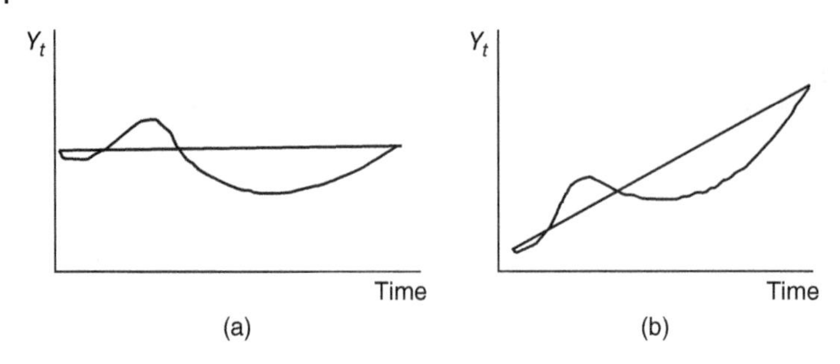

Figure 4.6 Cyclical component (imposed on the underlying trend). The horizontal axis is time t and the vertical axis is the time series Y_t.

4.3.4 Random

This corresponds to the error term E component of a time series. This component is unpredictable, and every time series has some unpredictable components that make it a random variable. In prediction, the objective is to "model" all the components to the point that the only component that remains unexplained is the random component.

4.4 Transformations to Achieve Stationarity

Nonstationary time series data used in models produces unreliable and false results and leads to poor understanding and forecasting. Therefore in order to obtain reliable and meaningful results, we need to transform the time series data so that it becomes stationary. We can often transform a nonstationary time series into a stationary series with one of the following techniques:

1. We can difference the data. Differencing helps to stabilize the mean. The new series is the difference between two consecutive observations in the original series. Let Z_t be a time series, ∇ be a difference operator and $\nabla^d Z_t$ denote the dth difference of Z_t for all t, then we define the first difference as

$$\nabla Z_t \equiv \nabla^1 Z_t = (1 - B)Z_t = Z_t - Z_{t-1}, \tag{4.1}$$

where B is the backward shift operator. The operator B is a useful notation when working with time series lags. By definition $BZ_t = Z_{t-1}$.
We note that the first differenced series will have only $T - 1$ values since it is not possible to calculate a difference ∇Z_t for the first observation.
Sometimes the differenced data will not be stationary, and it may be necessary to difference the data a second time, and we define the second difference as

follows:

$$\nabla^2 Z_t = (1 - B)^2 Z_t = \nabla(\nabla Z_t) = \nabla(Z_t - Z_{t-1}) \\ = \nabla Z_t - \nabla Z_{t-1} = Z_t - 2Z_{t-1} + Z_{t-2}. \tag{4.2}$$

Similarly, the second differenced series will have only $T - 2$ values since it is not possible to calculate a difference $\nabla^2 Z_t$ for the first two observations.
For the purpose of this text, we stop at the second difference since in practice it is almost never necessary to go beyond second-order differences.

Example 4.1 (Linear trend) Define a nonstationary process as

$$Z_t = bt + S_t, \tag{4.3}$$

where S_t is a stationary process. Then we can transform (4.3) to be a stationary process by differencing as follows:

$$\begin{aligned} W_t = \nabla Z_t &= Z_t - Z_{t-1} \\ &= bt + S_t - b(t-1) - S_{t-1} \\ &= b + (S_t - S_{t-1}). \end{aligned}$$

Thus, W_t is stationary because S_t is stationary.

Example 4.2 (Quadratic trend) Define a nonstationary process as

$$Z_t = bt^2 + S_t, \tag{4.4}$$

where S_t is a stationary process. In this example, a first difference will not transform the process to be stationary; however, a second difference will. We can transform (4.4) to be a stationary process by differencing as follows:

$$\begin{aligned} W_t = \nabla^2 Z_t &= Z_t - 2Z_{t-1} + Z_{t-2} \\ &= b[t^2 - 2(t-1)^2 + (t-2)^2] + S_t - 2S_{t-1} + S_{t-2} \\ &= 2b + (S_t - 2S_{t-1} + S_{t-2}). \end{aligned}$$

Therefore, W_t is stationary because S_t is stationary.
We remark that if the trend is a polynomial of order m, then $W_t = \nabla^m Z_t$ is stationary. This is because differencing m times eliminates polynomial trend of order m.

2. For nonconstant variance, taking the logarithm or square root of the series may stabilize the variance. For negative data, you can add a suitable constant to make all the data positive before applying the transformation. This constant

can then be subtracted from the model to obtain predicted (or the fitted) values and forecasts for future points.

3. If the data contains a trend, we can fit some type of curve to the data and then model the residuals from that fit. Since the purpose of the fit is to simply remove long-term trend, a simple fit, such as a straight line, is typically used.

We recall that the random walk model defined as $Y_t = Y_{t-1} + Z_t$, where Z_t is a stochastic component (white noise) distributed with mean 0 and variance σ^2 assumes that in each period, the variable Y_t takes a random step away from its previous value, and the steps are independently and identically distributed in size ("i.i.d."). This is equivalent to saying that the first difference of the variable is a series to which the mean model should be applied.

4.5 Elimination of Seasonality via Differencing

A seasonal difference is the difference between an observation and the corresponding observation from the previous year. It is mathematically defined as follows:

$$\nabla Z_t = Z_t - Z_{t-m},$$

where m is the number of seasons. For monthly data $m = 12$, for quarterly data $m = 4$, and so on. We remark that a seasonally differenced series is closer to being stationary. In fact the remaining nonstationarity can be removed with further first difference as defined in (4.1). For example if $\nabla_t^Z = Z_t - Z_{t-12}$ denotes monthly seasonally differenced series, then twice-differenced series is

$$\begin{aligned}\nabla^2 Z_t &= \nabla Z_t - \nabla Z_{t-1} \\ &= (Z_t - Z_{t-12}) - (Z_{t-1} - Z_{t-13}) \\ &= Z_t - Z_{t-1} - Z_{t-12} + Z_{t-13}.\end{aligned}$$

In practice, when both seasonal and first differences are applied, it makes no difference which is done first since the result will be the same. However, if seasonality is strong, it is recommended that seasonal differencing be done first because sometimes the resulting series will be stationary, and there will be no need for further first difference.

4.6 Additive and Multiplicative Models

The stable, predictable component of a timeseries is the specific combination of trend, cyclical, and seasonal components that characterize the particular time series. Two models accounts for the underlying pattern, an additive model and a multiplicative model. The **additive model** expresses the time series data Z_t as the sum of the trend (T_t), cyclical (C_t), seasonal (S_t), and error (E_t) components, i.e.

$$Z_t = T_t + C_t + S_t + E_t.$$

The cyclical, seasonal, and error components are each a fixed number of units above or below the underlying trend.

The **multiplicative model** accounts for the value of Y at time t as a product of the individual components, i.e.

$$Z_t = T_t \cdot C_t \cdot S_t \cdot E_t.$$

The multiplicative model expresses the cyclical, seasonal, and error components as percentages above or below the underlying trend. Each percentage is a ratio above or below the baseline of 1. The trend component T_t expressed in the original units of Y_t is multiplied by each of the C_t, S_t, and E_t ratios to yield the actual data.

With the multiplicative model, larger trends magnify the influence of the remaining components. For the additive model, the influence of a given value of C_t, S_t, and E_t is the same for all values of T_t.

In the collection of data taken over time, there is always some form of random variations or fluctuations. There exist methods for reducing or canceling the effect due to random variation. An often-used technique is called "smoothing." This technique, when properly applied, reveals more clearly the underlying trend, seasonal and cyclic components of the time series. There are two different groups of smoothing methods namely, **averaging methods** and **exponential smoothing methods**. Before we describe the two smoothing techniques, we need to discuss ways to compare different time series techniques for a given data set.

4.7 Measuring Accuracy of Different Time Series Techniques

By measuring forecast accuracy, the forecaster can carry out a validation study. In other words, the forecaster can try out a number of different forecasting models on some historical data, in order to see how each of these models would have worked had it been used in the past. The purpose of measuring forecast accuracy is to

- produce a metric or a single measure of a models usefulness or reliability;
- compare the accuracy of two forecasting models;
- search for an optimal model.

Four techniques that are commonly used to compare different time series techniques for a given data set are as follows: mean absolute deviation, mean absolute percent error, mean square error, and root mean square error. We discuss each technique in detail.

4.7.1 Mean Absolute Deviation

The mean absolute deviation (MAD) of a time series data sets is the average distance between each observation and the forecast. The distance between the each actual observation and the forecast value is known as the forecast error. Mathematically, the MAD is defined as follows:

$$\text{MAD} = \frac{1}{n}\sum_{i=1}^{n}|Y_i - \hat{Y}_i|,$$

where Y_i are the actual observations (values) and $\hat{Y}_i$ are the forecasted values data, and n is the number of observations.

Example 4.3 The table below displays a time series and its forecast.

Time (T)	11	12	13	14	15
Observation (Y_i)	3	14	14	20	7
Forecast ($\hat{Y}_i$)	11.7	11.6	11.4	11.6	11.1
Absolute error $\lvert Y_i - \hat{Y}_i\rvert$	8.7	2.4	2.6	8.4	4.1

From the data, we compute the MAD for the five observations as follows:

$$\begin{aligned}\text{MAD} &= \frac{1}{n}\sum_{i=1}^{n}|Y_i - \hat{Y}_i| \\ &= (8.7 + 2.4 + 2.6 + 8.4 + 4.1)/5 = 5.24.\end{aligned}$$

4.7.2 Mean Absolute Percent Error

The mean absolute percentage error (MAPE) is a statistical measure of how accurate a forecast system is. It measures this accuracy as a percentage and can be calculated as the average absolute percent error for each time period minus actual values divided by actual values. Mathematically, it is defined as follows:

$$\text{MAPE} = \frac{1}{n}\sum_{i=1}^{n}\left|\frac{Y_i - \hat{Y}_i}{Y_i}\right|.$$

In this example, we use the data from Example 4.3.

Example 4.4

Time (T)	11	12	13	14	15
Observation (Y_i)	3	14	14	20	7
Forecast ($\hat{Y}_i$)	11.7	11.6	11.4	11.6	11.1
$\left\lvert\frac{Y_i-\hat{Y}_i}{Y_i}\right\rvert$	2.9	0.171	0.186	0.42	0.586

From the data, we compute the MAPE for the five observations as follows:

$$\text{MAPE} = \frac{1}{n}\sum_{i=1}^{n}\left|\frac{Y_i - \hat{Y}_i}{Y_i}\right|$$

$$= (2.9 + 0.171 + 0.186 + 0.42 + 0.586)/5 = 0.8526.$$

The MAPE is the most common measure used to forecast error, and works best if there are no extremes to the data (and no zeros).

4.7.3 Mean Square Error

The mean square error (MSE) of a time series data sets is the average squared distance between the original time series and the forecast observations. Squaring the forecast error values forces them to be positive. It has the effect of putting more weight on large errors. Very large or outlier forecast errors are squared, which in turn has the effect of dragging the mean of the squared forecast errors out resulting in a larger mean squared error score. In effect, the score gives worse performance to those models that make large wrong forecasts.

Mathematically, it is defined as follows:

$$\text{MSE} = \frac{1}{n}\sum_{i=1}^{n}(Y_i - \hat{Y}_i)^2.$$

Example 4.5 The table below displays a time series and its forecast:

Time (T)	11	12	13	14	15
Observation (Y_i)	3	14	14	20	7
Forecast ($\hat{Y}_i$)	11.7	11.6	11.4	11.6	11.1
Squared error $(Y_i - \hat{Y}_i)^2$	75.69	5.76	6.76	70.56	16.81

From the data, we compute the MSE for the five observations as follows:

$$\text{MSE} = \frac{1}{n}\sum_{i=1}^{n}(Y_i - \hat{Y}_i)^2$$

$$= (75.69 + 5.76 + 6.76 + 70.56 + 16.81)/5 = 35.116.$$

The error values are in squared units of the forecasted values. A mean squared error of zero indicates no error.

4.7.4 Root Mean Square Error

The mean squared error described in Section 4.7.3 is the squared units of the predictions. It can be transformed back into the original units of the predictions by taking the square root of the mean squared error score. This is called the root mean squared error (RMSE). Mathematically, it is defined as follows:

$$\text{RMSE} = \sqrt{\frac{1}{n}\sum_{i=1}^{n}(Y_i - \hat{Y}_i)^2}.$$

Example 4.6 The table below displays a time series and its forecast:

Time (T)	11	12	13	14	15
Observation (Y_i)	3	14	14	20	7
Forecast ($\hat{Y}_i$)	11.7	11.6	11.4	11.6	11.1
Squared error $(Y_i - \hat{Y}_i)^2$	75.69	5.76	6.76	70.56	16.81

From the data, we compute the RMSE for the five observations as follows:

$$\text{RMSE} = \sqrt{\frac{1}{n}\sum_{i=1}^{n}(Y_i - \hat{Y}_i)^2}$$

$$= \sqrt{(75.69 + 5.76 + 6.76 + 70.56 + 16.81)/5}$$

$$\sqrt{35.116} = 5.926.$$

The RMSE error values are in the same units as the predictions. As with the mean squared error, an RMSE of zero indicates no error.

4.8 Averaging and Exponential Smoothing Forecasting Methods

In this section, we will present two simple time series forecasting techniques: averaging methods and exponential smoothing methods. Two versions of each approach are presented here: simple moving averages and weighted moving averages for the averaging methods and simple exponential smoothing and adjusted exponential smoothing for the exponential smoothing methods. Please see Chartfiled (2003) and Yaffee and McGee (1999) and references therein for other averaging and exponential smoothing forecasting approaches.

4.8.1 Averaging Methods

The main characteristic of the method of moving averages is that it generates a forecast for a particular time period by averaging the observed data values for the most recent n time periods. As each time period evolves, the observed data value for the new period is added to the average and the observed data value for the oldest period is subtracted from the average giving a new average value.

Two versions of this method are presented namely, simple moving averages and weighted moving averages.

4.8.1.1 Simple Moving Averages

The method of simple (equally weighted) moving averages smoothes out random fluctuations of data. This method is best used for short-term forecasts in the absence of seasonal or cyclical variations. On the other hand, this method is not particularly good in situations where the series has a trend.

The forecast for the value of Y at time $t+1$ that is made at time t equals the simple average of the most recent n observations:

$$\hat{Y}_{t+1} = \frac{Y_t + Y_{t-1} + \cdots + Y_{t-n+1}}{n}, \tag{4.5}$$

where $\hat{Y}_t$ is the forecast of the time series Y at time t and n is the number of actual values included in the average. In general, no method exists for determining n. One must try out several n values to see what works best.

This average is centered at period $t-(n+1)/2$, which implies that the estimate of the local mean will tend to lag behind the true value of the local mean by about $(n+1)/2$ periods. Thus, we say the average age of the data in the simple moving average is $(n+1)/2$ relative to the period for which the forecast is computed: this is the amount of time by which forecasts will tend to lag behind turning points in the data. For example, if we are averaging the last seven values, the forecasts will be about four periods late in responding to turning points.

Example 4.7 Consider the following example (Table 4.1) showing the volume of sales of a product over a time period of six hours. To calculate the three-hour moving average requires first that we sum the first three observations (130, 70, and 140). This three-hour total is then divided by 3 to obtain 113.33, as shown in the fourth row of column 3 in Table 4.2. This smoothed number, 113.33, becomes the forecast for hour 4, displayed in the fourth row, column 3 of forecast $\hat{Y}_t$.

By the same reasoning, we can obtain the forecast for hour 5 by moving one hour ahead and dropping the most distant hour. That is,

$$\begin{aligned}\hat{Y}_{t+1} &= (Y_t + Y_{t-1} + Y_{t-2})/3 \\ &= \frac{70+140+150}{3} = 120.\end{aligned}$$

Table 4.1 Time series data of the volume of sales of over a six hour period.

Hour	Sales(Y_t)
1	130
2	70
3	140
4	150
5	90
6	180

Table 4.2 Simple moving average forecasts.

Hour	Sales(Y_t)	Forecast ($\hat{Y}_t$)
1	130	—
2	70	—
3	140	—
4	150	113.33
5	90	120.0
6	180	126.67
7		140

Then a forecast for hour 6 is produced simply by working out the average actual volume of sales for hours 3, 4, and 5. The simple moving average forecasts are shown in Table 4.2.

We note that in Table 4.2, $\hat{Y}_t$ denotes the forecast value for period t. Thus, the forecast for period 4 will therefore be 113.33 and the forecast for period 7 will be 140. The difference between the Y_t value and the $\hat{Y}_t$ value is that the first refers to data that has occurred, whereas the second indicates a forecasted result.

The moving-average method provides an efficient mechanism for obtaining a value for forecasting stationary time series. The technique is simply an arithmetic average as time passes, with some lag-length determined optimally by an underlying cycle present in the time series data. Thus, in finance, moving-averages are frequently derived by financial analysts to generate market expectations, one of the most important input variables used by fund managers to allocate portfolios. The difficulty in using moving averages is their inability to capture the peaks and troughs, i.e. lowest level of the time series. When the market data are falling persistently, the moving average forecast tends to produce overpredicted values, while when the market is rising continually, the moving-average forecast will underpredict the market. In addition, since all the data points in the moving-average process are given equal weight, this approach fails to reflect the importance of time ordering with respect to the observations. For this reason, a weighted moving-average method has been suggested, where different weights are given to the data points. We briefly discuss the weighted moving-average method in Section 4.8.1.2.

4.8.1.2 Weighted Moving Averages

The method of weighted moving averages is another averaging time series forecasting method that smoothes out random fluctuations of data. The weighted moving averages imposes different weights on the observations being used for forecasting. This method is also best used for short-term forecasts in the absence of seasonal or cyclical variations. The only difference between simple moving average and weighted moving averages is that the latter uses weights in order to vary the effect of past data. This method assigns weights to each observed data point and works out a weighted mean as the forecast value for the next time period. This can be shown as follows:

$$\hat{Y}_{t+1} = w_1 Y_t + w_2 Y_{t-1} + \cdots + w_n Y_{t-n+1}, \tag{4.6}$$

where $0 \leq w_i \leq 1$, $\sum_{i=1}^{n} w_i = 1$, $\hat{Y}_t$ is the forecast of the time series Y at time t and n is the number of actual values included in the average.

Example 4.8 Consider the data used in Example 4.6. To calculate the three-hour moving average requires first that we sum the first three observations (130, 70, and 140) (Table 4.3). A possible set of weights could then be 0.35, 0.15, 0.50. We note the weights sum to one. In general we assign greater weight to recent data points. The forecast for periods 4–7 have been produced in Table 4.4 as follows:

$$\hat{Y}_4 = ((0.35 \times 130) + (0.15 \times 70) + (0.50 \times 140)) = 126,$$

$$\hat{Y}_5 = ((0.35 \times 70) + (0.15 \times 140) + (0.50 \times 150)) = 120.5,$$

$$\hat{Y}_6 = ((0.35 \times 140) + (0.15 \times 150) + (0.50 \times 90)) = 116.5,$$

$$\hat{Y}_7 = ((0.35 \times 150) + (0.15 \times 90) + (0.50 \times 180)) = 156.$$

Table 4.3 Time series data used in Example 4.6.

Hour	Sales(Y_t)
1	130
2	70
3	140
4	150
5	90
6	180

The weighted moving average forecasts are shown in Table 4.4:

Table 4.4 Weighted moving average forecasts.

Hour	Sales(Y_t)	Forecast ($\hat{Y}_t$)
1	130	—
2	70	—
3	140	—
4	150	126
5	90	120.5
6	180	116.5
7		156

We remark that in a series of n observations, the method of moving averages will not be able to generate a forecast for more than one period ahead. One way to produce forecasts for more than one period ahead would be using the technique of **trend projection**. To do this, we simply need to calculate the increment of the projection, given in relation (4.7), and add this to the last forecast value. This will produce a linear projection with increment defined as follows:

$$\text{Increment} = \frac{\text{Last } \hat{Y}_t - \text{First } \hat{Y}_t}{n-1}, \tag{4.7}$$

where Last $\hat{Y}_t$ is the last forecast value produced by the moving average model, First $\hat{Y}_t$ is the first forecast value produced by the moving average model, and n is the number of forecast values produced by the moving average models.

Table 4.5 Trend projection of weighted moving average forecasts.

Hour	Sales(Y_t)	Forecast ($\hat{Y}_t$)
1	130	—
2	70	—
3	140	—
4	150	126
5	90	120.5
6	180	116.5
7		156
8		166
9		176
10		186

The trend projection technique could then be used to produce forecasts for more periods ahead for the weighted moving average. In this example, $n = 4$ and the increment is calculated as follows:

$$\begin{aligned}\text{Increment} &= \frac{\text{Last } \hat{Y}_7 - \text{First } \hat{Y}_1}{n-1} \\ &= (156 - 126)/3 = 10.\end{aligned}$$

The forecast for periods 8–10 in Table 4.5 is computed as follows:

$$\begin{aligned}\hat{Y}_8 &= 156 + 10 = 166, \\ \hat{Y}_9 &= 166 + 10 = 176, \\ \hat{Y}_{10} &= 176 + 10 = 186.\end{aligned}$$

Averaging forecasting methods generally have one shortcoming. If n data points are to be included in the average, then $n - 1$ pieces of past data must be brought forward, in order to be combined with the current (the nth) observation. The past data must be stored in some way in order to produce the forecast. The problem of storing all the required data becomes expensive when a large number of forecasts are required. For example, if an organization is using a 10-point simple moving average model to forecast the demand for 6000 small parts, then for each part nine pieces of data must be stored for each forecast assuming that the current (10th) data value is available and does not need to be stored. If nine pieces of data are required for each forecast, then the forecaster will need 54 000 pieces of data

(9 × 6000) to be stored, in order to compute a single moving average forecast for every part, and this is computationally expensive.

4.8.2 Exponential Smoothing Methods

Exponential smoothing methods are averaging methods. In fact, exponential smoothing is a short name for an exponentially weighted moving average that requires only three pieces of data: the forecast for the most recent time period ($\hat{Y}_t$), the actual value for that time period (Y_t), and the value of the smoothing constant denoted by α.

Two versions of this method are presented namely, simple exponential smoothing and adjusted exponential smoothing.

4.8.2.1 Simple Exponential Smoothing

Simple exponential smoothing (usually referred to as exponential smoothing) is a time series forecasting method that smoothes out random fluctuations of data. It is best used for short-term forecasts in the absence of seasonal or cyclical variations. Exponential smoothing does not work very well if the series has a trend.

The exponential smoothing technique weights past data with weights that decrease exponentially with time, thus adjusting for previous inaccuracies in forecasts. It uses a weighting factor (known as the smoothing constant), which reflects the weight given to the most recent data values.

Exponential smoothing produces forecasts by using the following relations:

$$\hat{Y}_{t+1} = \hat{Y}_t + \alpha(Y_t - \hat{Y}_t), \tag{4.8}$$

where $\hat{Y}_{t+1}$ is the forecast value of the variable for period $t + 1$, $\hat{Y}_t$ is the forecast value of the variable for period t, Y_t is the actual value of the variable for period t, and α is the value of the smoothing constant which lies between 0 and 1.

The value α determines the degree of smoothing and how responsive the model is to fluctuations in the data. The larger the value given to α, the more strongly the model reacts to the most recent data. When the value of α is close to 1, the new forecast will include a substantial adjustment for any error that occurred in the preceding forecast. On the other hand, when the value of α is close to 0, the new forecast will be very similar to the old one. In general, if the time series is fluctuating irregularly, as a result of random variability, we choose a small value of α. On the other hand, if the series is more stable and shows little random fluctuation, the forecaster should choose a larger value of α.

From (4.8), we observe that exponential smoothing is simply the old forecast $\hat{Y}_t$ adjusted by α times the error $Y_t - \hat{Y}_t$ in the old forecast.

As an example, consider the sales example used earlier on in Example 4.6.

Table 4.6 Exponential smoothing forecasts of volume of sales.

Hour	Sales(Y_t)	Forecast ($\hat{Y}_t$)
1	130	130
2	70	130.00
3	140	106.00
4	150	119.60
5	90	131.76
6	180	115.06
7		141.04

Example 4.9 Let us use an exponential smoothing model with $\alpha = 0.4$ and use it to predict the volume of sales for the above time periods. Using relation (4.8), we have the following forecasts (Table 4.6):

$$\hat{Y}_2 = 130 + 0.4(130 - 130) = 130.00,$$
$$\hat{Y}_3 = 130 + 0.4(70 - 130) = 106,$$
$$\hat{Y}_4 = 106 + 0.4(140 - 106) = 119.60,$$
$$\hat{Y}_5 = 119.6 + 0.4(150 - 119.6) = 131.76,$$
$$\hat{Y}_6 = 131.76 + 0.4(90 - 131.76) = 115.06,$$
$$\hat{Y}_7 = 115.06 + 0.4(180 - 115.06) = 141.04.$$

Just like the moving averages techniques the exponential smoothing can only produce a forecast for one period ahead. The trend projection technique could then be used to generate forecasts for more periods ahead.

One of the disadvantages of exponential smoothing is the fact that it tends to produce forecasts that lag behind the actual trend. However, the adjusted exponential smoothing method can adjust exponentially smoothed forecasts to correct for a lag trend.

4.8.2.2 Adjusted Exponential Smoothing

The adjusted exponential smoothing just like the exponential smoothing technique is also used for short-term forecasts in the absence of seasonal or cyclical variations. In order to use the adjusted exponential smoothing method, one should have first produced a forecast for a particular time period. An adjusted forecast is then produced for that time period by using the following relation:

$$\hat{Y}_{t+1} = S_{t+1} + [(1 - \beta)/\beta]T_{t+1}, \quad (4.9)$$

Table 4.7 Exponential smoothing forecasts from Example 4.9.

Hour	Sales(Y_t)	Forecast (S_t)
1	130	130
2	70	130.00
3	140	106.00
4	150	119.60
5	90	131.76
6	180	115.06
7		141.04

where $\hat{Y}_{t+1}$ is the adjusted forecast value of the variable for period $t + 1$, S_{t+1} is the unadjusted forecast value of the variable for period $t + 1$, T_{t+1} is the trend factor for the period $t + 1$ and β is the value of the smoothing constant which lies between 0 and 1. The smoothing constant β has exactly the same meaning as the smoothing constant α used in the simple exponential smoothing approach.

The trend factor T_{t+1} is an exponentially smoothed factor that is used to convert the initial unadjusted exponential smoothing forecast to an adjusted exponential smoothing forecast. The trend factor for the period $t + 1$ is mathematically defined as follows:

$$T_{t+1} = \beta(S_{t+1} - S_t) + (1 - \beta)T_t, \tag{4.10}$$

where S_{t+1} is the unadjusted forecast value of the variable for period $t + 1$, S_t is the unadjusted forecast value of the variable for period t, T_t is the trend factor for period t and β is the value of the smoothing constant.

In the next example, an adjusted exponential smoothing model with a β value of 0.4 is used to adjust the forecasts produced by the exponential smoothing model developed in Example 4.9 (see Table 4.7).

Example 4.10 Let us use an exponential smoothing model with $\beta = 0.4$ and use it to predict the volume of sales for the above time periods.

Note that the trend factor for period 1 is always set to zero, i.e. $T_1 = 0$. The trend factor values for the other time periods are computed using relation (4.10) as follows:

$$T_2 = 0.4(130 - 130) + (1 - 0.4)0.00 = 0,$$

$$T_3 = 0.4(106.00 - 130) + (1 - 0.4)0.00 = -9.60,$$

$$T_4 = 0.4(119.6 - 106.00) + (1 - 0.4)0.00 = -0.32,$$

Table 4.8 Adjusted exponential smoothing forecasts.

Hour	Sales(Y_t)	Forecast (S_t)	Trend (T_t)	Adjusted ($\hat{Y}_t$)
1	130	130.00	0.00	—
2	70	130.00	0.00	130
3	140	106.00	−9.60	91.6
4	150	119.60	−0.32	119.12
5	90	131.76	4.67	138.77
6	180	115.06	−3.88	109.24
7		141.04	8.06	153.13

$$T_5 = 0.4(131.76 - 119.6) + (1 - 0.4)0.00 = 4.67,$$
$$T_6 = 0.4(115.06 - 131.76) + (1 - 0.4)0.00 = -3.88,$$
$$T_7 = 0.4(141.04 - 115.06) + (1 - 0.4)0.00 = 8.06.$$

Next, we calculate the adjusted $\hat{Y}_{t+1}$ using relation (4.9) as follows:

$$\text{adjusted } \hat{Y}_2 = 130 + 1.5(0) = 130,$$
$$\text{adjusted } \hat{Y}_3 = 106.00 + 1.5(-9.60) = 91.6,$$
$$\text{adjusted } \hat{Y}_4 = 119.6 + 1.5(-0.32) = 119.12,$$
$$\text{adjusted } \hat{Y}_5 = 131.76 + 1.5(4.67) = 138.77,$$
$$\text{adjusted } \hat{Y}_6 = 115.06 + 1.5(-3.88) = 109.24,$$
$$\text{adjusted } \hat{Y}_7 = 141.04 + 1.5(8.06) = 153.13.$$

The result is displayed in Table 4.8.

The trend projection technique discussed in Section 4.8.1.1 can be used to generate a forecast for more periods ahead.

The choice of values for α and β are subjective decision which are based on trial-and-error experimentation. The forecaster adjust the values of the parameters until a reasonable fit is obtained.

4.9 Problems

1 Let $\{S_t\}$ be a stationary process with mean zero and let a and b be constants.

$$Z_t = a + bt^2 + S_t. \tag{4.11}$$

Transform (4.11) to be stationary by differencing.

2 Let $\{Y_t\}$ be a stationary process with mean zero and let a and b be constants. If $Z_t = a + bt + s_t + Y_t$, where s_t is a seasonal component with period 12, show that

$$\nabla\nabla_{12}Z_t = (1 - B)(1 - B^{12})Z_t$$

is stationary.

3 Show that the autocovariance function can be written as follows:

$$\gamma(a, b) = E[(x_a - \mu_a)(x_b - \mu_b)] = E(x_a x_b) - \mu_a \mu_b,$$

where $E[x_a] = \mu_a$.

4 Let x_t and y_t be two time series, where

$$x_t = w_t,$$
$$y_t = w_t - \theta w_{t-1} + u_t,$$

where w_t and u_t are independent white noise series with variances σ_ω^2 and σ_u^2, respectively, and θ is an unspecified constant. Express the ACF, $\rho_y(h)$, for $h = 0, \pm 1, \pm 2, \ldots$ of the series y_t as a function of σ_ω^2, σ_u^2 and θ.

5 Show that $x_t = h_t$ and $y_t = h_t - \theta h_{t-1} + \epsilon_t$ are jointly stationary, where h_t and ϵ_t are independent white noise series with variances σ_h^2 and σ_ϵ^2, respectively, and θ is an unspecified constant.

6 The following data shows the number of liters of kerosene sold by TK LLC over the first eight months of the past year:

Month	Sales(1000's of liters)
January	22
February	24
March	30
April	33
May	36
June	47
July	52
August	65

(a) Using a two-point and a three-point simple moving average models produce forecasts for the period of March to September.

(b) Use the trend projection technique to generate a forecast for the period of October to December.
(c) Measure the accuracy of your forecasts using the four tests introduced in this chapter.
(d) Which model has produced more accurate forecasts?

7 Refer to the kerosene sales data given in Problem 3.
(a) Formulate two different weighted moving average models with weights of your choice to produce forecasts for March to September.
(b) Use the trend projection technique to generate a forecast for the period of October to December.
(c) Measure the accuracy of your forecasts using the four tests introduced in this chapter.
(d) Which model has produced more accurate forecasts?

8 Refer to the kerosene sales data given in Problem 3.
(a) Formulate two exponential smoothing models with smoothing constant values of your choice to produce forecasts for February to September.
(b) Use the trend projection technique to generate a forecast for the period of October to December.
(c) Measure the accuracy of your forecasts using the four tests introduced in this chapter.
(d) Which model has produced more accurate forecasts?
(e) How do these forecasts compare to the ones generated by your averaging models in Problems 3 and 4?

9 Refer to the kerosene sales data given in Problem 3.
(a) Formulate an adjusted exponential smoothing model with a smoothing constant value of your choice to adjust the forecasts produced by the exponential smoothing model that has given the best forecasts in Problem 5.
(b) Use the trend projection technique to generate a forecast for the period of October to December.
(c) Measure the accuracy of your forecasts using the four tests introduced in this chapter.
(d) How do these forecasts compare to the ones generated by your averaging models in Problems 3–5?

10 Sales of TK air conditioners have grown steadily during the past five years, as shown in the following table:

Year	Sales
1	450
2	495
3	518
4	563
5	584

(a) Using simple exponential smoothing constants of 0.3, 0.6, and 0.9, develop forecasts for years 2 through 6 for each constant. The sales manager had predicted, before the business started, that first year sales would be 410 air conditioners. Which smoothing constant gives the most accurate forecast?

(b) Using linear trend analysis, develop a forecasting model for the sales of Cool-Man air conditioners.

5
Introduction to R

5.1 Introduction

R is a programming language that is widely used by researchers and practitioners in different fields. R is a dynamically typed interpreted language and is typically used interactively. It has many built-in functions and libraries and is extensible, allowing users to define their own functions and procedures using C, C++, or Fortran. It also has a simple object system. R works well with data, making it a great language for anyone interested in statistics, data science, data analysis, and data visualization.

R is an integrated suite of software facilities for data manipulation, calculation, and graphical display. The following are some advantages of the R programming language:

- R is free and open-source. Open source means that people can modify and share because its design is publicly accessible.
- R is a large, coherent, integrated collection of intermediate tools for data analysis.
- R is an effective data handling and storage facility.
- R has graphical facilities for data analysis and display.

Technically, R is an expression language with a very simple syntax. It is case sensitive, so the letter "A" and "a" are different symbols and would refer to different variables. The set of symbols which can be used in R names depends on the operating system and country within which R is being run. With regards to the operating system, the very popular ones are the Microsoft windows system, Apple macOS system, and Linux operating system (i.e. Ubuntu, Fedora, Debian, etc.)

In this chapter, we will give an introduction to R and discuss some basics which are fundamental to some of the programs that will be used in this text. More details of R can be found in Paradis (2002) and references therein.

Data Science in Theory and Practice: Techniques for Big Data Analytics and Complex Data Sets, First Edition. Maria Cristina Mariani, Osei Kofi Tweneboah, and Maria Pia Beccar-Varela.

5.2 Basic Data Types

There are several basic data types in R which are of frequent occurrence in coding R calculations and programs. In this section, we will briefly review all of the different forms of data types. The list of all the data types provided by R are as follows:

- Numeric.
- Integer.
- Complex.
- Logical.
- Character or Strings.

5.2.1 Numeric Data Type

Decimal values are referred as numeric data types in R. This is the default working out data type. If you assign a decimal value for any variable x in the example

```
> x <- 2.0 # assign a decimal value
> x # print the value of x
[1] 2
> class(x) # print the class name of x
[1] "numeric"
```

5.2.2 Integer Data Type

In order to create an integer variable in R, we invoke the integer function. We can be sure that x is indeed an integer by applying the *is.integer* function. In the example below, we show how to create an integer variable in R:

```
> x = as.integer(2)
> y                   # print the value of x
[1] 2
> class(x)            # print the class name of x
[1] "integer"
> is.integer(x)       # is x an integer?
[1] TRUE
```

We can also coerce a numeric value into an integer with the *as.integer* function.

```
> as.integer(3.25)      # coerce a numeric value
[1] 3
```

5.2.3 Character

A character object is used to represent string values in R. We convert objects into character values with the function *as.character*(). A string is specified by using quotes. Both single (') and double (") quotes will work:

```
> a <- "hello"
> class(a)
[1] "character"
```

We can also coerce a numeric value into a character with the *as.character* function.

```
> x <- as.character(2.25)    # coerce a numeric value
> x                         # print the character string
[1] "2.25"
> class(x)                  # print the class name of x
[1] "character"
```

5.2.4 Complex Data Types

A complex value in R is defined via the pure imaginary value i as shown in the example below:

```
> z <- 2 + 3i              # create a complex number
> z                        # print the value of z
[1] 2+3i
> class(z)                 # print the class name of z
[1] "complex"
```

The following R code gives an error since R does not recognize -2 as complex value.

```
> sqrt(-2)
[1] NaN
Warning message:
In sqrt(-2) : NaNs produced
```

Instead, we have to rewrite -2 as $-2 + 0i$, before R recognizes it as a complex value.

```
 sqrt(-2+ 0i)
[1] 0+1.414214i
```

An alternative solution is to coerce -2 into a complex value by using the R code below:

```
> sqrt(as.complex(-2))
[1] 0+1.414214i
```

5.2.5 Logical Data Types

A logical value is often created via comparison between variables. The standard logical operations are "&" (and), "|" (or), and "!" (negation). We show an example of how logical values are created.

```
> x <- 2; y <- 4        # sample values
> z <- x > y            # is x larger than y?
> z                     # print the logical value
[1] FALSE
> class(z)              # print the class name of z
[1] "logical"
```

In this example display, the truth values of a statement using "&" (and), "|" (or), and "!" (negation).

```
> a  <- TRUE; b  <-FALSE
> a & b             # a AND b
[1] FALSE
> a | b             # a OR b
[1] TRUE
> !b                # negation of b
[1] TRUE
```

5.3 Simple Manipulations – Numbers and Vectors

5.3.1 Vectors and Assignment

R operates on named data structures. The simplest of such structure is the numeric vector, which is a single entity consisting of an ordered collection of numbers. Members in a vector are officially called components. To set up a vector named x, consisting of four numbers, namely 10.0, 11.1, 12.4, and 7.2, we use the R command:

```
> x <- c(10.0, 11.1, 12.4, 7.2)
```

This is an assignment statement using the function *c()* which in this context can take an arbitrary number of vector arguments and whose value is a vector obtained by concatenating its arguments end to end.

The sign "< –" is the assignment operator normally used in R. However, in context the operator "=" can be used as an alternative. Assignment can also be made using the function *assign*(). An equivalent way of making the same assignment as above is with

```
>  assign ("x", c(10.0, 11.1, 12.4, 7.2))
```

Assignments can also be made in the other direction, using the change in the assignment operator. So the same assignment could be made using for example

```
>  c(10.0, 11.1, 12.4, 7.2) -> x
```

5.3.2 Vector Arithmetic

Vectors can be used in arithmetic expressions, in which case the operations are performed element by element or elementwise. Vectors occurring in the same expression need not all be of the same length.

For example, suppose we have two vectors *x* and *y* which consist of four numbers defined below:

```
> x <- c(1, 2, 3, 4)
> y <- c(1, 3, 4, 5)
```

Then, if we multiply *x* by 4, we would get a vector with each of its members multiplied by 4 as shown below:

```
> 4 * x
[1]  4  8 12 16
```

And if we add *x* and *y* together, the sum would be a vector whose members are the sum of the corresponding members from *x* and *y* as shown below:

```
> x + y
[1] 2 5 7 9
```

Similarly, for subtraction, multiplication, and division, we get new vectors via elementwise operations as shown below:

```
> x - y
[1]  0 -1 -1 -1

> x * y
[1]  1  6 12 20

> x / y
[1] 1.0000000 0.6666667 0.7500000 0.8000000
```

We remark in the multiplication and division of vectors, the new vectors are obtained via elementwise operations. In addition, if two vectors are of unequal length, the shorter one will be recycled in order to match the longer vector. For example, the following vectors a and b have different lengths, and their sum is computed by recycling values of the shorter vector a. The shorter vectors in the expression are recycled until they match the length of the longest vector. In particular, a constant is simply repeated. A warning message will appear as displayed in the example below.

```
> a <- c(2, 3, 4)
> b <- c(2, 3, 5, 6, 6)
> a + b
[1] 4 6 9 8 9
Warning message:
In a + b : longer object length is not a multiple of
  shorter object length
```

The elementary arithmetic operators are the usual $+, -, *$ and $\wedge$ for raising to a power. In addition, all of the common arithmetic functions are available: log, exp, sin, cos, tan, sqrt, and several others, all have their usual meaning. The functions max () and min () select the largest and smallest elements of a vector, respectively. *range* is a function whose value is a vector of length two, namely c(min (x), max (x)). The function length(x) gives the number of elements in x and the function sum(x) gives the total of the elements in x, and the function prod(x) gives the product of all the elements in x.

Two statistical functions are mean(x) which calculates the sample mean (average), which is the same as sum(x)/length(x), and sample variance, i.e. var(x) which is defined as follows:

```
sum((x-mean(x))^2)/(length(x)-1)
```

5.3.3 Vector Index

We can retrieve values in a vector by declaring an index inside a single square bracket "[]" operator.

For example, the following shows how to retrieve a vector member. Since the vector index is 1– based, we use the index position 4 for retrieving the fourth member.

```
>  u <- c("dog", "cat", "lion", "elephant", "tiger")
> u[4]
[1] "elephant"
```

In the example above, $u[4]$ retrieved the fourth member in the vector u which is "elephant" in this case. Unlike other programming languages, the square bracket operator returns more than just individual members. In fact, the result of the square bracket operator is another vector, and *u*[4] is a vector slice containing a single member "elephant."

If the index is negative, it would strip the member whose position has the same absolute value as the negative index. For example, the following creates a vector slice with the fourth member removed.

```
>  u <- c("dog", "cat", "lion", "elephant")
> u[-4]
[1] "dog"  "cat"  "lion"
```

If an index is out-of-range, a missing value will be reported via the symbol NA as shown below:

```
>  u <- c("dog", "cat", "lion", "elephant")
> u[8]
[1] NA
```

A new vector can be sliced from a given vector with a **numeric index vector**, which consists of member positions of the original vector to be retrieved. In the next example, we show how to retrieve a vector slice containing the first and second members of a given vector *u*.

```
> u <- c("dog", "cat", "lion", "elephant")
> u[c(1, 2)]
[1] "dog" "cat"
```

In order to produce a vector slice between two indexes, we can use the colon operator ":." This can be convenient for situations involving large vectors. For example,

```
>  u <- c("dog", "cat", "lion", "elephant")
> u[1:3]
[1] "dog"  "cat"  "lion"
```

5.3.4 Logical Vectors

R allows manipulation of logical quantities. The elements of a logical vector can have the values TRUE, FALSE, and NA. The first two are often abbreviated as T and F, respectively. Note however that T and F are just variables that are set to TRUE and FALSE by default, but are not reserved words and hence can be overwritten by the user. Hence, you should always use TRUE and FALSE.

Some common logical operators in R are listed below:

- < (less than)
- > (greater than)

- <= (less than or equal)
- >= (greater than or equal)
- == (equal to)
- ! = (not equal to)
- | (entry wise or)
- || (or)
- ! (not)
- & (entry wise and)
- && (and)
- xor (a,b) (exclusive or)

For example, if $a1$ and $a2$ are logical expressions, then $a1\&a2$ is their intersection ("and"), $a1|a2$ is their union ("or"), and $!a1$ is the negation of $a1$. We provide an example as follows. Let a1 and a2 be vectors defined below:

```
> a1 = c(TRUE,FALSE)
> a2 = c(FALSE,FALSE)
> a1|a2  # union (entry wise or)
[1]  TRUE FALSE
> a1||a2 # or
[1] TRUE
> xor(a1,a2) # exclusive or
[1]  TRUE FALSE
```

5.3.5 Missing Values

In some instances, the components of a vector may not be completely known. When an element or value is "not available" or a "missing value" in the statistical sense, a place within a vector may be reserved for it by assigning it a special value. In R, missing values are represented by the symbol "NA" (not available). Impossible values (e.g. dividing by zero) are represented by the symbol "NaN" (not a number).

In general, any operation on an "NA" becomes an "NA." The motivation for this rule is simply that if the specification of an operation is incomplete, the result cannot be known and hence is not available. We present an example below:

```
> u <- c(1:5,NA)
> is.na(u)
[1] FALSE FALSE FALSE FALSE FALSE  TRUE
```

The function *is.na*(x) gives a logical vector of the same size as x with value TRUE if and only if the corresponding element in x is NA. Missing values are sometimes printed as $< NA >$ when character vectors are printed without quotes.

5.3.6 Index Vectors

Given a vector, one common task is to isolate particular entries that meet some criteria, i.e. subsetting. Subsets of the elements of a vector may be selected by appending to the name of the vector an index vector in square brackets. We will briefly show how to use R's indexing notation to pick out specific items within a vector. Such index vectors can be any of the four distinct types.

5.3.6.1 Indexing with Logicals

In this case, the index vector is recycled to the same length as the vector from which elements are to be selected. Values corresponding to TRUE in the index vector are selected and those corresponding to FALSE are omitted as shown in the example below.

```
> a1 <- c(1,2,3,4,5)
> a2 <- c(TRUE,FALSE,FALSE,TRUE,FALSE)
> a1[a2]
[1] 1 4
```

5.3.6.2 A Vector of Positive Integral Quantities

In this case, the values in the index vector must lie in the set $\{1, 2, \dots, \text{length}(x)\}$. The corresponding elements of the vector are selected and concatenated, in that order, in the result. The index vector can be of any length, and the result is of the same length as the index vector. For example,

```
> a1 <- c(1:15)        # vector of length 15
> a2 <- a1[1:10]       # selects first 10 elements of a1
> a2
 [1]  1  2  3  4  5  6  7  8  9 10
```

5.3.6.3 A Vector of Negative Integral Quantities

Such an index vector specifies the values to be excluded rather than included. For example,

```
> a1 <- c(1:15)        # vector of length 15
> a2 <- a1[-(1:10)]     # excludes first 10 elements of a1
> a2
 [1] 11 12 13 14 15
```

In this example, the vector a2 now consist of numbers 11, 12, 13, 14, and 15

5.3.6.4 Named Indexing

This case only applies where an object has a names attribute to identify its components. A subvector of the names vector may be used in the same way as the positive integral labels discussed above. In particular we assign names to reference members instead of numeric indexes. We present an example below.

```
> fruit <- c(2, 5, 8, 10)
> names(fruit) <- c("orange", "banana", "apple", "peach")
> breakfast <- fruit[c("apple","peach")]
> breakfast
apple peach
    8    10
```

The advantage is that alphanumeric names are often easier to remember than numeric indices. This option is particularly useful in connection with data frames, as we shall discuss later in this chapter.

An indexed expression can also appear on the receiving end of an assignment, in which case the assignment operation is performed only on those elements of the vector. The expression must be of the form *vector[index_vector]* as having an arbitrary expression in place of the vector name. For example,

```
> a1 <- c(1:5,NA)  # vector of length  6
> a1[is.na(a1)]  <- 0  # replaces any missing values in a1 by zeros
> a1
[1] 1 2 3 4 5 0
```

Thus, in the example, we replaced the NA in the vector a1, with the number 0.

5.3.7 Other Types of Objects

Vectors are very important type of object in R; however, there are several others which we will discuss in the subsequent subsections. We begin with the object matrices.

5.3.7.1 Matrices

We recall from Section 2.2.2 that matrices are rectangular array of real numbers. Matrices are the multidimensional generalizations of vectors. In fact, they are vectors that can be indexed by two or more indices and are printed in special ways. In R, the data elements of matrices must be of the same basic type. The following is an example of a matrix with two rows and three columns.

$$A_{2\times 3} = \begin{pmatrix} 1 & 4 & 8 \\ 0 & 4 & 9 \end{pmatrix}.$$

We reproduce a memory representation of the $A_{2\times 3}$ matrix in R with the matrix function in the next example:

```
> A = matrix(
+ c(1, 4, 8, 0, 4, 9),    # the data elements
```

```
+ nrow=2,                    # number of rows
+ ncol=3,                    # number of columns
+ byrow = TRUE)            # fill matrix by rows
> A
     [,1] [,2] [,3]
[1,]    1    4    8
[2,]    0    4    9
```

An element at the mth row and nth column of matrix A can be accessed by the expression $A[m, n]$. For example we can select the first row and third column element using the expression $A[1, 3]$ as shown below:

```
> A
     [,1] [,2] [,3]
[1,]    1    4    8
[2,]    0    4    9
> A[1,3]    # element at 1st row, 3rd column
[1] 8
```

The entire m-th row and entire nth column A can be extracted by the expression $A[m,]$ and $A[, n]$ respectively. For example

```
> A
     [,1] [,2] [,3]
[1,]    1    4    8
[2,]    0    4    9
> A[2,]  # extract the 2nd row
[1] 0 4 9
> A[,3]  # extract the 3rd column
[1] 8 9
```

We can construct the **transpose** of a matrix in R by interchanging its columns and rows with the function *t*().

```
> A
     [,1] [,2] [,3]
[1,]    1    4    8
[2,]    0    4    9
> t(A)   # transpose of A
     [,1] [,2]
[1,]    1    0
[2,]    4    4
[3,]    8    9
```

We can **deconstruct** a matrix by applying the *c* function, which combines all column vectors into one.

```
> A
     [,1] [,2] [,3]
[1,]    1    4    8
[2,]    0    4    9
> c(A)
[1] 1 0 4 4 8 9
```

5.3.7.2 List

Lists are a general form of vector in which the different elements need not be of the same type and are often themselves vectors or lists. Lists provide a convenient way to return the results of a statistical computation. For example, the following variable *x* is a list containing copies of three vectors *n*, *s*, *b*, and a numeric value 3.

```
> n = c(1, 2, 3)
> s = c("aa", "bb", "cc", "dd")
> b = c(TRUE, FALSE, TRUE, FALSE, FALSE)
> x = list(n, s, b, 3)     # x contains copies of n, s, b
```

We retrieve a list slice with the single square bracket "[]" operator. The following is a slice containing the second member of *x*, which is a copy of *s*.

```
> x[2]
[[1]]
[1] "aa" "bb" "cc" "dd"
```

With an index vector, we can retrieve a slice with multiple members. Here a slice containing the second and fourth members of x produces the following output.

```
> x[c(1, 2)]
[[1]]
[1] 1 2 3

[[2]]
[1] "aa" "bb" "cc" "dd"
```

We can assign names to list members, and reference them by names instead of numeric indexes.

For example, in the following, *v* is a list of two members, named "Christiana" and "Nadia."

```
> v = list(Christiana=c(1, 2, 3), Nadia=c("aa", "bb"))
> v
```

```
$Christiana
[1] 1 2 3

$Nadia
[1] "aa" "bb"
```

In order to reference a list member directly, we have to use the double square bracket "[[]]" operator. The following references a member of *v* by name:

```
> v[["Nadia"]]
[1] "aa" "bb"
```

A named list member can also be referenced directly with the "$" operator instead of the double square bracket operator.

```
> v$Nadia
[1] "aa" "bb"
```

We can attach a list to the R search path and access its members without explicitly mentioning the list. It should to be detached for cleanup.

```
> attach(v)
> Nadia
[1] "aa" "bb"
> detach(v)
```

5.3.7.3 Factor

A factor is a vector object used to specify a discrete classification of the components of other vectors of the same length. R provides both ordered and unordered factors. One of the most important uses of factors is in statistical modeling; since categorical variables enter into statistical models differently than continuous variables, storing data as factors insures that the modeling functions will treat such data correctly.

Factors in R are stored as a vector of integer values with a corresponding set of character values to use when the factor is displayed. The *factor*() function is used to create a factor. Both numeric and character variables can be made into factors, but a factor's levels will always be character values. You can see the possible levels for a factor through the levels command.

To change the order in which the levels will be displayed from their default sorted order, the *levels* = argument can be given a vector of all the possible values of the variable in the order you desire. If the ordering should also be used when performing comparisons, use the optional *ordered* = *TRUE* argument. In this case, the factor is known as an ordered factor. The levels of a factor are used when displaying the factor's values. The levels can be changed at the time you create a factor by passing a vector with the new values through the *labels* = argument.

To illustrate this point, consider a factor taking on integer values which we want to display as roman numerals.

```
> mydata <- c(1,2,2,3,1,2,3,3)
> fmydata <- factor(mydata)
> fmydata
[1] 1 2 2 3 1 2 3 3
Levels: 1 2 3
```

To convert the default factor fmydata to roman numerals, we use the assignment form of the levels function:

```
> levels(fmydata) = c('I','II','III')
> fmydata
[1] I   II  II  III I   II  III III
Levels: I II III
```

Factors represent a very efficient way to store character values, because each unique character value is stored only once, and the data itself is stored as a vector of integers. In the example below we use the table function to tabulate the number of months listed in the variable Months.

```
> Months = c("August","April","January","November","January",
+ "September","October","September","November","August",
+  "January","November","December","February","May","August",
+ "July","December","February","April")
> Months <- factor(Months)
> table(Months)
Months
    April  August  December  February  January  July  May  November
        2       3         2         2        3     1    1         3
  October  September
        1          2
```

Although the months clearly have an ordering, this is not reflected in the output of the table function. Additionally, comparison operators are not supported for unordered factors. Creating an ordered factor solves this issue.

```
> Months = c("August","April","January","November","January",
+ "September","October","September","November","August",
+  "January","November","December","February","May","August",
+ "July","December","February","April")
> Months <- factor(Months,levels=c("January","February","March",
+ "April","May","June","July","August","September","October","November",
  "December"),ordered=TRUE)
> table(Months)
Months
  January  February March  Apri  May  June July  August  September  October
        3         2     0     2    1     0    1       3          2        1
 November  December
        3         2
```

5.3.7.4 Data Frames

Data frames are matrix-like structures, in which the columns can be of different types. Data frames can be thought of as "data matrices" with one row per observational unit but with both numerical and categorical variables. Most experiments are best described by data frames where the treatments are categorical, but the response is numeric. Vector structures appearing as variables of data frames must all have the same length, and matrix structures must all have the same row size.

In R we can create a data frame using the *data.frame*() function. We show how to create a data frame.

In this example, *x* can be considered as a list of three components with each component having a two element vector.

```
> x <- data.frame("Index" = 1:2, "Age" = c(27,31),
  "Name" = c("John","Christie"))
> str(x)
'data.frame':  2 obs. of  3 variables:
 $ Index: int  1 2
 $ Age  : num  27 31
 $ Name : Factor w/ 2 levels "Christie","John": 2 1
```

We notice that in the third column, Name is of type factor, instead of a character vector. By default, *data.frame*() function converts character vector into factor. To suppress this behavior, we can pass the argument stringsAsFactors = FALSE.

```
> x <- data.frame("Index" = 1:2, "Age" = c(27,31),
 "Name" = c("John","Christie"), stringsAsFactors = FALSE)
> str(x)
'data.frame':  2 obs. of  3 variables:
 $ Index: int  1 2
 $ Age  : num  27 31
 $ Name : chr  "John" "Christie"
```

We can use either [], [[]] or $ operator to access columns of data frame.

```
> x["Age"]
  Age
1  27
2  31
> x[[2]]
[1] 27 31
> x$Age
[1] 27 31
```

We can also retrieve rows with a logical index vector. In the following example, for the vector *L*, the member value is TRUE if the Age is greater than 20, and FALSE if otherwise.

```
> L <- x$Age > 20
> L
[1] TRUE TRUE
```

5.3.8 Data Import

Most often, before we can start working, it is often necessary to import data into R prior to use. In this section, we will show how to import various files of data. We begin with Excel file.

5.3.8.1 Excel File

For this, we can use the function *read.xls* from the "gdata" package. It reads from an Excel spreadsheet and returns a data frame. The following shows how to load an Excel spreadsheet named "mydata.xls."

```
> library(gdata)                      # load gdata package
> help(read.xls)                      # documentation
> mydata = read.xls("mydata.xls")    # read from first sheet
```

Alternatively, we can use the function *loadWorkbook* from the XLConnect package to read the entire workbook, and then load the worksheets with readWorksheet. The XLConnect package requires Java to be preinstalled. Java is also a popular programming language, created in 1995. We present an example of how to read an excel file using the loadWorkbook from the XLConnect package.

```
> library(XLConnect)                  # load XLConnect package
> wk = loadWorkbook("mydata.xls")
> mydata= readWorksheet(wk, sheet="Sheet1")
```

5.3.8.2 CSV File

CSV is the short hand for comma separated values. Each cell inside such data file is separated by a special character, which usually is a comma, although other characters can be used as well.

The first row of the data file should contain the column names instead of the actual data. Here is a sample of the expected format.

```
Col1,Col2,Col3
10,a1,a2
20,b1,b2
30,c1,c2
```

After we copy and paste the data above in a file named "mydata.csv" with a text editor, we can read the data with the function read.csv.

```
> mydata = read.csv("mydata.csv")  # read csv file
> mydata
  Col1 Col2 Col3
1  10   a1   a2
2  20   b1   b2
3  30   c1   c2
```

5.3.8.3 Table File

A data table can reside in a text file. The cells inside the table are separated by blank characters. Here is an example of a table with three rows and three columns.

```
10  a1    a2
20  b1    b2
30  c1    c2
```

Now copy and paste the table above in a file named "mydata.txt" with a text editor. Then load the data into the workspace with the function *read.table.*

```
> mydata = read.table("mydata.txt")  # read text file
> mydata                             # print data frame
  V1   V2   V3
1  10  a1   a2
1  20  b1   b2
3  30  c1   c2
```

For further details of the function read.table, please consult the R documentation.

5.3.8.4 Minitab File

If the data file is in Minitab Portable Worksheet format, it can be opened with the function *read.mtp* from the foreign package. It returns a list of components in the Minitab worksheet.

```
> library(foreign)                 # load the foreign package
> help(read.mtp)                   # documentation
> mydata = read.mtp("mydata.mtp")  # read from .mtp file
```

5.3.8.5 SPSS File

For the data files in SPSS format, it can be opened with the function *read.spss* also from the foreign package. There is a "to.data.frame" option for choosing whether a data frame is to be returned. By default, it returns a list of components instead.

```
> library(foreign)                  # load the foreign package
> help(read.spss)                   # documentation
> mydata = read.spss("myfile", to.data.frame=TRUE)
```

We remark that the code samples above assume that the data files are located in the R **working directory**, which can be found with the function *getwd*().

```
> getwd()                # get current working directory
```

You can select a different working directory with the function *setwd*(), and thus avoid entering the full path of the data files.

5.4 Problems

1 Create a vector of coefficients for a quadratic equation, using the *sample* function. Here, we draw a sample of size 3 from $-10, -9, \dots, 9, 10$ with replacement.

2 Using the preamble in Problem 1,
(a) Determine the class of the object coeffs.
(b) Determine the length of the object coeffs.
(c) Determine the names associated with the vector.

3 Without using R, determine the result of the following computation

```
x <- c(1,2,3)
x[1]/x[2]^3-1+2*x[3]-x[2-1]
```

4 Given the logical vector

```
x <- seq(-2,2,length=10) > 0
```

(a) Negate this vector.
(b) Compute the truth table for logical *AND*.
(c) Compute the truth table for logical *OR*.

5 Generate a sample of random normal variable x, and a sample of random exponential variable y and perform the following computations:
(a) Compute the mean of x.
(b) Compute the mean of y.
(c) Compute the standard deviation of x.
(d) Compute the standard deviation of y.

(e) Compute the correlation between x and y.

6 Compound interest can be computed using the formula:

$$A = P\left(1 + \frac{r}{n}\right)^{nt},$$

where P is the principal amount, A is the accrued amount, r is the interest rate, n is the number of compoundings a year, and t is the total number of years.
Let the interest rate r be 3%, compounded monthly, and let the initial investment amount be $1250. Write R code to calculate the compound-interest, for an investment period of t years, where t changes from 1 to 10 in yearly increments.

7 Write a simple R program to display the sum of two variables: x and y.

8 Write a simple R program to sort a column of a matrix.

9 Using R, create a variable called z, assign $x + y$ to it, and display the result.

10 Write a R program that uses a for loop to print the decimal representations of $1/2, 1/3, \ldots, 1/10$, one on each line.

6

Introduction to Python

6.1 Introduction

Python is a high-level, interpreted scripting language developed in the late 1980s by Guido van Rossum at the National Research Institute for Mathematics and Computer Science in the Netherlands. It was officially released in 1991. It was developed for emphasis on code readability, and its syntax allows programmers to express concepts in fewer lines of code. Python has many uses, and mentioned below are some of the main task Python can accomplish:

- Python can handle big data and perform complex computations.
- Python can be used for production-ready software development.
- Python can be used in conjunction with other software to create workflows.
- Python can connect to database systems to read and modify files.
- Python can be used on a web server to create applications.

Just like R, Python works on several platforms including Microsoft windows system, Apple macOS system, and Linux operating system (i.e. Ubuntu, Fedora, Debian, etc.). One advantage of Python is the fact that it has syntax that allows one to write programs with fewer lines than some other programming languages. There are several versions of Python but as of 2021, the recent major version of Python is Python 3.9, which we shall be using in this book.

One unique feature of Python is that it uses new lines to complete a command, as opposed to other programming languages which often use semicolons or parentheses. In addition, Python depends on indentation, using whitespace, to define scope, such as the scope of loops, functions, and classes. Other programming languages such as C and C++ use curly brackets for this purpose.

Data Science in Theory and Practice: Techniques for Big Data Analytics and Complex Data Sets, First Edition. Maria Cristina Mariani, Osei Kofi Tweneboah, and Maria Pia Beccar-Varela.

In this chapter, we will give an introduction to Python and discuss some basics which are fundamental to some of the programs that will be used in this text. More details of Python can be found in Sweigart (2015) and references therein.

6.2 Basic Data Types

In this section, we will discuss all the basic data types built into Python. The lists of all the data types provided by Python are as follows:

- Number
  - Integer
  - Floating-point numbers
  - Complex
- Strings
- List
- Tuple
- Dictionary

We will also briefly present an overview of Python's built-in functions. These are pre-defined functions one can call to do useful things. As of now, the latest version of Python 3.9 has 69 built-in functions.

6.2.1 Number Data Type

The number data types store numeric values. Number objects are created when you assign a value to them. For example:

```
var1 = 2
var2 = 20
```

6.2.1.1 Integer

In Python, there is no limit to how long an integer value can even though it is constrained by the amount of memory your system has, but beyond that an integer can be as long as you need it to be. Python supports two types of integers:

- *int (signed integers)*: They are often called integers or ints and are positive or negative whole numbers with no decimal point. They represent positive and negative integers from -2^n to 2^{n-1}, where n is the number of bits.
- *long (long integers)*: Longs are integers of unlimited size. They are like integers and followed by an uppercase or lowercase L.

Python allows one to use a lowercase l with long, but it is recommended that we use only an uppercase L to avoid confusion with the number 1. Python displays

Table 6.1 Numbers.

int	long
10	51924361L
100	−0x19323L
−786	0122L
080	0xDEFABCECBDAECBFBAEL
0 × 69	−4721885298529L

long integers with an uppercase L. Table 6.1 displays some examples of numbers in int and long:

The underlying type of a Python integer, irrespective of the base used to specify it is called int:

```
>>> type(20)
<class 'int'>
>>> type(0x20)
<class 'int'>
```

6.2.1.2 Floating-Point Numbers

The float type in Python assigns a floating-point number. Float values are specified with a decimal point. Usually, the character *e* or *E* followed by a positive or negative integer may be appended to specify scientific notation as shown below:

```
>>> type(6.2)
<class 'float'>
>>> type(.6e7)
<class 'float'>

>>> 1.79e308
1.79e+308
>>> 1.8e308
inf
```

Almost all platforms represent Python float values as 64-bit "double-precision" values, according to the *IEEE*754 standard. In that case, the maximum value a floating-point number can be is approximately 1.8×10^{308}. Python will indicate a number greater than that by the string inf which denotes infinity as shown in the example above. The closest a nonzero number can be to zero is approximately 5.0×10^{-324}. Anything closer to zero than that is effectively zero:

```
>>> 5e-324
5e-324
>>> 1e-325
0.0
```

Floating point numbers are represented internally as binary (base-2) fractions. Most decimal fractions cannot be represented exactly as binary fractions, so in most cases, the internal representation of a floating-point number is an approximation of the actual value.

6.2.1.3 Complex Numbers

Complex numbers are specified as $<$ real part $>$ + $<$ imaginary part $>$ j. The $<$ real part $>$ and $<$ imaginary part $>$ are real numbers, and j is an in indeterminate satisfying $j^2 = -1$. For example

```
>>> 1+5j
(1+5j)
>>> type(1+5j)
<class 'complex'>
```

6.2.2 Strings

Strings are sequences of character data. The string type in Python is denoted str. Strings in Python are identified as a contiguous set of characters represented in quotation marks. String literals may be delimited using either single or double quotes. All the characters between the opening delimiter and matching closing delimiter are part of the string. This is illustrated below:

```
>>> print(" I love data science.")
 I love data science.
>>> type(" I love data science.")
<class 'str'>
```

A string in Python can contain as many characters as you wish. The only limitation is your machine's memory resources. A string can also be empty.

Subsets of strings can be taken using the slice operators ([ ] and [:]) with indexes starting at 0 in the beginning of the string and working their way from −1 at the end. −1 refers to the last character of the string.

The plus (+) sign is the string concatenation operator and the asterisk (*) is the repetition operator. For example

```
str = 'Hello Data Science!'

print(str)          # Prints complete string
```

```
print(str[0])        # Prints first character of the string
print(str[2:5])      # Prints characters starting from 3rd to 5th
print(str[2:])       # Prints string starting from 3rd character
print(str * 2)       # Prints string two times
print(str + "TEST") # Prints concatenated string
```

This will produce the following result:

```
Hello Data Science!
H
llo
llo Data Science!
Hello Data Science!Hello Data Science!
Hello Data Science!TEST
```

6.2.3 Lists

A list is a collection of objects which are ordered and changeable. The objects in a list are separated by commas and enclosed within square brackets ([]). They are the most versatile of Python's various data types. To some extent, lists are similar to arrays. The main difference between a list and an array is that all the items belonging to a list can be of different data type.

Just like strings, the values stored in a list can be accessed using the slice operators ([] and [:]) with indexes starting at 0 in the beginning of the list and working their way to end. Negative indexing means beginning from the end. For example, −1 refers to the last item, −2 refers to the second last item etc. The plus (+) sign is the list concatenation operator, and the asterisk (*) is the repetition operator. We present an example below:

```
list = [ 'abcde', 888 , 2.25, 'Maria', 90.00 ]
tinylist = [1234, 'Maria']
print(list)           # Prints complete string
print(list[0])        # Prints first character of the string
print(list[2:5])      # Prints characters starting from 3rd to 5th
print(list[2:])       # Prints string starting from 3rd character
print(tinylist * 2)       # Prints string two times
print(list + tinylist) # Prints concatenated string
```

This produces the following result:

```
['abcde', 888, 2.25, 'Maria', 90.0]
abcde
[2.25, 'Maria', 90.0]
[2.25, 'Maria', 90.0]
[1234, 'Maria', 1234, 'Maria']
['abcde', 888, 2.25, 'Maria', 90.0, 1234, 'Maria']
```

6.2.4 Tuples

Tuples is another sequence data type that is similar to a list. Tuples are collection of objects which is ordered and unchangeable. They consist of a number of values separated by commas. Unlike lists, tuples are enclosed within parentheses.

The main differences between lists and tuples are the following: Lists are enclosed in brackets i.e. [] and their elements and size can be changed, while tuples are enclosed in parentheses i.e. () and cannot be updated. Tuples can be thought of as read-only lists. For example,

```
tuple = ( 'abcde', 888 , 2.25, 'Maria', 90.00)
tinytuple = (1234, 'Maria')
print(tuple)            # Prints complete string
print(tuple[0])         # Prints first character of the string
print(tuple[2:5])       # Prints characters starting from 3rd to 5th
print(tuple[2:])        # Prints string starting from 3rd character
print(tinytuple * 2)       # Prints string two times
```

This produces the following result:

```
('abcde', 888, 2.25, 'Maria', 90.0)
abcde
(2.25, 'Maria', 90.0)
(2.25, 'Maria', 90.0)
(1234, 'Maria', 1234, 'Maria')
```

The following code is invalid with tuple, because we attempted to update a tuple, which is not allowed. Similar case is possible with lists.

```
list = [ 'abcde', 888, 2.25, 'Maria', 90.00]
tuple = ( 'abcde', 888, 2.25, 'Maria', 90.00)
tuple[3] = 3000    # Invalid syntax with tuple
list[3] = 3000     # Valid syntax with list
```

6.2.5 Dictionaries

A dictionary is a collection of objects which are unordered, changeable, and indexed. In Python, dictionaries are written with curly brackets () and values can be assigned and accessed using square braces []. They work like associative arrays and consist of key-value pairs. A dictionary key can be almost any Python type, but are usually numbers or strings. Values, on the other hand, can be any arbitrary Python object. The example below creates and prints a dictionary.

```
thisdict = {
  "brand": "Dodge",
  "model": "Stratus",
```

```
  "year": 2003
}
print(thisdict) # Prints complete dictionary
print(thisdict.keys())   # Prints all the keys
print(thisdict.values()) # Prints all the values
```

This produces the following result:

```
{'brand': 'Dodge', 'model': 'Stratus', 'year': 2003}
dict_keys(['brand', 'model', 'year'])
dict_values(['Dodge', 'Stratus', 2003])
```

6.3 Number Type Conversion

Python converts numbers internally in an expression containing mixed types to a common type for evaluation. But sometimes, you need to coerce a value explicitly from one type to another to satisfy the requirements of an operator or a function parameter. For example

- int(x) converts the number x to a plain integer.
- long(x) converts the number x to a long integer.
- float(x) converts the number x to a floating-point number.
- complex(x) converts the number x to a complex number with real part x and imaginary part zero.
- complex(x, y) converts the number x and y to a complex number with real part x and imaginary part y. x and y are numeric expressions.

6.4 Python Conditions

Python allows manipulation of logical quantities. Just like R, the elements of a logical vector in Python can have the values TRUE, FALSE, and NA. The logical operators are the following:

- < (less than)
- > (greater than)
- <= (less than or equal)
- >= (greater than or equal)
- == (equal to)
- ! = (not equal to)

The python logical operators can be used in several ways, most commonly in "if statements" and loops.

6.4.1 If Statements

In its simplest form, the if statement looks like this:

```
if <expr>:
<statement>
```

where `<expr>` is an expression that is evaluated in the Boolean context and `<statement>` represents a valid Python statement, which must be indented.

If the expression i.e. `<expr>` is true (evaluates to a value that is true), then the python statement i.e. `<statement>` is executed. If expression is false, then the python statement is skipped over and not executed.

Note that the colon (:) following `<expr>` is required. Some programming languages for example C and C++ requires `<expr>` to be enclosed in parentheses, but Python does not. We present an example of this type of if statement:

```
a = 20
b = 500
if b > a:
  print("b is greater than a")
```

This produces the following result:

```
b is greater than a
```

In this example we use two variables, *a* and *b*, which are used as part of the if statement to test whether *b* is greater than *a*. Since *a* is 20, and *b* is 500, we know that 500 is greater than 20, and so we print to screen that "*b* is greater than *a*."

We remark that Python relies on indentation (i.e. whitespace at the beginning of a line) to define scope in the code. Other programming languages such as C and C++ often use curly brackets for this purpose. An if statement without indentation will raise an error. For example,

```
a = 20
b = 500
if b > a:
print("b is greater than a") # you will get an error
```

This produces the following error:

```
print("b is greater than a")
        ^

IndentationError: expected an indented block
```

The correct code should be

```
a = 20
b = 500
if b > a:
  print("b is greater than a") # you will get an error
```

which produces the output:

```
b is greater than a
```

6.4.2 The Else and Elif Clauses

Occasionally you may want to evaluate a condition and take one path if it is true and specify an alternative path if it is not. This can be accomplished with an else clause:

```
if <expr>:
<statement(s)>
else:
<statement(s)>
```

If the expression i.e. `<expr>` is true, the first suite is executed, and the second is skipped. On the other hand, if `<expr>` is false, the first suite is skipped and the second is executed. Either way, execution then resumes after the second suite. Both suites are defined by indentation, as described above.

We present an example below:

```
a = 2000
b = 500
if b > a:
    print("b is greater than a")
else:
    print("b is not greater than a")
```

This produces the following output:

```
b is not greater than a
```

There are syntaxes for branching execution based on several alternatives. For these instances one may use single or several elif (short for else if) clauses. Python evaluates each `<expr>` in turn and executes the suite corresponding to the first that is true. If none of the expressions are true, and an else clause is specified, then its suite is executed:

```
if <expr>:
<statement(s)>
```

```
elif <expr>:
<statement(s)>
elif <expr>:
<statement(s)>
    ...
else:
<statement(s)>
```

We present an example below:

```
a = 2000
b = 2000
if b > a:
  print("b is greater than a")
elif a == b:
  print("a and b are equal")
else:
  print("a is greater than b")
```

This produces the following result:

```
a and b are equal
```

6.4.3 The While Loop

In this section, we learn how Pythons while statement is used to construct loops. With the while loop, we can execute a set of statements as long as a condition is true. The format of a while loop is

```
while <expr>:
<statement(s)>
```

Here, `<statement(s)>` represents the block of codes or instructions to be repeatedly executed, often referred to as the body of the loop. This is denoted with indentation, just as in an if statement. We present as example below:

```
i = 2
while i < 7:
  print(i)
  i += 1
```

This produces the following result:

```
2
3
4
```

```
5
6
```

The while loop requires relevant variables to be ready, in this example we need to define an indexing variable, *i*, which we set to 2.

6.4.3.1 The Break Statement

The break statement is inserted into a while loop. With the break statement we can stop the loop even if the while condition is true. For example

```
i = 2
while i < 7:
  print(i)
  if i == 4:
      break
  i += 1
```

This produces the following result:

```
2
3
4
```

6.4.3.2 The Continue Statement

With the continue statement, we can stop the current iteration, and continue with the next iteration. For example:

```
i = 1
while i < 7:
  i += 1
  if i == 4:
      continue
  print(i)
```

This produces the following result:

```
2
3
5
6
7
```

6.4.4 For Loops

A Python for loop is used for iterating over a sequence (that is either a list, a tuple, a dictionary, or a string). With the for loop, we can execute a set of statements, once for each item in a list, tuple, dictionary, etc. The format of Pythons for loop looks like this:

```
for <var> in <iterable>:
<statement(s)>
```

where `<iterable>` is a collection of objects, for example, a list or tuple. The `<statement(s)>` in the loop body are denoted by indentation, as with all Python control structures, and are executed once for each item in `<iterable>`. The loop variable `<var>` takes on the value of the next element in `<iterable>` each time through the loop. Below is an example:

```
fruits = ["orange", "apple", "pineapple"]
for x in fruits:
  print(x)
```

This produces the following output:

```
orange
apple
pineapple
```

The for loop unlike the while loop does not require an indexing variable to set beforehand.

6.4.4.1 Nested Loops

A nested loop is defined as a loop inside a loop. The "inner loop" will be executed one time for each iteration of the "outer loop." The following example prints each adjective for every fruit:

```
adjective = ["yellow", "small", "fresh"]
fruits = ["orange", "apple", "pineapple"]
for x in adjective:
    for y in fruits:
        print(x,y)
```

This produces the following result:

```
yellow orange
yellow apple
yellow pineapple
small orange
small apple
small pineapple
fresh orange
fresh apple
fresh pineapple
```

6.5 Python File Handling: Open, Read, and Close

Python provides an inbuilt function for opening, writing, and reading files.

Suppose we have the following file, saved datascience.txt located in the same folder as Python:

```
Hello! Welcome to datascience
This file is for testing purposes.
Good Luck!
```

To open the file, use the built-in open() function. The open() function returns a file object, which has a read() method for reading the content of the file. By default, the read() method returns the whole text, but you can also specify how many characters you want to return. For example you can return one line by using the readline() method. We remark that it is always a good practice to always close the file when you are done reading its content. We close the file using the close() function. Table 6.2 shows the various files mode in python.

6.6 Python Functions

Python functions are groups of related statements that perform a specific task. Functions help break programs into smaller and modular chunks. As our program grows larger and larger, functions make it more organized and manageable. One advantage of functions is the fact that it avoids repetition and makes code reusable. Basically, we can divide functions into the following two types: Built-in functions, i.e. functions that are built into Python and user-defined functions, i.e. functions defined by the users themselves.

Table 6.2 Files mode in Python.

Mode	Description
"r"	This is the default mode. It Opens file for reading
"w"	This Mode Opens file for writing. If file does not exist, it creates a new file. If file exists, it truncates it
"x"	Creates a new file. If file already exists, the operation fails
"a"	Open file in append mode. If file does not exist, it creates a new file
"t"	This is the default mode. It opens in text mode
"b"	This opens in binary mode
"+"	This will open a file for reading and writing (updating)

A Python function has the following syntax:

```
def function_name(parameters):
"""docstring"""
statement(s)
```

Here, the keyword def marks the start of function header. The parameters defines arguments through which we pass values to a function. The parameters are optional. The string(docstring) is used to explain in brief, what a function does. One or more valid python statements make up the function body. Statements must have same indentation level. Below is an example of a function in Python:

```
def greet(name):
"""This function greets
the person passed in as
parameter"""
print("Hello, " + name + ". Good evening!")
```

6.6.1 Calling a Function in Python

After defining a function, we can call it from another function, program, or even the Python prompt. To call a function, we simply type the function name with appropriate parameters. For example,

```
greet('Maria')
```

This produces the following output:

```
Hello, Maria. Good evening!
```

6.6.2 Scope and Lifetime of Variables

The scope of a variable is the portion of a program where the variable is recognized. Parameters and variables defined inside a function is not visible from outside. Hence, they are known as a local scope. Global scope refers to variables which can be accessed inside or outside of the function.

The lifetime of a variable is the period throughout which the variable exits in memory. The lifespan of variables inside a function is as long as the function executes. They are immediately destroyed once we return from the function. Thus a function does not remember the value of a variable from its previous calls.

Here is an example to illustrate the scope of a variable inside a function.

```
def my_func():
x = 40
```

```
print("Value inside function:",x)

x = 90
my_func()
print("Value outside function:",x)
```

This produces the following output:

```
Value inside function: 40
Value outside function: 90
```

Here, we can see that the value of x is 90 initially. Even though the function `my_func()` changed the value of x to 40, it did not affect the value outside the function. This is because the variable x inside the function is different (local to the function) from the one outside. Although they have the same names, they are two different variables with different scopes.

6.7 Problems

1 Using Python create a variable named carname and assign the value Toyota to it.

2 Write a simple Python program to display the sum of two variables: x and y.

3 Write a simple Python program to display the percentage of missing values of rows and columns, respectively.

4 Using Python, create a variable called z, assign $x + y$ to it, and display the result.

5 Write a simple Python program to estimate the coefficients of linear regression.

6 Write correct and efficient Python definitions for the functions given below.
(a) $2x^5 5x^3 + 1.9x^2 7.1$.
(b) $g(x, y) = 3x^2 - 2y^2 + 8$.
(c) $x\sqrt{\frac{x-3}{x+2}}$.

7 Write a Python program that uses a for loop to print the decimal representations of $1/2, 1/3, \ldots, 1/10$, one on each line.

8 Without using Python, determine the result of the following computation:

(a)
```
num = 12
while num > 4:
    print(num)
    num = num - 1
```

(b)
```
count = 0
for letter in 'Snow!':
    print('Letter #', count, 'is', letter)
    count += 1
```

9 Using Python generate a sample of random normal deviates x, and a sample of random exponential deviates y and perform the following computations:
(a) Compute the mean of x.
(b) Compute the mean of y.
(c) Compute the standard deviation of x.
(d) Compute the standard deviation of y.
(e) Compute the correlation between x and y.

10 Compound interest can be computed using the formula:

$$A = P\left(1 + \frac{r}{n}\right)^{nt},$$

where P is the principal amount, A is the accrued amount, r is the interest rate, n is the number of compoundings a year, and t is the total number of years.
Let the interest rate r be 4%, compounded monthly, and let the initial investment amount be $1350. Write a Python program to calculate the compound interest, for an investment period of t years, where t changes from 1 to 10 in yearly increments.

7

Algorithms

7.1 Introduction

An algorithm is a sequence of instructions, typically to solve a class of problems or perform a computation. Algorithms are specifications for performing calculation, data processing, automated reasoning, and other tasks. Algorithms help us complete tasks and, when they are written in code, are what make our computers work. Algorithms are different from computer codes. Algorithms are systematic logical approach, which is a well-defined, step-by-step procedure that allows a computer to solve a problem. A computer code or program code is the set of instructions forming a computer program, which is executed by a computer. It uses short phrases to write code for a program before it is implemented in a specific programming language. A code is a series of steps that machines can execute.

In this chapter, we will identify examples of algorithmic thinking and write simple algorithms in pseudocode.

7.2 Algorithm – Definition

Algorithm is a procedure that describes the exact steps needed for the computer to solve a problem or reach a goal. It describes how to do something, and your computer will do it exactly that way every time. The ingredients are the inputs, and the results are the outputs. Algorithms can be expressed using natural language, flowcharts, etc. The following are some common usage of algorithms:

1. It is used for data processing, calculation, and other related computer and mathematical operations.
2. It is used to manipulate data in various ways, such as inserting a new data item, searching for a particular item, or sorting an item, etc.

Data Science in Theory and Practice: Techniques for Big Data Analytics and Complex Data Sets, First Edition. Maria Cristina Mariani, Osei Kofi Tweneboah, and Maria Pia Beccar-Varela.

Not all procedures can be called an algorithm. To have a good algorithm, it should exhibit the following characteristics:

1. Input and output should be defined precisely.
2. Each steps in algorithm should be clear and defined.
3. Algorithm should be most effective among many different ways to solve a problem.
4. An algorithm should not have program code. Instead, the algorithm should be written in such a way that, it can be used in similar programming languages.
5. Algorithms must terminate after a finite number of steps.
6. Algorithms should be achievable with the available resources. It should not contain any unnecessary or redundant steps.

7.3 How to Write an Algorithm

When writing algorithms, we do not have well-defined standards. However, it is problem and resource dependent, i.e. algorithms should be accomplished with the available resources. Algorithms are not implemented to support a particular programming code. As mentioned in Section 7.2, they should be written in such a way that, it can be used in similar programming languages.

All programming languages share basic code constructs such as loops (do, for, while), flow-control (if-else), etc. These common constructs can be used to write an algorithm.

Algorithm writing is a process and is implemented after the problem domain is well-defined. That is, we should have a good knowledge of the problem domain, for which we are creating a solution. Algorithms tell the programmers the steps to code the program. A computer programmer, sometimes called more recently a coder is a person who writes code or program for many kinds of computer software.

Let us learn how to write an algorithm by using an example. In this example, we will design an algorithm to add two numbers entered by the user and displays the result.

Example 7.1

```
Step 1: Start
Step 2: Declare variables number1, number2 and sum.
Step 3: Read values number1 and number2.
Step 4: Add number1 and number2 and assign the result to sum.
          sum <-- number1 + number2
Step 5: Display sum
Step 6: Stop
```

In design and analysis of algorithms, Example 7.1 is used to describe an algorithm. It makes it easy for the analyst to analyze the algorithm ignoring all undesired definitions. The analyst can also observe what operations are being used and how the process is flowing. In general, writing the step numbers is optional. Algorithms are designed to get a solution to a specific problem. However, a problem can be solved in more than one ways. In fact several solution algorithms can be derived for the same given problem. The subsequent step is to analyze those proposed solution algorithms and implement the best suitable solution. This explains the characteristics of an algorithm mentioned in the previous section, which stated that an algorithm should be most effective among many different ways to solve a problem.

7.3.1 Algorithm Analysis

Algorithmic **efficiency** is the characteristics of an algorithm which relates to the number of computational resources used by the algorithm. An algorithm must be analyzed to determine its resource usage, and the efficiency of an algorithm can be quantified based on usage of different resources. For greatest efficiency, we have to reduce resource usage. However, different resources such as time and space complexity cannot be compared directly. It's efficiency usually depends on which measure of efficiency is considered most important.

Efficiency of an algorithm can be analyzed at two different stages, before implementation and after implementation. They are the following:

- *A priori analysis*: This refers to the theoretical analysis of an algorithm. Efficiency of an algorithm is quantified by assuming that all other factors, such as, processor speed, are constant and have no effect on the execution.
- *A posterior analysis*: This refers to the empirical analysis of an algorithm. The selected algorithm is implemented using programming language. This is then implemented on target computer machine. In this analysis, real statistics for example running time and space required are collected.

In general, algorithm analysis deals with the execution or running time of various operations involved. The running time of an operation can be defined as the number of computer instructions executed per operation.

7.3.2 Algorithm Complexity

Suppose **A** is an algorithm and m is the size of input data, the time and space used by the algorithm **A** are the two main factors, which decide the efficiency of **A**.

- *Time factor*: Time is measured by counting the number of key operations such as comparisons in the sorting algorithm. This an algorithm to sort items in a certain order.
- *Space factor*: The space factor is quantified by counting the maximum memory space required by the algorithm.

The complexity of an algorithm $f(m)$ shows the running time and/or the storage space needed by the algorithm in terms of m as the size of input data.

7.3.3 Space Complexity

Space complexity of an algorithm describes the amount of memory space needed by the algorithm in its life cycle. That means how much memory, in the worst case, is needed at any point in the algorithm. The space required by an algorithm is equal to the sum of the following two components i.e. the fixed part and variable part.

- A fixed part is the space required to store certain data and variables, that are independent of the size of the problem. For example, simple variables and constants used, program size, etc.
- A variable part is the space required by variables whose size depends on the size of the problem. For example dynamic memory allocation. There are two ways that memory gets allocated for data storage:
  - *Compile time (or static) allocation*: Memory for named variables is allocated by the compiler. A compiler is a computer program that translates computer code written in one programming language into another language that is understood by the computer.
  - *Dynamic memory allocation*: Allocate storage space while the program is running.

The space complexity $S(A)$ of any algorithm A is $S(A) = C + S(I)$, where C is the fixed part and $S(I)$ is the variable part of the algorithm, which depends on instance characteristic I. We illustrate this with an example:

```
Algorithm: Sum(a, b)
Step 1: Start
Step 2: c <-- a + b + 3
Step 3: Stop
```

In this example, we have three variables a, b, and c and one constant, i.e. 3. Hence,the space complexity $S(A) = 1 + 3$. Space depends on data types of the given variables and constant types.

7.3.4 Time Complexity

Time complexity of an algorithm represents the amount of time required by the algorithm to run to completion. Time requirements can be defined as a numerical function $T(n)$, where $T(n)$ can be measured as the number of steps, provided each step consumes constant time.

For example, addition of two n-bit integers takes n steps. Consequently, the total computational time is $T(n) = c * n$, where c is the time taken for the addition of two bits. Here, we observe that $T(n)$ grows linearly as the input size increases.

7.4 Asymptotic Analysis of an Algorithm

We recall in Section 7.3 that many solution algorithms can be derived for a given problem. Therefore, the question that normally arises is given two algorithms for example, algorithm A and algorithm B for a task, how do we find out which one is better? One solution is to implement both algorithms and run the two codes on your computer for different inputs and see which one takes less time. There are two drawbacks with this approach for analysis of algorithms.

1. It might be possible that for some inputs, algorithm A performs better than the algorithm B and vice versa.
2. It might also be possible that for some inputs, algorithm A performs better on one machine and algorithm B works better on other machine for some other inputs.

Asymptotic analysis handles the above issues when analyzing algorithms.

Usually, the time required by an algorithm falls under three types namely; best case scenario, average case scenario and worst case scenario. We explain all the three cases below:

- *Best case*: This is the minimum amount of time required for program execution.
- *Average case*: This is the average amount of time required for program execution.
- *Worst case*: This is the maximum amount of time required for program execution.

Using asymptotic analysis, we conclude the best case, average case, and worst case scenario of an algorithm. In asymptotic analysis, we calculate the performance of an algorithm in terms of its input size. We evaluate how the time (or space) taken by an algorithm increases with the input size.

In fact asymptotic analysis is input bound, i.e. if there is no input to the algorithm, it is concluded to work in a constant time. We remark that other than the "input," all other factors are considered constant.

The asymptotic analysis of an algorithm defines the mathematical framing of its run-time performance in mathematical units of computation. For example, the running time of one operation is computed as $f(n)$ and for another operation it is computed as $g(n^2)$. This means the first operation running time will increase linearly with an increase in n and the running time of the second operation will increase quadratically when n increases. Similarly, the running time of both operations will be nearly the same if n is significantly small.

7.4.1 Asymptotic Notations

The commonly used asymptotic notations to calculate the running time complexity of an algorithm are as follows: big O notation, Ω notation, and Θ notation. We now explain each in more detail.

7.4.1.1 Big O Notation

The notation $O(n)$ is the formal way to express the upper bound of an algorithm's running time. It measures the worst case time complexity or the longest amount of time an algorithm can possibly take to execute.

If the running time is $O(f(n))$, then for large n, the running time is at most $c \cdot f(n)$ for some constant c. This reasoning is shown in Figure 7.1.

7.4.1.2 The Omega Notation, Ω

The notation $\Omega(n)$ is the formal way to express the lower bound of an algorithm's running time. It measures the best case time complexity or the best amount of time an algorithm can possibly take to execute.

If the running time is $\Omega(f(n))$, then for large n, the running time is at least $c \cdot f(n)$ for some constant c. This is shown in Figure 7.2.

7.4.1.3 The Θ Notation

The Θ notation is the formal way to express both the lower bound and the upper bound of an algorithm's running time.

If the running time is $\Theta(f(n))$, it implies that once n gets large enough, the running time is at least $c_1 \cdot f(n)$ and at most $c_2 \cdot f(n)$ for some constants c_1 and c_2. This is shown in Figure 7.3. Table 7.1 list some common asymptotic notations.

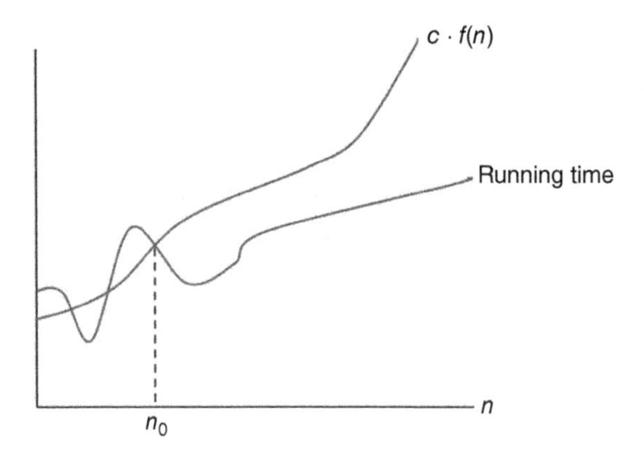

Figure 7.1 The big O notation.

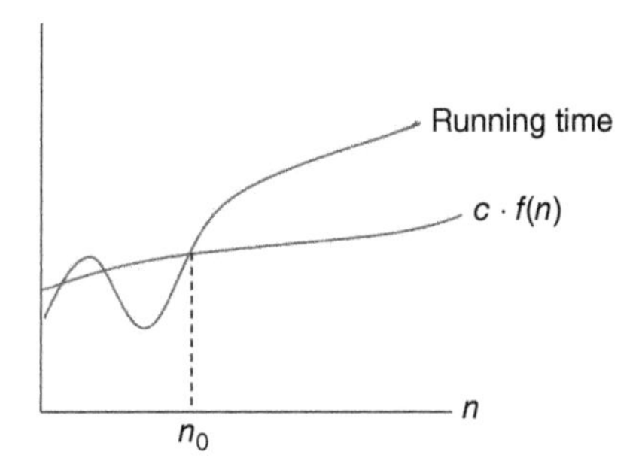

Figure 7.2 The Ω notation.

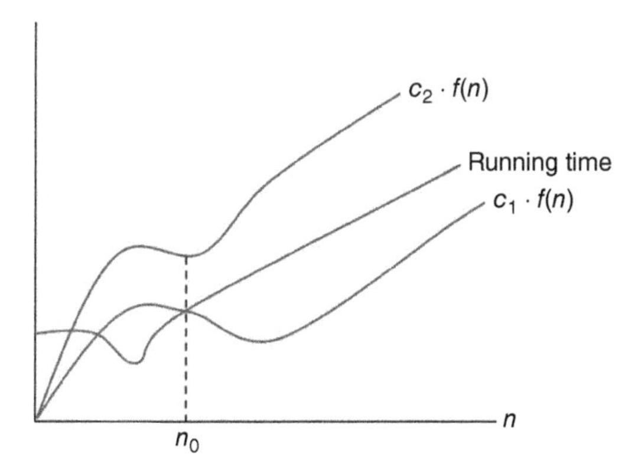

Figure 7.3 The Θ notation.

Table 7.1 Common asymptotic notations.

Name	Notation
Constant	O()
Logarithmic	$O(\log n)$
Linear	$O(n)$
$n \log n$	$O(n \log n)$
Quadratic	$O(n^2)$
Cubic	$O(n^3)$
Exponential	$2^{O(n)}$

7.5 Examples of Algorithms

Next we present two examples of algorithms. In the first example, we write an algorithm to find all roots of a quadratic equation $ax^2 + bx + c = 0$:

```
Step 1: Start
Step 2: Declare variables a, b, c, D, r1, r2, real_p and imaginary_p;
Step 3: Calculate discriminant
          D <-- b^2-4ac
Step 4: If D >= 0
               r1 <-- (-b+ D)/2a
               r2 <-- (-b- D)/2a
               Display r1 and r2 as roots.
        Else
               Calculate real part and imaginary part
               real_p <-- b/2a
               imaginary_p <--  (-D)/2a
               Display real_p +i(imaginary_p) and real_p - i(imaginary_p)  as
                  roots
Step 5: Stop
```

In the second example, we design an algorithm to find the largest value in an array:

```
Step 1: Start
Step 2: Take an array A and define its values
Step 3: Declare variable large_num
Step 4: large_num <-- 0
Step 5: Loop for each value of A
Step 6: If A[n] > large_num,
large_num <--  A[n]
Step 7: After loop finishes,
Display large_num
Step 8: Stop
```

We remark that algorithm is not a computer code. Algorithms are the instructions which gives idea on how to write the computer code.

7.6 Flowchart

Flowchart is a diagrammatic representation of an algorithm. Flowchart is very helpful in writing program and explaining program to others. Different symbols are used for different states in flowchart. For example: Input/Output and decision-making has different symbols. Figure 7.4 describes all the symbols that are used in making flowcharts:

Symbol	Purpose	Description
 →	Flow line	Used to indicate the flow of logic by connecting symbols
	Terminal(Stop/Start)	Used to represent start and end of flowchart
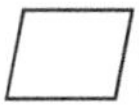	Input/Output	Used for input and output operation
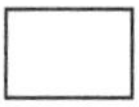	Processing	Used for arithmetic operations and data-manipulations
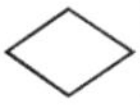	Decision	Used to represent the operation in which there are two alternatives, true and false
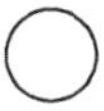	On-page connector	Used to join different flowline
	Off-page connector	Used to connect flowchart portion on different page
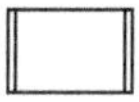	Predefined process/Function	Used to represent a group of statements performing one processing task

Figure 7.4 Symbols used in flowchart.

In Figures 7.5 and 7.6, we draw the flowchart to add two numbers entered by the user and to find all roots of a quadratic equation $ax^2 + bx + c = 0$ respectively.

7.7 Problems

1 Given that $f(x) = 2x^4 - 3x^2 + 2$, simplify $f(x)$ using the big O notation to describe its growth rate as x approaches infinity.

2 Suppose $f(n) = 2x^2 + 200x + 100$. Using the Θ notation, what is the running time?

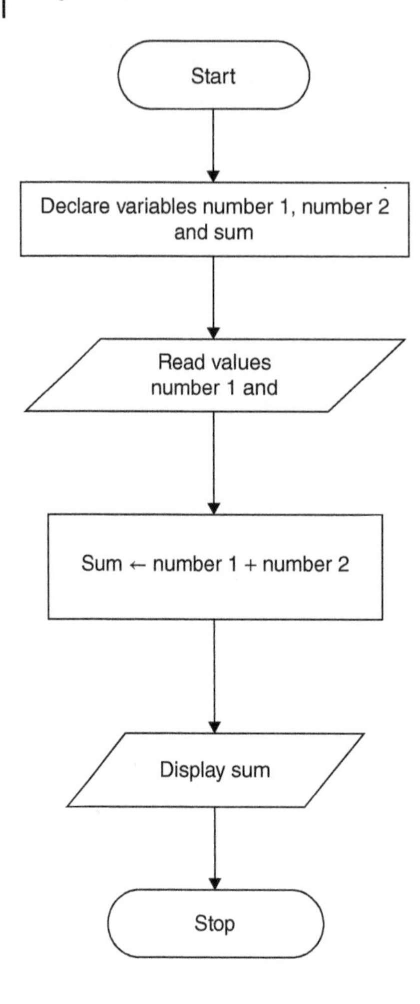

Figure 7.5 Flowchart to add two numbers entered by user.

3 Write an algorithm and draw a flowchart to find the largest of two numbers.

4 Design an algorithm and draw a flowchart to find even numbers between 1 and 100.

5 Using flowcharts, write an algorithm to read 100 numbers and then display the sum and average.

6 Write an algorithm and draw a flowchart to find the prime numbers between 1 and 100.

7 Discuss four some advantages of using flowcharts.

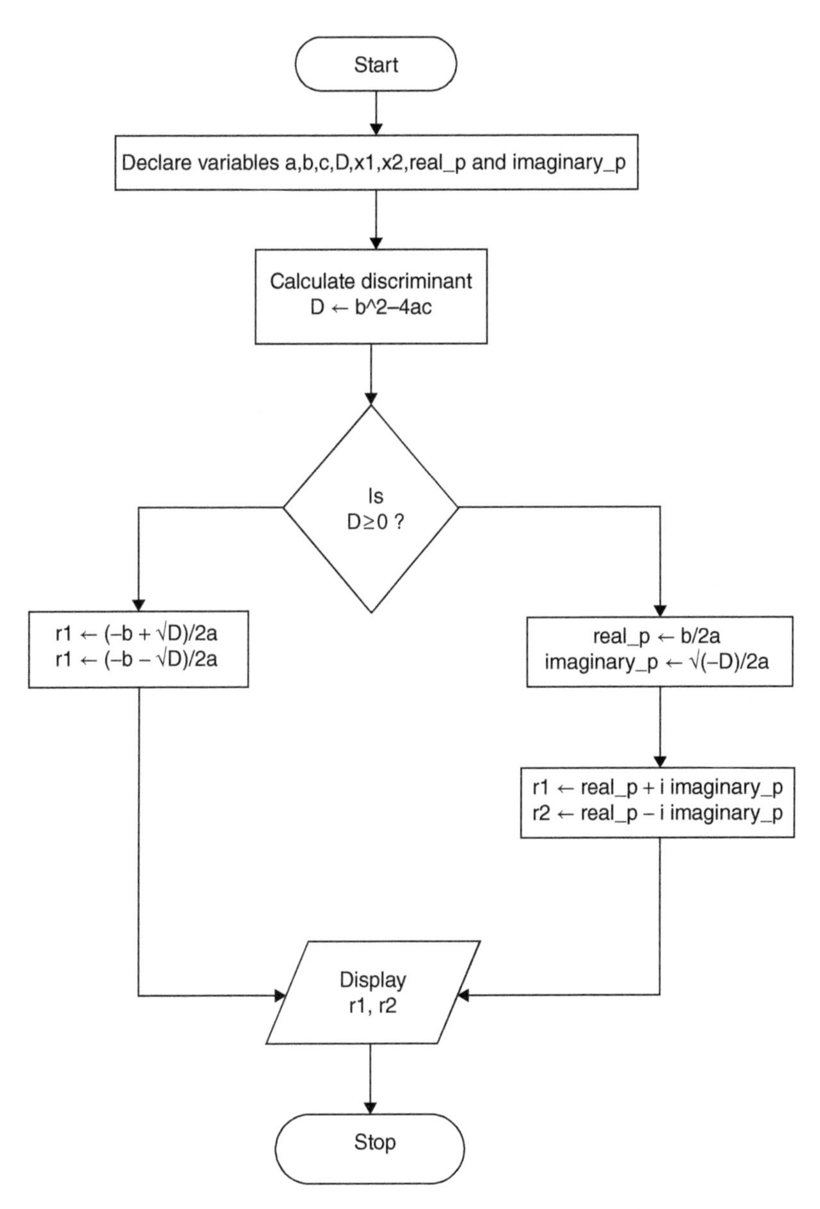

Figure 7.6 Flowchart to find all roots of a quadratic equation $ax^2 + bx + c = 0$.

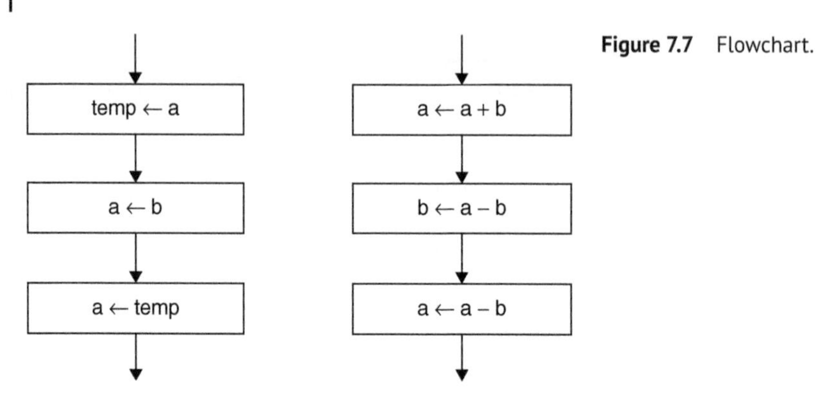

Figure 7.7 Flowchart.

8 Discuss the difference between an algorithm and a pseudocode.

9 Discuss the difference between an algorithm and a computer program.

10 Trace each of the following flowcharts (Figure 7.7) using the initial values $a = 10$ and $b = 6$. What useful computation are both of these flowcharts performing?

8

Data Preprocessing and Data Validations

8.1 Introduction

In general, real-world data is always incomplete and hence cannot be processed by a model unless the raw data has been transformed into an understandable format. That is why we need to preprocess and validate data before sending through a model. In this chapter, we will discuss various techniques for preprocessing and validating data before applying any technique to analyze and describe it. We will begin this chapter with a discussion on data preprocessing.

8.2 Definition – Data Preprocessing

Data preprocessing is a technique that involves transforming raw data into an understandable format. After the transformation, the preprocess data can be sent through a model. Real-world data is generally incomplete, i.e. lacking attribute values, lacking certain attributes of interest, or containing only aggregate data, inconsistent, i.e. containing discrepancies in names, noisy, i.e. containing errors or outliers and/or lacking in certain behaviors or trends, and is likely to contain many errors. Data preprocessing is a proven method for resolving such issues. There are several steps in data preprocessing. We outline them below and explain each in more details in the subsequent sections. The tasks in data preprocessing are as follows:

- *Data cleaning*: Fill in missing values, smooth noisy data, identify or remove outliers, and resolve inconsistencies.
- *Data transformation*: Normalization.
- *Data reduction*: Reducing the volume but producing the same or similar analytical results.

Data Science in Theory and Practice: Techniques for Big Data Analytics and Complex Data Sets, First Edition. Maria Cristina Mariani, Osei Kofi Tweneboah, and Maria Pia Beccar-Varela.

8.3 Data Cleaning

Data cleaning is defined as the process of identifying and correcting erroneous entries from a dataframe, table, or database and refers to detecting incomplete, or irrelevant parts of the data and then replacing, modifying, or deleting that data. After cleaning, a data set should be consistent with other similar data sets in the system. The steps and techniques for data cleaning will vary from dataset to dataset. However, the approach outlined in this book is applicable to most datasets.

8.3.1 Handling Missing Data

Most algorithms do not accept missing values and so we need to find a systematic way to handle such instances. Missing data can produce biased results, leading to invalid conclusions. Please see Kang (2013) for a comprehensive discussions of the prevention and handling of the missing data.

Missing data (or missing values) is defined as the data value that is not stored for a variable in the observation of interest. As mentioned earlier, missing data can have a significant effect on the conclusions that can be drawn from the data. Thus, there is the need to handle missing data before analyzing the data.

8.3.2 Types of Missing Data

In general, there are three types of missing data according to the mechanisms of missingness namely, missing completely at random (MCAR), missing at random (MAR), and missing not at random (MNAR). We explain each in more details.

8.3.2.1 Missing Completely at Random

A data set is said to be Missing Completely at Random (MCAR) if the probability that the data are missing is not related to either the specific value which is supposed to be obtained or the set of observed responses. If data are missing by design because of an equipment failure or because the samples are lost in transit or technically unsatisfactory, such data are regarded as being MCAR. The advantage of data that are MCAR is that the results from the analysis remain unbiased.

8.3.2.2 Missing at Random

A data set is said to be Missing at Random (MAR) if the probability that the responses are missing depends on the set of observed responses, but is not related to the specific missing values which are expected to be obtained.

In most cases, we may assume that MAR does not present a problem since its produced at random therefore it is unbiased. However, MAR does not mean that the missing data can be ignored. If a missing data is MAR, we may expect that the probability of the missing data in each case is conditionally independent of the variable, which is obtained currently and expected to be obtained in the future, given the history of the obtained variable prior to that case.

8.3.2.3 Missing Not at Random

Missing Not at Random (MNAR) is when the missing data does not follow the characteristics of the MCAR or MAR. In fact the cases of MNAR data are serious. The only way to obtain good results is to model the missing data. The model may then be incorporated into a more complex one for estimating the missing values.

8.3.3 Techniques for Handling the Missing Data

The best solution to handling missing data is to prevent it from occuring by well-planning the study and using effective methods for the data collection. We discuss below a systematic approach for handling missing data.

8.3.3.1 Listwise Deletion

The commonly used approach for handling missing data is to simply delete those instances with the missing data and analyze the remaining data. This approach is known as the complete case analysis or listwise deletion. Listwise deletion is the most frequently used method for handling missing data, and thus has become the default option for analysis in most statistical software packages. Some practitioners and researchers insist that it may introduce bias in the estimation of the parameters. However, if the assumption of MCAR is satisfied, a listwise deletion is known to produce unbiased estimates and conservative results.

If the sample size is large enough, and the assumption of MCAR is satisfied, the listwise deletion may be a reasonable strategy. However, when there is no large sample, or the assumption of MCAR is not satisfied, the listwise deletion is not the optimal strategy.

8.3.3.2 Pairwise Deletion

Pairwise deletion removes information only when the datapoint needed to test a particular assumption is missing. If there are missing data elsewhere in the data set, the existing values are used in the statistical testing. Since a pairwise deletion uses all information observed, it preserves more information than the listwise deletion, which may delete the case with any missing data.

In general pairwise deletion is known to be less biased for the MCAR or MAR data, and the appropriate mechanisms are included as covariates.

8.3.3.3 Mean Substitution

Mean substitution is when the mean value of a variable is used in place of the missing data value for that same variable. This helps to utilize the collected data in an incomplete dataset. The theoretical background of the mean substitution is that the mean is a reasonable estimate for a randomly selected observation from a normal distribution. However, if the missing values are not strictly random, especially in the case that the distribution of the data is highly skewed, the mean substitution method may not be appropriate and will lead to inconsistent bias. In this case, the median substitution will be more appropriate. Median substitution is when the median value of a variable is used in place of the missing data value of the same variable.

8.3.3.4 Regression Imputation

Imputation is the process of replacing the missing data with estimated values. Instead of deleting any case that has any missing value, this approach preserves all cases by replacing the missing data with a probable value estimated by other available information. After all missing values have been replaced by this approach, the data set is analyzed using the standard techniques for a complete data.

In regression imputation, the existing variables are used to make a prediction, and then the predicted value is substituted as if an actual obtained value. This technique has a number of advantages because the imputation retains a great deal of data over the listwise or pairwise deletion and avoids significantly altering the standard deviation or the shape of the distribution. However, as in a mean or median substitution, while a regression imputation substitutes a value that is predicted from other variables, no novel information is added, while the sample size has been increased and the standard error is reduced.

8.3.3.5 Multiple Imputation

Multiple imputation is another technique for handling missing data. In a multiple imputation, instead of substituting a single value for each missing data, the missing values are replaced with a set of reasonable values which contain the variations of the correct values. This technique begins with a prediction of the missing data using the existing data from other variables. The missing values are then replaced with the predicted values, and a full data set called the imputed data set is created.

An advantage of multiple imputation is that in addition to restoring the natural variability of the missing values, it incorporates the uncertainty due to the missing data, which results in a valid statistical inference. Restoring the variations and uncertainties of the missing data can be achieved by replacing the missing data with the imputed values which are forecasted using the variables correlated with the missing data. Please refer to Kang (2013) for more details of the multiple imputation method for imputing data.

8.3.4 Identifying Outliers and Noisy Data

Noise is defined as a random error or variance in a measured variable. Given a numeric attribute such as age, height or weight, how can we "smooth" out the data to remove the noise? We present below a technique for smoothing out noise.

8.3.4.1 Binning

Binning methods smooths out a sorted data value by consulting the values around it. The sorted values are distributed into a number of "buckets," or bins. Because binning methods take into account the neighborhood of values, they perform local smoothing. The smoothing can be performed by bin means, bin median, or bin boundaries.

We can also correct inconsistent data by using domain knowledge or expert decision. Other basic statistical techniques such as box plots can be used to identify outliers, which may represent noise. We briefly explain the box plot and present an example.

8.3.4.2 Box and Whisker plot

The box and whisker plot (box plot) displays the distribution of data based on the five number summary: minimum, first quartile, median, third quartile, and maximum. In the simplest box plot, the central rectangle spans the first quartile to the third quartile (the interquartile range or IQR). A segment inside the rectangle shows the median, and "whiskers" above and below the box show the locations of the minimum and maximum. The above descriptions are shown in Figure 8.1.

The five-number summary divides the data into sections that each contain approximately 25%, percent of the data in that set. This simplest possible box plot

Figure 8.1 The box plot.

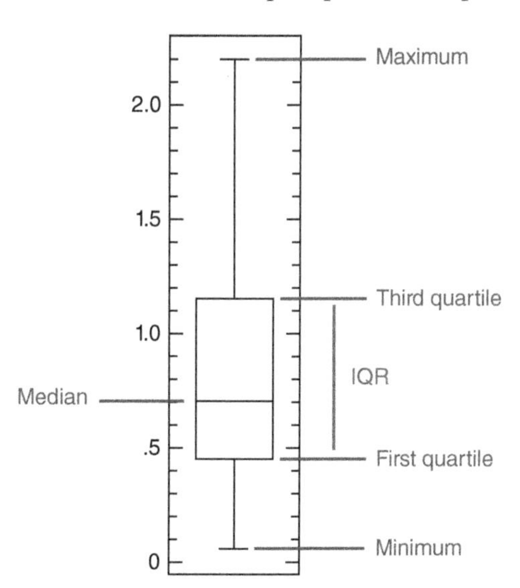

displays the full range of variation, the likely range of variation (i.e. the IQR), and the median. Most often, real datasets will display surprisingly high maximums or surprisingly low minimums called outliers. An outlier is an observation that lies outside the overall pattern of a distribution. Usually, the presence of an outlier indicates some sort of problem. This can be a case which does not fit the model under study, or an error in measurement. A convenient definition of an outlier is a point which falls more than 1.5 times the IQR above the third quartile or below the first quartile.

We present an example as follows:

Example 8.1 Given the following data sets, determine the outliers if any exist. Let the data range be 199, 201, 236, 250, 271, 278, 283, 287, 301, 303, and 341. We begin by computing the Median.

$$\text{Lower quartile (Q1)} = \frac{1}{4}(11 + 1)\text{th term= third term,}$$
$$\text{Q1} = 236,$$
$$\text{Median (Q2)} = \frac{1}{2}(11 + 1)\text{th term= sixth term,}$$
$$\text{Q2} = 278,$$
$$\text{Upper quartile (Q3)} = \frac{3}{4}(11 + 1)\text{th term= ninth term,}$$
$$\text{Q3} = 301,$$
$$\text{Inter quartile range (IQR)} = Q3 - Q1 = 301 - 278,$$
$$\text{IQR} = 23,$$
$$\text{Lower limit} = Q1 - 1.5\text{IQR} = 236 - 1.5(23),$$
$$\text{Lower limit} = 201.5,$$
$$\text{Upper limit} = Q3 + 1.5\text{IQR} = 301 + 1.5(23),$$
$$\text{Upper limit} = 335.5.$$

Hence, it is clear that any data point lying above 333.5 or below 201.5 is an outlier. Therefore, the outliers in the above data sets are 199, 201, and 341. The box plot of Example 8.1 is presented in Figure 8.2.

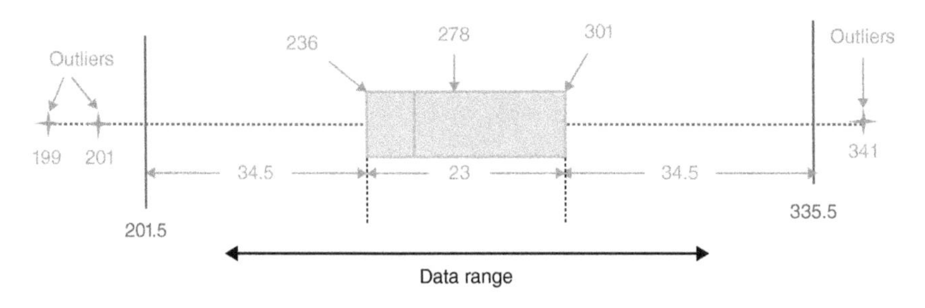

Figure 8.2 Box plot example.

8.4 Data Transformations

Many algorithms attempt to find trends in the data by comparing features of data points. However, there is an issue when the features are on different scales. A way to overcome this issue is by transforming the data. Data transformation is the process of converting data from one format or structure into another format or structure. Data transformation can be simple or complex based on the required changes to the data between the source (initial) data and the target (final) data. The goal of normalization is to make every datapoint have the same scale so each feature is equally important. In this section, we describe two normalization techniques namely, min–max normalization and Z-score normalization.

8.4.1 Min–Max Normalization

The min–max normalization is one of the popular ways to normalize data. For every in the data set to be normalized, the minimum value of that data set gets transformed into a zero and the maximum value gets transformed into a one, and every other value gets transformed into a decimal between zero and one. The formula for the min–max normalization is given as:

$$v' = \frac{v - \min(A)}{\max(A) - \min(A)}(\text{newmax}(A) - \text{newmin}(A)) + \text{newmin}(A), \quad (8.1)$$

where v' is the new value of each entry in data, v is the old value of each entry in data, A is the attribute data, $\min(A)$ and $\max(A)$ are the minimum and maximum absolute value of A, respectively, and newmax(A), newmin(A) is the max and min value of the range (i.e. boundary value of range required), respectively. We present an example below.

Suppose that the minimum and maximum values of the future income of Mogital analytics are \$10 000 and \$95 000, respectively. We would like to map income to the range [0.0, 1.0]. By using the min–max normalization, a value of \$75 000 for income is transformed as follows:

$$v' = \frac{75\,000 - 10\,000}{95\,000 - 10\,000}(1.0 - 0.0) + 0.0 = 0.765.$$

This means that on the range [0.0, 1.0], \$75 000 is equivalent to 0.765.

8.4.2 *Z*-score Normalization

Z-score normalization is another technique for normalizing data. The data values are normalized based on mean and standard deviation of the data A. The formula used is:

$$v' = \frac{v - \bar{A}}{\sigma_A}, \quad (8.2)$$

where v', v are the new and old entry of the data, respectively, $\bar{A}$ is the mean of A and σ_A is the standard deviation of A. If a value is exactly equal to the mean of all the values of the feature, it will be normalized to 0. If it is below the mean, it will be a negative number, and if it is above the mean, it will be a positive number. The size of those negative and positive numbers is determined by the standard deviation of the original feature. If the unnormalized data had a large standard deviation, the normalized values will be closer to 0. We present an example below.

Suppose that the mean and standard deviation of the values for the future income of Mogital analytics are \$54 000 and \$16 000, respectively. With Z-score normalization, a value of \$75 000 for income is transformed to 1.3125 as illustrated below:

$$v' = \frac{75\,000 - 54\,000}{16\,000} = 1.3125.$$

8.5 Data Reduction

Data reduction is a technique that can be applied to obtain a reduced representation of a data set that is much smaller in volume compared to the original dataset but still contains critical information. We present below some data reduction strategies:

1. *Sampling*: Selecting a subset of the data objects to be analyzed.
2. *Feature selection*: Selecting a subset of the features to be analyzed.
3. *Dimensionality reduction*: Creating new features that are a combination of the old features.

Sampling is commonly used approach for selecting a subset of the data to be analyzed. It is typically used because it is too expensive or time-consuming to process all the data. The key idea in sampling is to obtain a representative sample of the data. Some sampling techniques include the following:

- *Systematic sampling*: Selecting instances from an ordered sampling window.
- *Simple random sampling*: Every possible sample of the same size has the same chance of being selected.
- *Stratified random sampling*: Dividing data into groups (strata) and select a random sample from each group.
- *Cluster sampling*: Dividing data into groups (clusters) and select all of the members in one or more, but not all, of the clusters.

For feature selection, we select a minimal set of features such that the probability distribution of the class is close to the one obtained by all the features. A good feature vector is defined by its capacity to discriminate between examples from different classes. For example maximize the inter-class separation and minimize the intra-class separation.

We will discuss more about data reduction techniques in Chapter 10.

8.6 Data Validations

Data validation is the process of ensuring that data have undergone data cleansing. It checks the accuracy and quality of data, and it is typically performed prior to importing and processing. Data validation ensures that data to be processed or analyzed is complete (i.e. no blank or null values), unique (i.e. contains distinct values that are not duplicated), and the range of values is consistent with what you expect. Very often, data validation is used as a part of processes such as ETL (Extract, Transform, and Load) where you move data from a source database to a target data warehouse so that you can join it with other data for analysis. A database is an organized collection of data, generally stored and accessed electronically from a computer system. A data warehouse is a system used for reporting and data analysis. Data validation helps ensure that when you perform any analysis, your results are accurate. We outline below the steps to data validation.

Step 1. Determining data sample: If you have a large volume of data, you will probably want to validate a sample of your data rather than the entire set. The researcher or practitioner needs to decide what volume of data to sample, and what error rate is acceptable to ensure the success of the project.

Step 2. Validating the database: Before transferring the data to the warehouse, we need to ensure that all the required data is present in the existing database. In addition, we need to determine the number of records and unique IDs, and compare the source and target data fields.

Step 3. Validating the data format: Determine the overall health of the data and the changes that will be required of the source data to match the types in the target. Then, search for incomplete data counts, duplicate data, incorrect formats, and null field values.

8.6.1 Methods for Data Validation

In this section, we discuss some techniques for data validation. We begin our discussion with the simple statistical criterion (SSC).

8.6.1.1 Simple Statistical Criterion

The SSC method consists of establishing upper and lower limits for data acceptance defined by a probability confidence interval based on the mean and standard deviation of a given dataset. The application of this technique consists of calculating the mean and standard deviation of the data subset, and eliminating any values outside of the confidence interval. The confidence interval is a range of values we are fairly sure our true value lies in. The confidence interval is arbitrarily defined as the mean plus or minus 3.6 times the standard deviation of the set of values (i.e. $\overline{X} \pm 3.6\text{SD}$ where $\overline{X}$ and SD are the mean and standard deviation, respectively) after observing that the number of eliminated values increased considerably when a narrower confidence interval was tested on some of the variables. We remark that the selected criterion excludes 1 in every 3143 data points, or 0.032% of data, in a set of normally distributed values. Although data may not be normally distributed, this method is used because of its simplicity. This method is the basis for most validation approaches in many different areas of research.

8.6.1.2 Fourier Series Modeling and SSC

This method consists of modeling the natural seasonal variations of variables using Fourier series in order to calculate the seasonal means and then to apply the SSC outlier criterion on the model residuals instead of the actual variable values. The incorporation of Fourier helps to detect possible seasonal variations. The Fourier series models are used to test for significant seasonal effects and are in the form:

$$y(t) = a_0 + \sum_{i=1}^{n} a_i \cos(i\omega t) + b_i \sin(i\omega t), \tag{8.3}$$

where y is the modeled variable, n is the model order, a_0 is the model constant, a_i and b_i are the model coefficients, t is the time unit, and ω is the angular velocity.

The least squares fitting method is then used to adjust the different Fourier models for each variable. Starting with a first-order ($n = 1$) Fourier series, the model order(n) is progressively increased, to a maximum of four or until one or both terms (sine and cosine) were not significant (i.e. the p-value > 0.05). The p-value or probability value is the probability of obtaining test results at least as extreme as the results actually observed during the test, assuming that the null hypothesis is correct. Model residuals are then calculated by subtracting the seasonal means estimated by the corresponding Fourier model from the measured values. Finally, outlier data is found by applying the SSC on the model residuals discussed in Section 8.6.1.1.

8.6.1.3 Principal Component Analysis and SSC

This method consists of transforming a pair of variables into principal components and then applying the SSC on the values of the second principal component. The principal components are linear combinations of the original variables weighted by their contribution to explaining the variance in a particular orthogonal dimension. The first principal component accounts for as much of the variability in the data as possible, and the deviation from this main variation is represented by the second principal component, which was used for outlier detection. We will discuss principal components in more detail in Chapter 11.

Please refer to S. A. Jimenez-Marquez and Thibau (2002) for more details of the data validation techniques described in this section.

8.7 Problems

1 Discuss the steps in data preprocessing.

2 Given the following 14 points,

$$5, 7, 10, 15, 19, 20, 21, 22, 23, 23, 23, 23, 24, 24$$

find

(a) Median.
(b) Quartiles (Q_1 and Q_3).
(c) Interquartile range.
(d) Calculate $1.5 \cdot$ IQR below the first quartile and check for low outliers.
(e) Calculate $1.5 \cdot$ IQR above the third quartile and check for high outliers.
(f) Display the five number summary by drawing a box plot.

3 Discuss the difference between missing completely at random and missing at random and present examples each.

4 Discuss the difference between parametric hypothesis test and nonparametric hypothesis test.

5 Discuss the multicollinearity issue of a predictive model.

6 Discuss the difference between listwise deletion and pairwise deletion.

7 Explain the bias–variance trade-off in a predictive model.

8 Take a complete dataset (with no missing values) of interest to you with two variables, x and y. Call this the "complete data." Write a program in R or Python to cause approximately half of the values of x to be missing. Design the missing values to be at random but not completely at random; that is, the probability that x is missing should depend on y. Call this new dataset, with the missing values in x, the "available data."

9 Given the following 10 points,

$$5, 7, 10, 15, 19, 20, 21, 22, 23, 24,$$

transform the data

(a) Using the min–max normalization to the range [0, 1].
(b) Using Z-score normalization.

10 Take a complete dataset and by using any of the methods discussed in Section 8.6.1, check the validity of the dataset.

9

Data Visualizations

9.1 Introduction

Before applying a mathematical or statistical model to analyze any data sets, it is always essential to perform an exploratory data analysis (EDA). Exploratory data analysis refers to a set of techniques to visualize and summarize data in such a way that interesting characteristics and or properties will become apparent. Unlike classical methods which usually begin with an assumed model for the data, EDA techniques are used to visualize the data to identify trends, pattern, or structures in the data before we apply an appropriate model. No matter what data you want to analyze, visualizing the data seems to be a necessary step. Data visualization is the mapping from data space to graphic space, i.e. we process and filter the data, transform it into an expressible visual form, and then render it into a user-visible view.

In this chapter, we will discuss the definition, concept, implementation process, and tools for data visualization.

9.2 Definition – Data Visualization

Data visualization is the visual representation of information and data using graphical elements such as charts, graphs, and maps. Data visualization tools provide an accessible way to see and understand trends, outliers, and complex patterns in data that go undetected. Data visualization uses human skills to enhance data processing and analysis. Visualization can help us extract more complex information from data.

To illustrate the usefulness of data visualization, we present an example as follows. An ice cream vendor keeps track of how much ice cream they sell versus the temperature on that day. Table 9.1 list the figures for the last 11 days. Figure 9.1 is a representation of the same data as a Scatter Plot.

Data Science in Theory and Practice: Techniques for Big Data Analytics and Complex Data Sets, First Edition. Maria Cristina Mariani, Osei Kofi Tweneboah, and Maria Pia Beccar-Varela.

Table 9.1 Temperature versus ice cream sales.

Day	Temperature (°C)	Ice cream sales
1	16.4	325
2	11.9	185
3	15.2	332
4	18.5	506
5	22.1	522
6	19.4	312
7	25.1	614
8	23.4	544
9	18.1	421
10	22.6	445
11	17.2	508

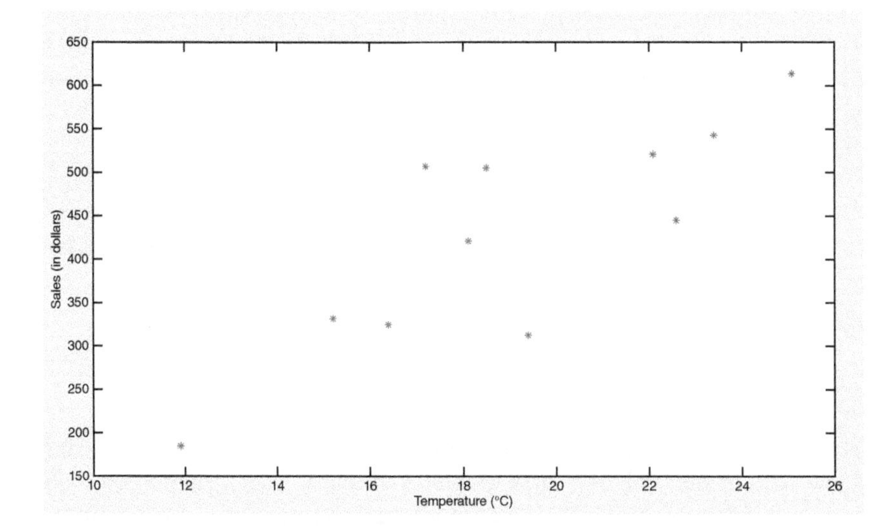

Figure 9.1 Scatter plot of temperature versus ice cream sales.

From Fig. 9.1, we observe that warmer weather leads to more sales, but the relationship is not perfect. Without the visualization, it would be difficult to identify this pattern in the data.

In general, researchers and practitioners do not need technical training to visualize data. However, to effectively visualize data and make insightful inferences, these are some skills that are highly recommended.

- *Basic mathematics*: Trigonometric function, linear algebra, and geometric algorithm.
- *Algorithms*: Basic algorithms, statistical algorithms, and common layout algorithms.
- *Data analysis*: Data cleaning, statistics, and data modeling.
- *Design aesthetics*: Design principles, aesthetic judgment, color, interaction, and cognition.
- *Visual basis*: Visual coding, visual analysis, and graphical interaction.
- *Visualization solutions*: Correct usage of charts and visualization of common business scenarios.

Data visualization comprises of three branches, namely scientific visualization, information visualization, and visual analytics. We will explain each branch in details as follows.

9.2.1 Scientific Visualization

Scientific visualization is a multidisciplinary research and application field in science, focusing on the visualization of three-dimensional phenomena, that is architecture, physical, or biological systems. Its purpose is to graphically illustrate scientific data, enabling scientists to understand, explain, and collect patterns from the data. For example, Figure 9.2 shows the visualization (heat map) of clustered handwritten digit data. From the heat map, we can observe 10 different patterns where each pattern corresponds to the digits 0–9. From the heat map we observe, a white vertical line for the 1st and 40th variable. This suggests that there are no observations recorded for the 1st and 40th variable. Without the heat map, it is very difficult to identify these patterns in the handwritten digit data.

9.2.2 Information Visualization

Information visualization is the study of the interactive graphical representations of abstract data to enhance human cognition. Abstract data includes both digital and nondigital data such as geographic information, and text. Visuals such as histograms, barcharts, trend graphs, flow charts, and tree diagrams all belong to information visualization, and the design of these graphics transforms abstract concepts into visual information.

As a practitioner, the process of creating information visualization starts with understanding the information needs of the target user group. Qualitative research methods such as user interviews can reveal how, when, and where the visualization will be used. Taking these insights, a practitioner can determine which form of data organization is needed and useful.

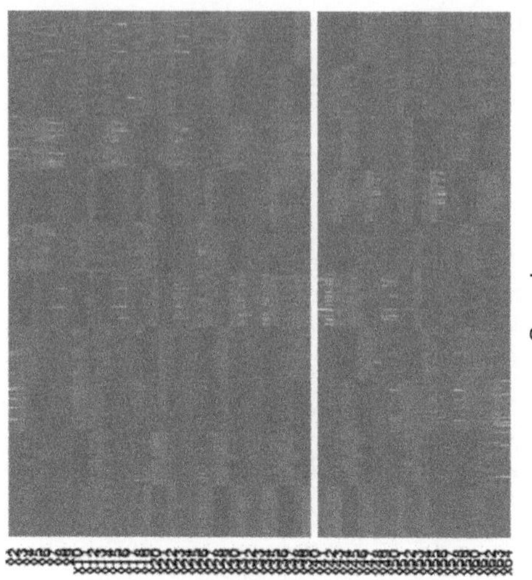

Figure 9.2 Heatmap of handwritten digit data.

Figure 9.3 displays the map of earthquake magnitudes recorded in the country Chile. The map displays the locations where earthquakes were recorded. In the figure, the circles correspond to the spatial distribution of earthquakes. This is an example of information visualization.

9.2.3 Visual Analytics

Visual analytics deals with combining interactive visual representations with the underlying analytical processes (e.g. statistical procedures, data mining techniques) such that high-level complex activities can be effectively performed (e.g. sense making, reasoning, decision-making).

In Figure 9.4, the map shows the spatial distribution of earthquakes along the tectonic boundary between the Nazca and the South American tectonic plates. The rectangles represent the four study regions. The white circles mark the location of the very large earthquakes ($Mw \geq 8$) in each of the four regions. The magnitudes are recorded as a function of time origin of earthquakes for the four study regions. The corresponding magnitudes and time origin of the earthquakes for the four study regions are also displayed on Fig. 9.4.

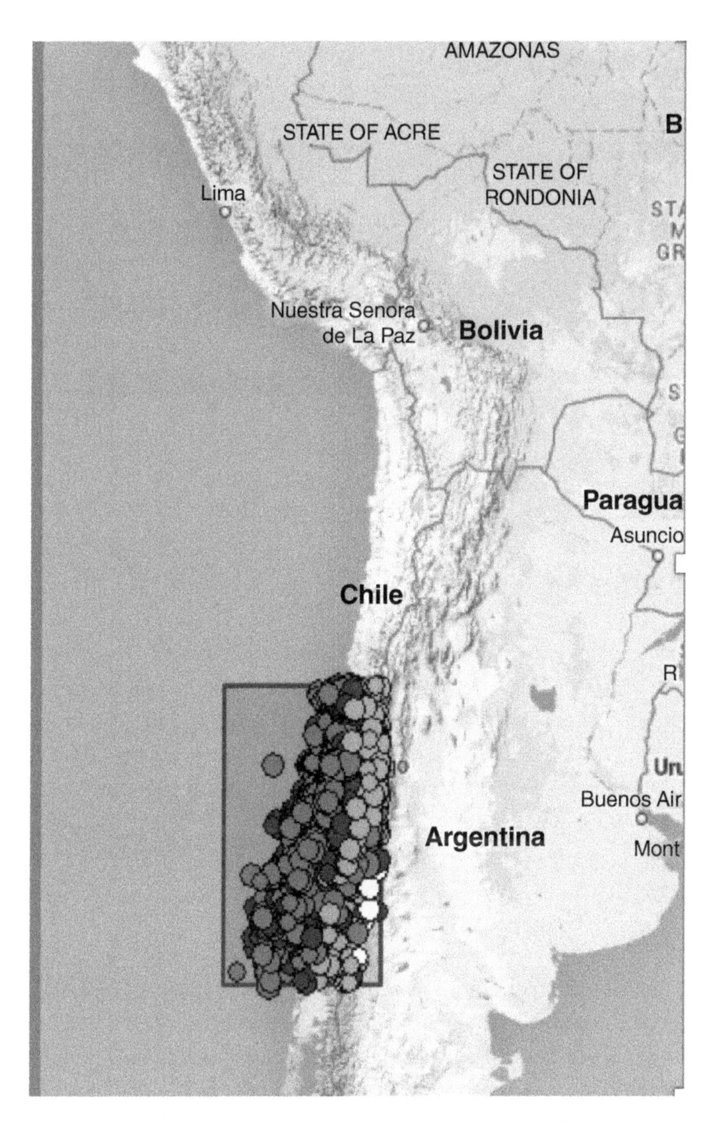

Figure 9.3 Map of earthquake magnitudes recorded in Chile.

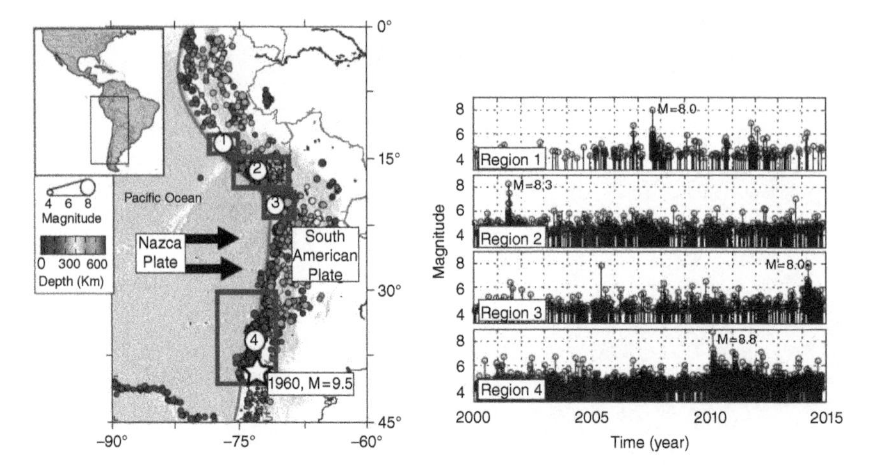

Figure 9.4 Spatial distribution of earthquake magnitudes (Mariani et al. 2016).

9.3 Data Visualization Techniques

Developing appropriate visuals to display and describe data requires a number of judgments. The practitioner must determine which questions to ask, identify the appropriate data, and select effective visual encodings to map data values to graphical features such as position, size, shape, color, etc. In this section, we will provide a brief discussion of techniques for visualizing and interacting with diverse data sets. We will focus on a few of the more sophisticated techniques that deal with complex data sets.

9.3.1 Time Series Data

Time series data is one of the most common forms of recorded data. As defined in Chapter 4, a time series data consists of data points indexed in time order. The time-varying phenomena are central to many domains such as finance (stock prices), science (temperatures, pollution levels, electric potentials), and public policy (crime rates). One often needs to compare a large number of time series simultaneously and choose from a number of visualizations. We can display time series data using line graphs or line charts, stacked graphs, horizon graphs, and small multiples.

A line graph (or chart) is used to show how data evolves over time. It provides quick ways to indicate trends when the data are in time order.

The next plot for displaying a time series data is by using a stacked bar. A stacked bar is a chart that uses bars to show comparisons between categories of data, but with the ability to break down and compare parts of a whole. Each bar in the graph represents a whole, and segments in the bar represent different parts or categories of that whole.

A horizon graph or chart is a space-efficient time series visualization technique across a range of chart that presents a time series data into a compact space thereby preserving resolution. Horizon graphs can meaningfully display 50 or more full sets of time-series data on a page in a way that supports comparisons among them.

Another type of graph for visualizing a time series data is small multiple chart (sometimes called trellis chart, lattice chart, grid chart, or panel chart). Multiple chart is a series of similar graphs or charts using the same scale and axes, allowing easy comparisons. It uses multiple views to show different partitions of a dataset.

9.3.2 Statistical Distributions

Other visualizations have been designed to reveal how a set of numbers is distributed and thus help a researcher or practitioner better understand the statistical properties of the data. One important use of visualizations is to understand how data is distributed to inform data transformation and modeling decisions. Common techniques include the histogram, which shows the frequency of values grouped into bins or "buckets", and the box-and-whisker plot, which can convey statistical features such as the mean, median, quartile boundaries, or extreme outliers. In addition, a number of other techniques exist for assessing a distribution and examining interactions between multiple dimensions. We explain them below.

9.3.2.1 Stem-and-Leaf Plots

Stem-and-leaf plots are for assessing a collection of numbers. It is an alternative to the histogram. It typically bins numbers according to the first significant digit, and then stacks the values within each bin by the second significant digit. This representation uses the data itself to create a frequency distribution, replacing the "information-empty" bars of a traditional histogram bar chart and allowing one to assess both the overall distribution and the contents of each bin.

Figure 9.5 displays the stem-and-leaf plot of the numbers of text messages sent in one day by 50 cell phone users.

From the display, we can see that more than 50% of the cell phone users sent between 20 and 50 text messages.

9.3.2.2 Q–Q Plots

The Q–Q (quantile–quantile) plot compares two probability distributions by graphing their quantiles against each other. If the two probability distributions are similar, the plotted values will lie roughly along the central diagonal in a

Stem	Leaf
1	6 9
2	0 3 4 6 6 8 9 9 9 9
3	0 0 2 3 3 3 3 4 8 9
4	0 1 1 3 8 9
5	2 3 6 8
6	6 7 9
7	2 6 6 6 8
8	0 0 6 8 9
9	9
10	2
11	5
12	2
13	
14	9

Key: 10|2 = 102

Figure 9.5 Number of text messages sent.

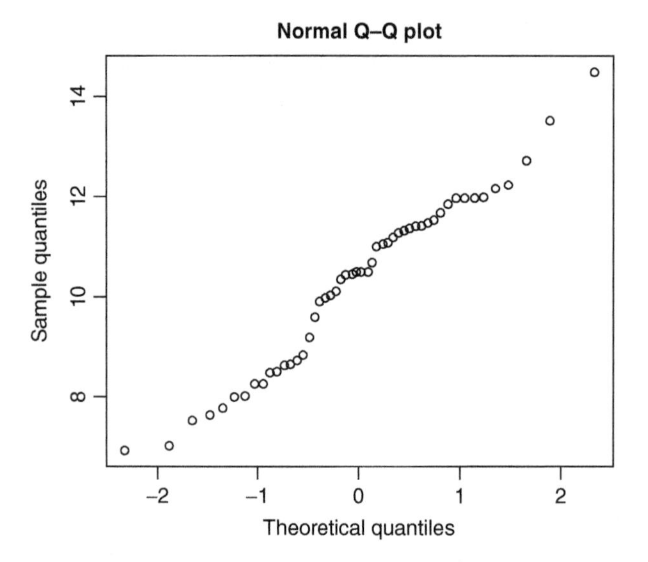

Figure 9.6 Normal Q–Q plot.

straight line. Quantiles are points in your data below which a certain proportion of your data fall. If the two probability distributions are linearly related, the plotted values will again lie along a line, though with varying slope and intercept.

Figure 9.6 is a normal Q–Q plot when both sets of quantiles truly come from the normal distributions.

In Section 9.4, we discuss some commonly used data visualization tools.

9.4 Data Visualization Tools

Data visualization tools provide researchers and practitioners with an easy way to design and create visual representations of data sets. When dealing with complex data sets that include several data points, automating the process of creating a visualization helps to identify complex patterns and trends in the data set. These data visualizations can then be used for a variety of purposes such as creating data dashboards, reports, sales and marketing materials, and anywhere else information needs to be interpreted immediately. Data visualization is about how to present your data, in order for the target audience to gain insights most effectively.

Several data visualization tools have two things in common. First, they are easy to use, and secondly, most have excellent documentation and tutorials and are designed in ways that feel intuitive to the user. A good visualization tool should be able to handle high dimensional data sets.

Next, we describe some popular data visualization tools. We begin with the Tableau.

9.4.1 Tableau

Currently, Tableau is one of the useful visualization tools available on the market. The Tableau software is an interactive data visualization software founded in January 2003 by Christian Chabot, Pat Hanrahan, and Chris Stolte, in Mountain View, California. Tableau has a variety of options available, including a desktop app, server and hosted online versions, and a free public option (Tableau public). There are hundreds of data import options available, from CSV files to Google Ads and Analytics data to Salesforce data.

The output options include multiple chart formats as well as mapping capabilities. That means users can create color-coded maps that showcase geospatial datasets in a format that is much easier to visualize than a table or chart.

The public version of Tableau is free to use for anyone looking for a powerful way to create data visualizations that can be used in a variety of settings. They have an extensive gallery of infographics and figures that have been created with the public version to serve as inspiration for those who are interested in creating their own.

Figure 9.7 displays a loan risk analysis data dashboard that analyzes bank loan data to assess the risk of loan default. The figure was generated using Tableau.

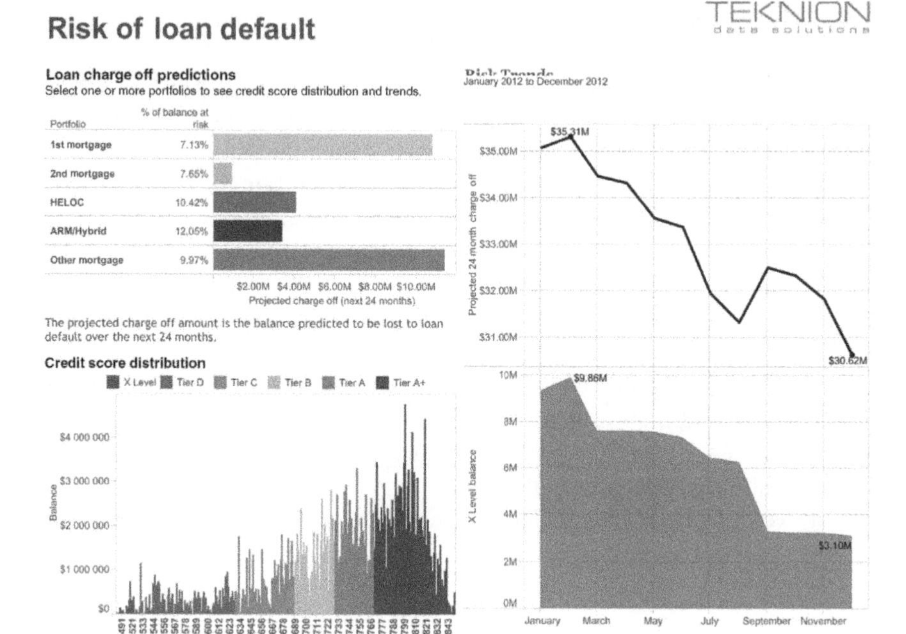

Figure 9.7 Risk of loan default. Source: Tableau Viz Gallery.

Tableau is a great visualization tool for those who need to create graphs in addition to other types of charts. Tableau Public is also an alternative for anyone who wants to create public-facing visualizations.

9.4.2 Infogram

Infogram is a fully featured visualization tool that allows users to create effective visualizations of data for reports, social media posts, maps, dashboards, and several others by clicking on a desired visual and dragging it to a location as required.

The finalized visualizations can be exported into a number of formats: .PNG, .JPG, .GIF, .PDF, and .HTML. Interactive visualizations are also possible, suitable for embedding into webpages or apps. Infogram also offers a WordPress plugin that makes embedding visualizations even easier for WordPress users.

Figures 9.8 and 9.9 display the top five publishing markets by revenue (mEur) in 2015 and high-yield defaulted issuer and volume trends, respectively. Both figures were generated using infogram.

The map is an excellent way to illustrate worldwide data, and the charts make data easier to compare. Infogram is easy to use and helps create professional-looking designs without a lot of visual design skill or technical background.

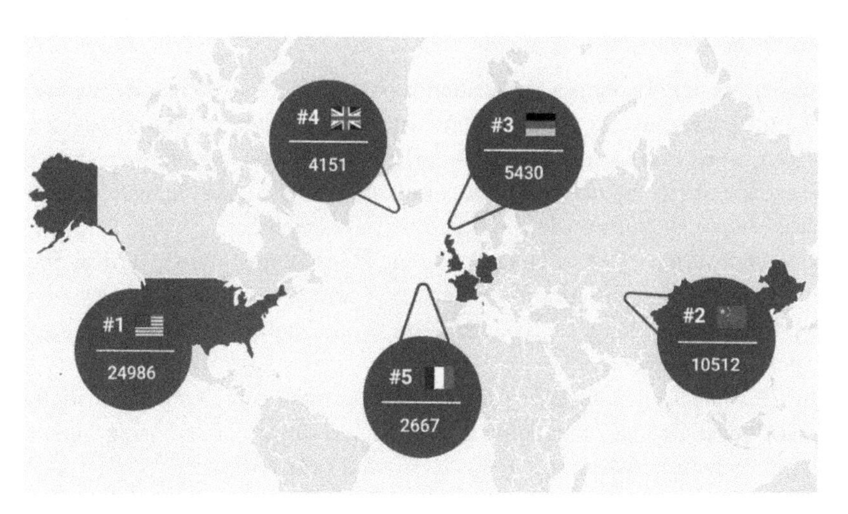

Figure 9.8 Top five publishing markets. Source: Modified from International Publishers Association – Annual Report.

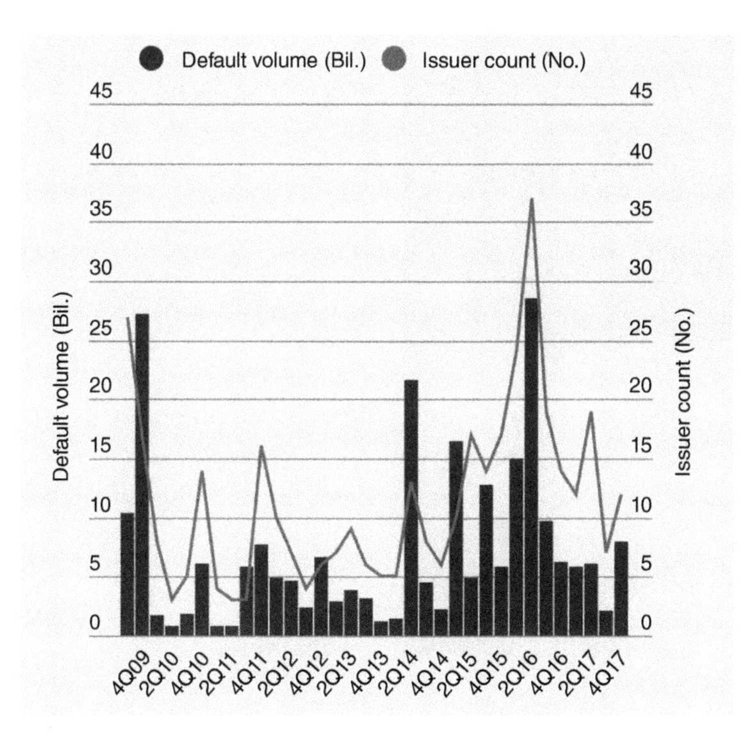

Figure 9.9 High yield defaulted issuer and volume trends. Source: Based on Fitch High Yield Default Index, Bloomberg.

9.4.3 Google Charts

Google Charts is a free data visualization tool that is used for creating interactive charts for embedding online. It works with dynamic data and the outputs are based purely on HTML5 and scalable vector graphics (SVG), therefore they work in browsers without the use of additional plugins. Data sources include Google Spreadsheets, Google Fusion Tables, and other SQL databases.

There are a variety of chart types, including maps, scatter charts, vertical and horizontal bar charts, histograms, pie charts, treemaps, timelines, gauges, and many others. These charts can be customized completely, via simple cascading style sheets (CSS) editing.

Figure 9.10 displays a statistics page for popular movies and cinema locations of a make-belief cinema chain company. The figure was generated using Google charts and it includes a Map and a stacked bar chart visualization.

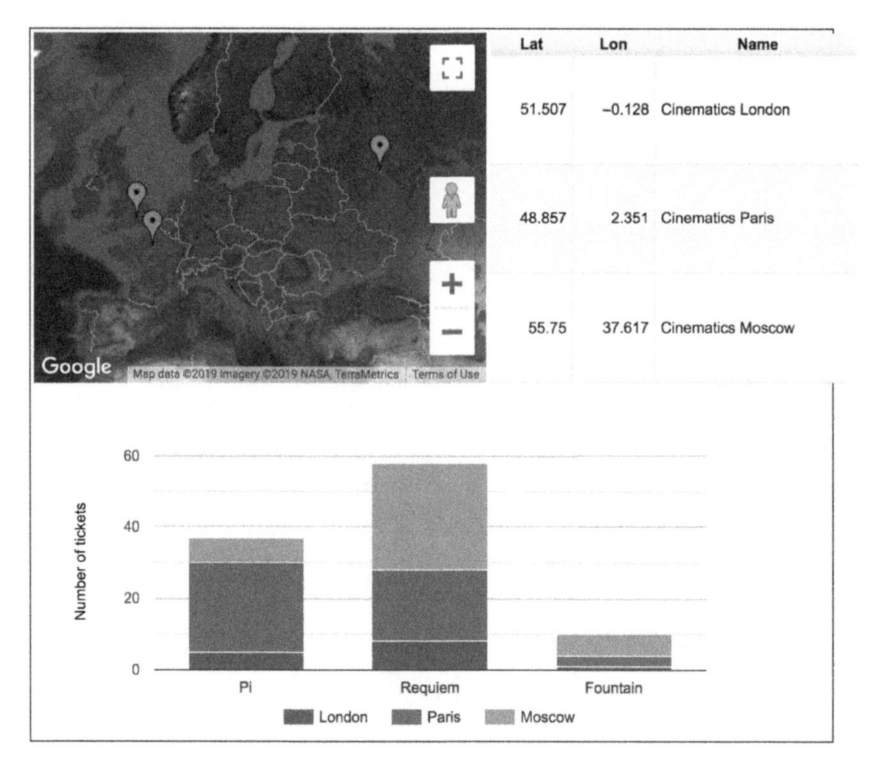

Figure 9.10 Statistics page for popular movies and cinema locations. Source: Google Charts.

There are several other data visualizations tools such as the Google data studio, R statistical software, Python, Microsoft Power BI and several others. The reader can consult (Chapman 2019) for more details of this topic.

9.5 Problems

1. Discuss why designers of visualizations need to understand human visual perception?
2. Define a time series data set and discuss various techniques to visualize time series data sets.
3. Choose an appropriate visualization for a statistical distribution and discuss the usefulness of that technique.
4. When is one visualization better than another? How can we measure it?
5. What is the most effective way to encode data with multiple visual variables?
6. How do you build tools that help people build sophisticated and accurate visualizations without programming?
7. Discuss the difference between scientific visualization and information visualization.
8. List and describe various ways of visualizing vector attributes.
9. Select any complete data sets that contains two variables. Visualize the data sets to identify any pattern or trend in the data.
10. What are the advantages of having a human involved in data analysis?

10

Binomial and Trinomial Trees

10.1 Introduction

In this chapter, we will learn methods to approximate stochastic processes using the binomial tree methodology (see Clewlow and Strickland (1998) and Hull (2008) for more details of the tree methodology). We begin this chapter with an introduction to the binomial tree methodology.

10.2 The Binomial Tree Method

The binomial tree was developed during the late 1970s by Cox et al. (1979) and perfected in the early 1980s. The binomial tree model generates a pricing tree in which every node represents the price of an underlying financial instrument at a given point in time. The methodology can be used to price options with non-standard features such as path dependence and barrier events.

Suppose we have a stochastic process S_t solving the SDE:

$$dS_t = \mu(S_t)dt + \sigma(S_t)dB_t, \tag{10.1}$$

where $\mu(\cdot)$ and $\sigma(\cdot)$ are known function. The theory presented at the end of this chapter applies to d-dimensional stochastic processes S_t not just one dimensional. Later in this chapter, we shall present an approximation to a stochastic volatility model which will involve a two-dimensional process.

The main idea behind the binomial tree is to construct a discrete version of the stock process S_t in (10.1).

The process in (10.1) is a continuous process. Furthermore, since it is stochastic every time one constructs a path, the path would look different. The present time of a stock price is known to be S_0. Therefore, the binomial tree is constructed so that for each path in the tree, there is a possible path for the process in (10.1) at the specified times. The specified times refer to the steps in the tree. Furthermore,

Data Science in Theory and Practice: Techniques for Big Data Analytics and Complex Data Sets, First Edition. Maria Cristina Mariani, Osei Kofi Tweneboah, and Maria Pia Beccar-Varela.

when the number of steps increase, the approximation improves and at the limit we obtain the same realization given by (10.1).

In the subsection 10.3.1, we consider a specific example of the binomial tree methodology which is the one step binomial tree.

10.2.1 One Step Binomial Tree

The one-step binomial tree method is a particular example of the binomial tree technique where we can only exercise the option in one step. For instance, consider a stock whose price is initially S_0. In this instance, we are interested in deriving the current price (f_0) of a European Call option on the stock. Suppose that the option lasts for time T and that during the lifespan of the option the stock price can either move up from its initial price S_0 to a new level, S_0u or down from S_0 to a new level, S_0d where ($u > 1$ and $d < 1$). Then, the movement of the stock price can be described by the one-step binomial tree. For details of the one-step tree, consult Sections 2.2 and 2.3 in Clewlow and Strickland (1998). More details about the construction of the one-step binomial tree can be found in Hull (2008).

It turns out that the stochastic equation of the return (i.e. $X_t = \log S_t$) is much simpler, and thus in practice is much easier to construct the binomial tree that simulates the evolution of the return instead of the stock. The resulting tree for the return X_t is called the additive tree, and the tree for the stock process S_t is called the multiplicative tree for reasons that are evident by just looking at the details of the constructed trees. For each choice of the martingale measure (probabilities in the tree), the two trees (multiplicative and additive) are perfectly equivalent. More specifically, suppose that we have constructed a tree for the return process X_t which has $x + \Delta x_u$ and $x + \Delta x_d$ additive steps up and down from x, respectively. Then, a multiplicative tree for S_t could be constructed immediately by taking the next steps from $S = e^x$ as $Su = e^{x+\Delta x_u} = Se^{\Delta x_u}$ and $Sd = e^{x+\Delta x_d} = Se^{\Delta x_u}$. Similarly, the additive tree is constructed from the multiplicative tree by taking $\Delta x_u = \log u$ and similarly for the down step.

In order to proceed, we state the Girsanov's theorem (Øksendal 2010a).

Theorem 10.1 **(Girsanov, one-dimensional)** Let $B(t), 0 \leq t \leq T$, be a Brownian motion on a probability space $(\Omega, \mathcal{F}, \mathbb{P})$. Let $\mathcal{F}(t), 0 \leq t \leq T$, be the accompanying filtration, and let $\theta(t), 0 \leq t \leq T$, be a process adapted to this filtration. For $0 \leq t \leq T$, define

$$\tilde{B}(t) = \int_0^t \theta(u)du + B(t),$$

$$Z(t) = \exp\left\{ -\int_0^t \theta(u)dB(u) - \frac{1}{2}\int_0^t \theta^2(u)du \right\},$$

and define a new probability measure by

$$\tilde{\mathbb{P}}(A) = \int_A Z(T) d\mathbb{P}, \quad \forall A \in \mathcal{F}.$$

Under $\tilde{\mathbb{P}}$, the process $\tilde{B}(t), 0 \leq t \leq T$, is a Brownian motion.

Applying Girsanov'stheorem to (10.1) and using the Itô's formula on the process $X_t = \log S_t$, we obtain the stochastic differential equation:

$$dX_t = \left(r - \frac{\sigma^2}{2}\right) dt + \sigma\, dW_t. \tag{10.2}$$

The Girsanov's theorem describes the change of measure for diffusion processes. The one-step tree for this process is presented in Figure 10.1. At any point in the tree, we can go up to $x + \Delta x_u$ with probability p_u or down to $x + \Delta x_d$ with probability $p_d = 1 - p_u$.

To make connection with the model in (10.2), we impose the conditions that the mean increase and variance increase from the tree and model should be equal.

This is a sufficient condition since the process X_t in (10.2) is a normal random variable for t fixed. We also know that a normal variable is entirely characterized by its mean and variance. Thus, equating the mean and variance increases for an infinitesimal time step will be sufficient to make sure that the binomial process will converge to the path given by the process (10.2). Then, the conditions we need to enforce are:

$$\begin{cases} p_u \Delta x_u + p_d \Delta x_d & = \left(r - \frac{\sigma^2}{2}\right) \Delta t, \\ p_u \Delta x_u^2 + p_d \Delta x_d^2 & = \sigma^2 \Delta t + \left(r - \frac{\sigma^2}{2}\right)^2 \Delta t^2, \\ p_u + p_d & = 1. \end{cases} \tag{10.3}$$

The system (10.3) has three equations and four unknowns i.e. p_u, p_d, Δx_u and Δx_d; thus, we have a choice of a parameter. Furthermore, note that for any choice of Δx_u and Δx_d, the tree recombines i.e. an upward move by the stock price followed by a downward move generates the same stock prices as a downward move

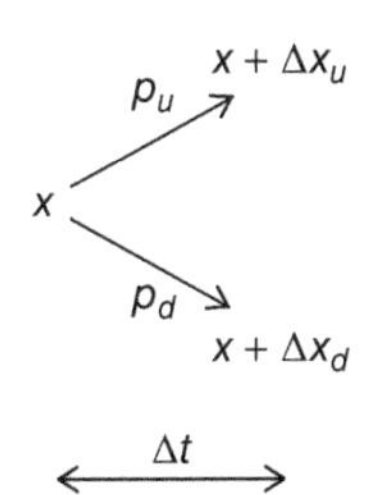

Figure 10.1 One-step binomial tree for the return process.

followed by an upward move. For example first step successors are $x + \Delta x_u$ and $x + \Delta x_d$. At the second step, their successors will be:

$$x + 2\Delta x_u \text{ and } x + \Delta x_u + \Delta x_d \text{ for the upper node,}$$
$$\text{and } x + \Delta x_d + \Delta x_u \text{and } x + 2\Delta x_d \text{ for the lower node.}$$

Clearly, one of the nodes is identical and the tree recombines. For example, consider a binomial tree with $n = 10$ steps that recombines and one that does not recombine. The recombining tree has a total number of nodes: $1 + 2 + 3 + \cdots + 11 = 11 \times 10/2 = 55$ nodes, whereas the one that does not recombine has $1 + 2 + 2^2 + 2^3 + \cdots + 2^{10} = 2^{11} - 1 = 2047$ nodes. The latter has a much bigger number of calculations that quickly becomes unmanageable.

All the trees constructed with any particular choice of parameters gives a good approximating tree. Some popular choices are obtained by taking $\Delta x_u = \Delta x_d$ in the system (10.3). This produces what is called the Trigeorgis tree (see Trigeorgis 1991) which supposedly has better approximation power. Solving the system (10.3) we fix Δ so by taking $\Delta x_u = \Delta x_d$ will yield:

$$\begin{cases} \Delta x &= \sqrt{\left(r - \frac{\sigma^2}{2}\right)^2 \Delta t^2 + \sigma^2 \Delta t}, \\ p_u &= \frac{1}{2} + \frac{1}{2}\frac{\left(r - \frac{\sigma^2}{2}\right)\Delta t}{\Delta x}. \end{cases} \tag{10.4}$$

The equivalent tree constructed for the stock is the Cox et al. (1979) (CRR) tree.

Another popular tree is obtained by setting the probabilities of jumps in (10.3) equal (i.e. $p_u = p_d = 1/2$). In this case, the tree for S is called Jarrow and Rudd (1983).

Remarks

- The system (10.3) could be further reduced by subtracting the constant term $\left(r - \frac{\sigma^2}{2}\right)\Delta t$ from each node, the resulting tree will be much simpler.
- Finding the probabilities, p_u and p_d is equivalent to finding the martingale measure under which the pricing is made. For each solution one can construct an additive tree (moving from x to $x + \Delta x$ and $x - \Delta x$ with probabilities p_u and p_d respectively) or a multiplicative tree (moving from S to S_u and S_d with probabilities p_u and p_d respectively).

Next, we discuss an application of the binomial tree methodology in option pricing. Options are contracts that give option buyers the right to buy or sell a security at an established price on or before a specified day. We will study the European options and American options. A European option is an options contract that limits its execution to its expiration date. An American option on the other hand can be exercised any time up to and including the date of expiration. We begin the application of the binomial tree with the European option.

10.2.2 Using the Tree to Price a European Option

After constructing the tree, we basically have possible paths of the process X_t or equivalently the process $S_t = \exp(X_t)$. Since these are possible paths we can use the binomial tree to calculate the present price of any path dependent option.

We begin this study by looking at the European Call. The European Call is written on a particular stock with current price S_0, and is characterized by maturity, T and strike price, K.

The first step is to divide the interval $[0, T]$ into $n+1$ equally spaced points, then we will construct our tree with n steps, $\Delta t = T/n$ in this case. The times in our tree will be: $t_0 = 0, t_1 = T/n, t_2 = 2T/n, \ldots, t_n = nT/n = T$.

Next we construct the return X_t tree as above starting with $X_0 = x_0 = \log(S_0)$, and the last branches of the tree ending in possible values for $X_{t_n} = X_T$. We remark that since we constructed the possible values at time T we can calculate for every terminal node in the tree the value of the Call option at that terminal node using the change of variables:

$$C(T) = (S_T - K)_+ = (e^{X_T} - K)_+. \tag{10.5}$$

Thus now we know the possible payoffs of the option at time T. Suppose we want to calculate the value of the option at time $t = 0$. Using the Girsanov theorem and the Harrison and Pliska result the discounted value of the option value specifically the process $\{e^{-rt}C(t)\}$ is a continuous time martingale. The Harrison and Pliska results asserts that a frictionless financial market is arbitrage free if and only if the price process is a martingale under a probability measure equivalent to the objective one. This results allows to relate the arbitrage free arguments with martingale theory. Therefore we may write that the value at time 0 must be:

$$C(0) = \mathbf{E}[e^{-rT}C(T)|\mathcal{F}_0].$$

Now we can use the basic properties of conditional expectation and the fact that $T/n = \Delta t$ or $e^{-rT} = (e^{-r\Delta T})^n := \delta^n$ to write:

$$\begin{aligned}C(0) &= \mathbf{E}[\delta^n C(T)|\mathcal{F}_0] = \mathbf{E}[\delta^{n-1}\mathbf{E}[\delta C(T)|\mathcal{F}_1]|\mathcal{F}_0] \\ &= \delta\mathbf{E}[\delta\mathbf{E}[\ldots\delta\mathbf{E}[\delta C(T)|\mathcal{F}_{n-1}]|\ldots\mathcal{F}_1]|\mathcal{F}_0],\end{aligned}$$

where $\delta = e^{-r\Delta t}$ is a discount factor.

This formula allows us to recursively go back in the tree toward the time $t = 0$. When this happens, we will eventually reach the first node of the tree at C_0 and that will give us the value of the option. More precisely since we know the probabilities of the down and up steps to be p_d and p_u respectively, and the possible Call values one step ahead to be C^1 and C^2, we can calculate the value of the option one step before in time as:

$$C = e^{-r\Delta t}(p_d C^1 + p_u C^2). \tag{10.6}$$

What is remarkable about the above construction is that we constructed the return tree just by calculating the value of the option at the final nodes. Note that when we stepped backwards in the tree we did not use the intermediate values of the stock, we only used the probabilities. This is due to the fact that we are pricing the European option which only depends on the final stock price. This situation will change when pricing any path dependent option such as the American, Asian, barrier, etc.

Furthermore, we do not actually need to go through all the values in the tree for this option. For instance, the uppermost value in the final step has probability p_u^n. The next one can only be reached by paths with probability $p_u^{n-1}p_d$ and there are $\binom{n}{1}$ of them.

10.2.3 Using the Tree to Price an American Option

The American option can be exercised at any time, thus we will need to calculate its value at $t = 0$ and the optimal exercise time called τ. Estimating τ is not a simple task, but in theory it can be done. τ is a random variable and therefore it has an expected value and a variance. The expected value can be calculated by looking at all the points in the tree where it is optimal to early exercise the option and the probabilities for all such points. Then we can calculate the expected value.

For the value of the American option we will proceed in similar way as for the European option. We will construct the tree in the same way we did for the European option and then we will calculate the value of the option at the terminal nodes. For example in an American Put option using the change of variable we have:

$$P(T) = (K - S_T)_+ = (K - e^{X_T})_+. \tag{10.7}$$

Then we recursively go back in the tree in a similar way as we did in the case of the European option. The only difference is that for the American option, we have the early exercise condition. So at every node we have to decide if it would have been optimal to exercise the option rather than to hold onto it. More precisely using the same notation as before in (10.6), we again calculate the value of the option today holding onto it as:

$$C = e^{-r\Delta t}(p_d C^1 + p_u C^2).$$

What actually happens if we exercised the option at that point in time, say t_i? Then we would obtain $(K - S_{t_i})_+$. Because we can only do one of the two, the value of the option at this particular node will be the maximum of the two values. There after, we recursively work the tree backward all the way to $C(0)$ and that will yield the value of the American Put option.

We now discuss a more general approach of pricing path dependent options using the binomial tree techniques.

10.2.4 Using the Tree to Price Any Path Dependent Option

From subsection 10.2.3, we see that the binomial tree could be applied to any path dependent option. The trick is to keep track of the value of the option across various paths of the tree.

The binomial tree is not appropriate for pricing options when the volatility varies. This is because varying volatility introduces another parameter in the model and so we will not have enough equations to solve it. A different tree construction method such as the trinomial tree is required to solve such problems. The trinomial tree model will be discussed in Section 10.4.

Binomial trees for dividend paying asset are very important, however their construction is similar to the regular binomial tree construction. The model developed so far can be modified easily to price options on underlying assets other than dividend and non-dividend-paying assets.

Another application of the binomial tree methodology is in the computation of hedge sensitivities. A brief introduction is presented below. Please refer to Clewlow and Strickland (1998) for more details.

In section 10.3, we will study the binomial discrete model that relates to binomial tree model.

10.3 Binomial Discrete Model

The binomial discrete model is based upon a simplification of the underlying financial instruments involved in option pricing, but its implications capture the essential features of more complicated continuous models. We first introduce the one-step binomial model and then discuss the more general multi-step model.

10.3.1 One-Step Method

Consider a European Call option (we recall that it is possible to exercise at time T) C, its strike price is K, and we have two possible states at time T, a price"up" S_u and a price "down," S_d

$$
\begin{array}{ccc}
 & & S_u \\
 & & C_u \\
 & \nearrow & \\
S & & \\
 & \searrow & S_d \\
 & & C_d
\end{array}
$$

where $C_T = \max\{S_T - K, 0\}$.

Suppose that we have a portfolio,

$$\pi = \begin{cases} \Delta, & \text{shares in long,} \\ 1, & \text{option in short,} \end{cases}$$

so $\pi = \Delta S - C$. Because of the non-arbitrage assumption, we fix Δ so that $\pi_u = \pi_d$. We deduce that Δ, the *hedging* coefficient, has to satisfy the equation:

$$\Delta S_u - C_u = \Delta S_d - C_d$$

and then,

$$\Delta = \frac{C_u - C_d}{S_u - S_d}. \tag{10.8}$$

Example 10.1 Consider a portfolio consisting of a long position in Δ shares and a short position in one Call option. We compute the value of Δ that makes the portfolio riskfree. If the stock price moves upward from \$30 to \$35, the value of the shares is 35Δ and the value of the option is one. Therefore the total value of the portfolio is $35\Delta - 1$. If the stock price moves downward from \$30 to \$25, the value of the shares becomes 25Δ and the value of the option is zero. Thus the total value of the portfolio is 25Δ. The portfolio is riskfree if the value of Δ is chosen so that the final value of the portfolio is the same for both cases i.e.

$$35\Delta - 1 = 25\Delta \implies \Delta = 0.1$$

A riskless portfolio is therefore

$$\pi = \begin{cases} 0.1 & \text{shares in long} \\ 1 & \text{option in short} \end{cases}$$

If the stock price moves up to \$35, the value of the portfolio is $35 \times 0.1 - 1 = 2.5$. On the other hand, if the stock price moves down to \$25, the value of the portfolio is $25 \times 0.1 = 2.5$. This means that regardless of whether the stock price moves up or down, the value of the portfolio stays the same i.e. 2.5 at the end of the life of the option.

Remarks 10.1

1. For the continuous model, the formula (10.8) becomes $\frac{\partial C}{\partial S}$.
2. Since the portfolio π is "risk-free" its expected value at time T is defined as:

$$E(\pi_T) = \pi_u = \pi_d.$$

   Therefore, because of the non-arbitrage condition, we can assume that:

$$\pi = \frac{1}{1+R}\pi_u.$$

Hence, from $\pi = \Delta S - C$ and $\pi_u = \Delta S_u - C_u$, if we solve in terms of C and make the relevant substitutions we have:

$$C = \frac{1}{1+R}(\Delta S(1+R) + C_u - \Delta S_u). \tag{10.9}$$

Replacing Δ with $\frac{C_u - C_d}{S_u - S_d}$ in (10.9) we obtain:

$$\begin{aligned} C &= \frac{1}{1+R}\left(\frac{C_u - C_d}{S_u - S_d}S(1+R) + C_u - \frac{C_u - C_d}{S_u - S_d}S_u\right) \\ &= \frac{1}{1+R}(pC_u + (1-p)C_d), \end{aligned}$$

where

$$p = \frac{S(1+R) - S_d}{S_u - S_d}.$$

The relation C is a multiple of a convex combination of C_u and C_d, with a factor $\frac{1}{1+R}$. At this point, two questions arise:

1. Does p represent a probability, i.e. $0 \leq p \leq 1$?
2. Do we have $p = \hat{p}$?, where $\hat{p}$ is an estimator.

To answer these questions, we have the following lemmas.

Lemma 10.1 Does p represent a probability, i.e. $0 \leq p \leq 1$?

Proof: We begin the proof by showing that there is no arbitrage if and only if

$$S_d < S(1+R) < S_u.$$

Equivalently, there is no arbitrage if and only if

$$0 \leq p \leq 1.$$

Next suppose that $S(1+R) > S_u$. Then it is possible to construct an arbitrage portfolio such that:

$$\pi = \begin{cases} X, & \text{bonds in long,} \\ 1, & \text{share in short,} \end{cases}$$

where,

$$\begin{aligned} \pi &= x\frac{1}{1+R} - S_d, \\ \pi_u &= x - S_u, \\ \pi_d &= x - S_d. \end{aligned}$$

Now fixing $x = S(1+R)$, we have that $\pi = 0$, $\pi_u > 0$ and $\pi_d > 0$, which is a contradiction. Therefore, $0 < p < 1$.

As a consequence, we write

$$C = \frac{1}{1+R}E(C_T),$$

where $C_T = \max\{S_T - K, 0\}$.

The reader is required to verify the other inequality $S_d < S(1+R)$. □

It is observed that even if we can assume that p "is a probability," *it is not* the probability of the up and down asset states.

Lemma 10.2 Do we have $p = \hat{p}$?

Proof: According to the proof of Lemma 10.1, if we write

$$S = p_1 S_u + p_2 S_d,$$
$$\frac{1}{1+R} = p_1 + p_2.$$

Then

$$p_1(1+R) = \frac{p_1}{p_1 + p_2} = \hat{p}_1$$

and

$$p_2(1+R) = \frac{p_1}{p_1 + p_2} = \hat{p}_2 = 1 - \hat{p}_1.$$

Hence

$$\begin{aligned}(1+R)S &= p_1(1+R)S_u + (1+R)p_2 S_d \\ &= \hat{p}_1 S_u + \hat{p}_2 S_d = \hat{p}_1 S_u + (1-\hat{p}_1)S_d \\ &= \hat{p}_1(S_u - S_d) + S_d\end{aligned}$$

and we can get:

$$\hat{p}_1 = \frac{(1+R)S - S_d}{S_u - S_d} = p.$$

Hence the proof. □

In subsection 10.3.2, we consider a generalization of the one-step method which is the multi-step method.

10.3.2 Multi-step Method

For the multi-step method, it is possible to exercise in multiple steps. Now we will consider "up" and "down" steps in different periods of time, and we will obtain a binomial tree. The notation for the value of the option is shown on the tree. For example, after two down movements the value of the option is S_{dd} and two up movements is denoted S_{uu}.

$$
\begin{array}{ccc}
 & & S_{uu} \quad \cdots \\
 & S_u & \\
 & & S_{ud} \quad \cdots \\
S & & \\
 & S_{du} & \\
 & S_d & \\
 & & S_{dd} \quad \cdots
\end{array}
$$

If there are no dividends, the expected values remain invariant. However, we recall that to drop the dividends is not realistic in the case in which the underlying asset is a share.

Assuming that $S_u = Su$ and $S_d = Sd$ where $u > 1 > d$ (factors producing the up and down of the original price respectively) then the tree will lose branches, because for example, $S_{ud} = S_{du}$.

$$
\begin{array}{ccc}
 & & S_{uu} \quad \cdots \\
 & S_u & \\
 & \cdots & \\
S & & S_{ud} \\
 & & \\
 & S_d & \\
 & & S_{dd} \quad \cdots
\end{array}
$$

Suppose that we divide the interval $[0, T]$ into n subintervals of length ΔT, i.e.

$$T = n\Delta T.$$

Then we can replace the *return* R by a risk-free rate r so that the factor $\frac{1}{1+R}$ becomes $e^{-r\Delta t}$.

Using the fact that $S_u = Su$ and $S_d = Sd$ the *non arbitrage condition* $S_u > S(1 + R) > S_d$ becomes

$$u > Se^{r\Delta t} > d.$$

From subsection 10.3.1, we know that

$$\frac{Se^{r\Delta t} - S_d}{S_u - S_d} = \frac{e^{r\Delta t} - d}{u - d} = p$$

does not depends on S. From this interesting property we can conclude that the probability p does not depend on the asset price, it is *an intrinsic parameter of the asset.*

10.3.2.1 Example: European Call Option

Suppose a strike price $K = 105$, $u = 1$, $d = 0.9$ and $r = 7\%$ (annual) with T semesters and $\Delta T = \frac{1}{2}$. Suppose that the original price of the asset is $S = 100$. In this case, we have the following tree:

		121
		(16)
	110	
	10,4764	
100		99
(6,8597)		(0)
	90	
	(0)	
		81
		(0)

The probability is then $p = 0.6781$. Next, when we consider a European Put option, the situation is different. In this case, we have:

		121
		(0)
	110	
	(1,865)	
100		99
(4,772)		(6)
	90	
	(11,3885)	
		81
		(24)

Remarks 10.2 We observe that there exists a relation between the values of the Put and the Call options:

$$C_t + Ke^{-r(T-t)} = P_t + S_t.$$

This relation is called the *Put–Call parity*.

The Put–Call parity is a principle that describes the relationship between the price of European Put options and European Call options of the same underlying asset, strike price, and expiration date.

10.4 Trinomial Tree Method

The trinomial tree is an alternate way to approximate the stock price model. The stock price once again follows the equation:

$$dS_t = rS_t\, dt + \sigma S_t\, dW_t. \tag{10.10}$$

In Clewlow and Strickland (1998) the authors worked on a continuously paying dividend asset and the drift in the equation is replaced by $r - \delta$. In this text, all the methods we will implement would require an input r. It is easy to regain the formulas for a continuously paying dividend asset by just replacing this parameter r with $r - \delta$. It is equivalent to work with the return $X_t = \log S_t$ instead of directly with the stock and we obtain:

$$dX_t = \nu\, dt + \sigma\, dW_t, \quad \text{where } \nu = r - \frac{1}{2}\sigma^2. \tag{10.11}$$

A one-step trinomial tree is presented below:

$$\begin{array}{ll} & S_u \\ S & S \\ & S_d \end{array}$$

Trinomial trees allows the option value to increase, decrease or remain stationary at every time step as illustrated above.

Once again we match the expectation and variance. In this case, the system contains three equations and three unknowns so we do not have a free choice as in the binomial tree case. In order to have a convergent tree, numerical experiments suggest that we impose the condition:

$$\Delta x \geq \sigma\sqrt{3\Delta t} \quad \text{Clewlow and Strickland (1998).} \tag{10.12}$$

For stability, there must be restrictions on the relative sizes of Δx and Δt. Equation (10.12) ensures that our method is stable and converges to the exact solution. Once the tree is constructed, we find an American or European option values by stepping back through the tree in a similar manner with what we did for the binomial tree. The only difference is that we calculate the discounted expectation of three node values instead of two as we did for the binomial tree. The main advantage of the trinomial tree over the binomial tree construction is the fact that the trinomial tree is appropriate for pricing options when the volatility varies. This is because when we vary the volatility, we introduce other parameters in the model that increases the number of equations which is easily solved by the trinomial tree construction method.

The trinomial tree produces more paths (3^n) than the binomial tree (2^n). Surprisingly, the order of convergence is not affected by this extra number. In both cases, the convergence of the option values is of the order $O(\Delta x^2 + \Delta t)$. Optimal convergence is always guaranteed due to the condition (10.12). This is as a results of numerical experiments.

Condition (10.12) makes a lot of difference when we deal with barrier options, i.e. options that are path dependent. This is due to the fact that the trinomial tree contains a larger number of possible nodes at each time in the tree. The trinomial tree is capable of dealing with the situations when the volatility changes over time, i.e. is a function of time. Please refer to Mariani and Florescu (2020) for more discussions of the tree methodology.

10.4.1 What is the Meaning of Little o and Big O?

Suppose we have two functions f and g. We say that f is of order little o of g at x_0:

$$f \sim o(g) \Leftrightarrow \lim_{x \to x_0} \frac{f(x)}{g(x)} = 0.$$

We say that f is of order big O of g at x_0:

$$f \sim O(g) \Leftrightarrow \lim_{x \to x_0} \frac{f(x)}{g(x)} = C,$$

where C is a constant.

In our context, if we calculate the price of an option using an approximation (e.g. trinomial tree) called $\hat{\Pi}$, and the real (unknown) price of the option is called Π, then we say that the approximation is of the order $O(\Delta x^2 + \Delta t)$ and we mean that:

$$|\hat{\Pi} - \Pi| = C(\Delta x^2 + \Delta t)$$

whenever Δx and Δt both go to zero for some constant C.

10.5 Problems

1 Construct a binomial tree $\left(u = \frac{1}{d}\right)$ with three quarters for an asset with present value \$95, if $r = 0.1$ and $\sigma = 0.4$. Using the tree compute the prices of:

(a) A European Call option with strike 105.

(b) A European Put option with strike 105.

(c) An American Put option with strike 105.
(d) A European Call option with strike 105, assuming that the underlying asset pays out a dividend equivalent to 1/5 of its value at the end of the second quarter.
(e) An American Call option with strike 105 for an underlying as in (d).
(f) An *asian option*, with strike equal to the average of the underlying asset values during the period.
(g) A *barrier option* (for different barriers).

2 The price of a share is 80. During the following six months, the price can go up or down in a 10% per month. If the risk-free interest rate is 8% per year, continuously compounded. What is the value of a European Call option expiring in one year with strike price 80?

3 In the binomial model, obtain the values of u, d, and p given the volatility σ and the risk-free interest rate r, for the following cases:
(a) $p = \frac{1}{2}$.
(b) $u = \frac{1}{d}$. **Hint:** $E(S^2_{t+\Delta t}) = S^2_t e^{(2r+\sigma^2)\Delta t}$.

4 Compute the value of an option with strike \$100 expiring in four months on underlying asset with present value is \$97, using the binomial model. The risk-free interest rate is 7% per year and the volatility is 20%. Assume that $u = \frac{1}{d}$.

5 In a binomial tree with n steps, let $f_j = f_{u\ldots ud\ldots d}$ (j times u and $n-j$ times d). For a European Call option expiring at $T = n\Delta t$ with strike K, show that

$$f_j = \max\{Su^j d^{n-j}, 0\}.$$

In particular, if $u = \frac{1}{d}$, we have that

$$f_j \geq 0 \Leftrightarrow j \geq \frac{1}{2}\left(n + \frac{\log(K/S)}{\log(u)}\right).$$

(a) Is it true that the probability of positive payoff is the probability of $S_T \geq K$? Justify your answer.
(b) Find a general formula for the present value of the option.

6 Explain the difference of pricing a European option by using a binomial tree with one period and risk neural valuation.

7 We know that the present value of a share is \$40, and that after one month it will be \$42 or \$38. The risk-free interest rate is 8% per year continuously compounded.

(a) What is the value of a European Call option that expires in one month, with strike price \$39?

(b) What is the value of a Put option with the same strike price?

8 Explain the difference between pricing a European option by using a binomial tree with one period and assuming no arbitrage, and by using risk neutral valuation.

9 The price of a share is \$100. During the following six months the price can go up or down in a 10% per month. If the risk-free interest rate is 8% per year, continuously compounded.

(a) what is the value of an European Call option expiring in one year with strike price \$100?

(b) Compare with the result obtained when the risk-free interest rate is monthly compounded.

10 The price of a share is \$40, and it is incremented in 6% or it goes down in 5% every three months. If the risk-free interest rate is 8% per year, continuously compounded, compute:

(a) The price a European Put option expiring in six months with a strike price of \$42.

(b) The price of an American Put option expiring in six months with a strike price of \$42.

11

Principal Component Analysis

11.1 Introduction

Principal component analysis (PCA) is a dimension-reduction technique that is used to reduce a large set of variables to a small set that still preserves most of the important information in the large set. PCA is popular because it can effectively find an optimal representation of a large data set with fewer dimensions. It is effective at filtering noise and decreasing redundancy in a data set. If you have a data set with many continuous variables, PCA can help select important features for your target variable. In addition, PCA is also popular for visualizing data sets with high dimensionality. We proceed with background of PCA in the next section.

11.2 Background of Principal Component Analysis

A multivariate analysis problem often starts out with data involving a considerable number of correlated variables. Let **X** be a data matrix with n rows and p columns. The p elements of each column are measurements on a subject such as height, weight, and age.

$$\mathbf{X}_{n\times p} = \begin{bmatrix} x_{1,1} & x_{1,2} & \cdots & x_{1,k} & \cdots & x_{1,p} \\ x_{2,1} & x_{2,2} & \cdots & x_{2,k} & \cdots & x_{2,p} \\ \vdots & \vdots & & \vdots & \vdots & \vdots \\ x_{j,1} & x_{j,2} & \cdots & x_{j,k} & \cdots & x_{j,p} \\ \vdots & \vdots & & \vdots & \vdots & \vdots \\ x_{n,1} & x_{n,2} & \cdots & x_{n,k} & \cdots & x_{n,p} \end{bmatrix}.$$

The central idea of PCA is to reduce the dimension of a $n \times p$ data matrix **X** consisting of a large number of correlated variables, while retaining as much as possible of the variation present in the data set. The term dimension reduction

Data Science in Theory and Practice: Techniques for Big Data Analytics and Complex Data Sets, First Edition. Maria Cristina Mariani, Osei Kofi Tweneboah, and Maria Pia Beccar-Varela.

come from the fact that although p components are required to reproduce the total system variability, much of which can be accounted for by a small number $k \leq p$ of the principal components (PCs).

The k principal components can then replace the initial p variables, and the original data set, consisting of n measurements on p variables, is reduced to a data set consisting of n measurements on k principal components. This is achieved by transforming to a new set of variables, the principal components, which are uncorrelated, and ordered so that the first few components retain most of the variation present in all of the original variables. The transformation must preserve the useful part of the data while discarding the noise.

In Section 11.3, we present some motivations for using PCA.

11.3 Motivation

One could think of many reasons where transforming a data set at hand to a low-dimensional space might be desirable, e.g. it makes the data easier to manipulate with and requires less computational resource.

11.3.1 Correlation and Redundancy

Environmental variables may be highly correlated or "redundant" with one another. Redundancy means that two or more variables might be providing the same information about the response variable thereby leading to unreliable coefficients of the predictors. For example soil pH, calcium, magnesium, and cation exchange capacity are usually very tightly correlated. One goal of PCA is to eliminate redundant and correlated variables in a data matrix since they directly impact model performances. The removal of redundant variables might assist in interpretability, but it must be noted that as long as the correlation between two variables is less than 1, then there is some variation in each variable which is not redundant with the other.

Suppose you did a survey on height and weight of students and observed that these two variables were roughly correlated in that heavier students tend to be taller and vice versa. This is observed in Figure 11.1.

So in this example, PCA will transform the set of correlated variables into a set of linearly uncorrelated variables called principal components. PCA aims to eliminate multicollinearity by transforming the original features in our data into new features without losing important traits of the original features. Multicollinearity is defined as a state of very high intercorrelations among the independent variables in a data matrix. Multicollinearity is a type of disruption in the data, and if present, the statistical inferences made about the data may not be reliable.

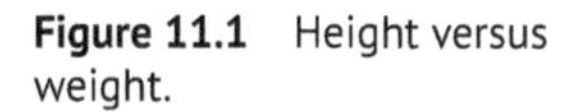
Figure 11.1 Height versus weight.

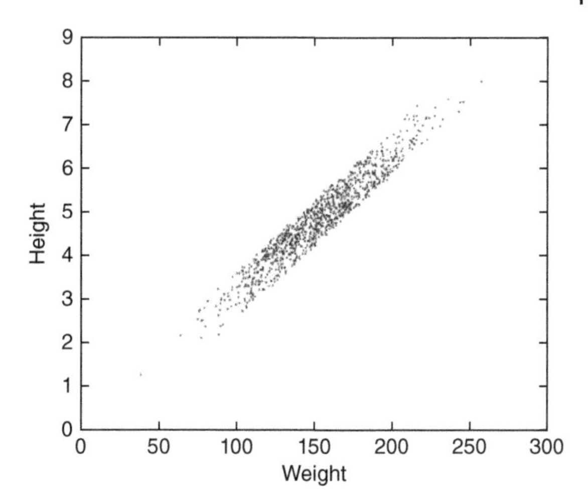

11.3.2 Visualization

We recall in Chapter 9 that before we analyze any data sets, it is always essential to visualize the data to identify trends, pattern or structures in the data before you apply an appropriate model. Visualizing four or more dimensional data is very challenging. However, with the help of PCA, we can reduce a four-dimensional data into two or three dimensions so that you can plot and hopefully understand the data better.

PCA also helps identify patterns or structures among objects which could not be visualized otherwise. For example looking at Figure 11.2, we observe that the graph of the plot of dimensions does not reveal any pattern in the data sets. However, after applying the PCA technique, we can observe a structure in the plot of the first and second PCA.

11.4 The Mathematics of PCA

PCA is a linear projection operator that maps a variable of interest to a new coordinate frame where the axes represent maximal variability. Expressed mathematically, PCA transforms an input data matrix $\mathbf{X}(n \times p)$ to an output $Y(n \times k),\ k \leq p$ via the following

$$Y = \mathbf{XP},$$

where $\mathbf{P}_{(p \times k)}(p \times k)$ is the projection matrix of which each column is a principal component that is, are unit vectors that bear orthogonal directions.

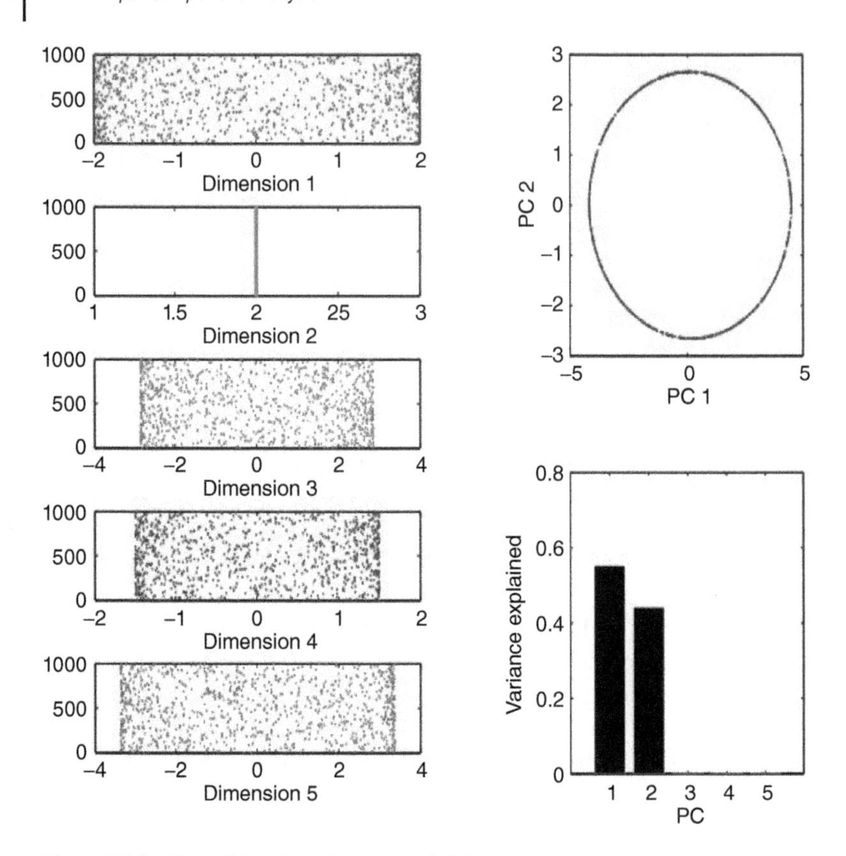

Figure 11.2 Visualizing low-dimensional data.

The principal components are linear combinations of the p random variables $X_1, X_2, \ldots, X_p$. Geometrically, these linear combinations represent the selection of a new coordinate system obtained by rotating the original system with $X_1, X_2, \ldots, X_p$ as the coordinate axes. The new axes represent the directions with maximum variability and provide a simpler and more parsimonious description of the covariance structure.

Let the random vector $\mathbf{X}' = [X_1, X_2, \ldots, X_p]$ have the covariance matrix Σ with eigenvalues $\lambda_1 \geq \lambda_2 \geq \cdots \geq \lambda_p \geq 0$.

Consider the linear combinations:

$$Y_1 = a'_1\mathbf{X} = a_{11}X_1 + a_{12}X_2 + \cdots + a_{1p}X_p$$
$$Y_2 = a'_2\mathbf{X} = a_{21}X_1 + a_{22}X_2 + \cdots + a_{2p}X_p$$
$$\vdots$$
$$Y_p = a'_p\mathbf{X} = a_{p1}X_1 + a_{p2}X_2 + \cdots + a_{pp}X_p.$$

Then we have,

$$\mathrm{Var}(Y_i) = a'_i \sum a_i, \quad i = 1, 2, \dots, p, \tag{11.1}$$

$$\mathrm{Cov}(Y_i, Y_k) = a'_i \sum a_k, \quad i, k = 1, 2, \dots, p. \tag{11.2}$$

The principal components are those uncorrelated linear combinations $Y_1, Y_2, \dots, Y_p$ whose variances in (11.1) are as large as possible.

The first principal component is the linear combination with maximum variance. That is, it maximizes $\mathrm{Var}(Y_1) = a'_1 \sum a_1$. We observe that $\mathrm{Var}(Y_1) = a'_1 \sum a_1$ can be increased by multiplying it by some constant. To eliminate this issue, it is convenient to restrict attention to coefficient vectors of unit length. We therefore define the first principal component to be the linear combination of $a'_1\mathbf{X}$ that maximizes the $\mathrm{Var}(a'_1\mathbf{X})$ subject to $a'_1a_1 = 1$. The second principal component is the linear combination of $a'_2\mathbf{X}$ that maximizes the $\mathrm{Var}(a'_2\mathbf{X})$ subject to $a'_2a_2 = 1$ and $\mathrm{Cov}(a'_1\mathbf{X}, a'_2\mathbf{X} = 0)$. At the ith step, the ith principal component is the linear combination of $a'_i\mathbf{X}$ that maximizes: $\mathrm{Var}(a'_i\mathbf{X})$ subject to $a'_ia_i = 1$ and $\mathrm{Cov}(a'_i\mathbf{X}, a'_k\mathbf{X} = 0)$ for $k < i$.

We state the following results without proof. Please refer to Johnson and Wichern (2014) for the proof.

Result 11.1 Let Σ be the covariance matrix associated with the random vector $\mathbf{X}' = [X_1, X_2, \dots, X_p]$. Let Σ have the eigenvalue–eigenvector pairs $(\lambda_1, e_1), (\lambda_2, e_2), \dots, (\lambda_p, e_p)$ where $\lambda_1 \geq \lambda_2 \geq \cdots \geq \lambda_p \geq 0$. Then the ith principal component is given by

$$Y_i = e'_i\mathbf{X} = e_{i1}X_1 + e_{i2}X_2 + \cdots + e_{ip}X_p, \quad i = 1, 2, \dots, p. \tag{11.3}$$

With these conditions,

$$\mathrm{Var}(Y_i) = e'_i \sum e_i = \lambda_i, \quad i = 1, 2, \dots, p, \tag{11.4}$$

$$\mathrm{Cov}(Y_i, Y_k) = e'_i \sum e_k = 0, \quad i \neq k. \tag{11.5}$$

If some λ_i are equal, the choices of the corresponding coefficient vectors, e_i, and hence Y_i are not unique.

Result 11.2 Let $\mathbf{X}' = [X_1, X_2, \dots, X_p]$ have covariance matrix associated with eigenvalue–eigenvector pairs $(\lambda_1, e_1), (\lambda_2, e_2), \dots, (\lambda_p, e_p)$ where $\lambda_1 \geq \lambda_2 \geq \cdots \geq \lambda_p \geq 0$. Let $Y_1 = e'_1\mathbf{X}, Y_2 = e'_2\mathbf{X}, \dots, Y_p = e'_p\mathbf{X}$ be the principal components. Then

$$\sigma_{11} + \sigma_{22} + \cdots + \sigma_{pp} = \sum_{i=1}^{p} \mathrm{Var}(X_i) = \lambda_1 + \lambda_2 + \cdots + \lambda_p = \sum_{i=1}^{p} \mathrm{Var}(Y_i). \tag{11.6}$$

We remark that the first PC is the direction which maximizes variability of the data when projected on that axis. The second PC is the direction, among those orthogonal to the first, maximizing variability. Each subsequent principal component is orthogonal to the previous ones, and points in the directions of the largest variance of the residual subspace.

In the example that follow, we present a heuristic approach to PCA. Before we present the example, we explain the concept of eigenvalues and eigenvectors.

11.4.1 The Eigenvalues and Eigenvectors

In the context of understanding PCA at a high level, all that we really need to know about eigenvectors and eigenvalues is that the eigenvectors of the covariance matrix are the axes of the principal components in a dataset. The eigenvectors define the directions of the principal components calculated by the PCA. The eigenvalues associated with the eigenvectors describe the magnitude of the eigenvector, or how far spread apart the observations (points) are along the new axis.

The first eigenvector will span the greatest variance (separation between points) found in the dataset, and all subsequent eigenvectors will be orthogonal to the one calculated before it. This is how we can know that each of the principal components will be uncorrelated with one another.

Each eigenvector found by PCA will pick up a combination of variance from the original variables in the data set. For symmetric matrices, eigenvectors for distinct eigenvalues are orthogonal, i.e.

$$Sv_{\{1,2\}} = \lambda_{\{1,2\}} v_{\{1,2\}} \quad \text{and} \quad \lambda_1 \neq \lambda_2 \to v_1 \cdot v_2 = 0$$

and all eigenvalues of a real symmetric matrix are real, i.e.

$$|S - \lambda I| = 0 \quad \text{and} \quad S = S^T \to \lambda \in \mathbb{R}.$$

In the example that follow, we consider the oval graph presented in Figure 11.3. The data points are displayed in $x - y$ axis. The x variable could be age and y the hours spent on the Internet. These are the two dimensions that the data set is currently being measured in. The principal component of the oval is a line splitting it longways as shown in Figure 11.4.

It turns out the other eigenvectors (remember there are only two of them as it is a 2-D problem) is perpendicular to the principal component. We recall that the eigenvectors have to be able to span the whole x–y area; in order to do this effectively, the two directions need to be orthogonal (i.e. perpendicular) to one another. This is why the x and y axis are orthogonal to each other in the first place. So the second eigenvector is shown in Figure 11.5.

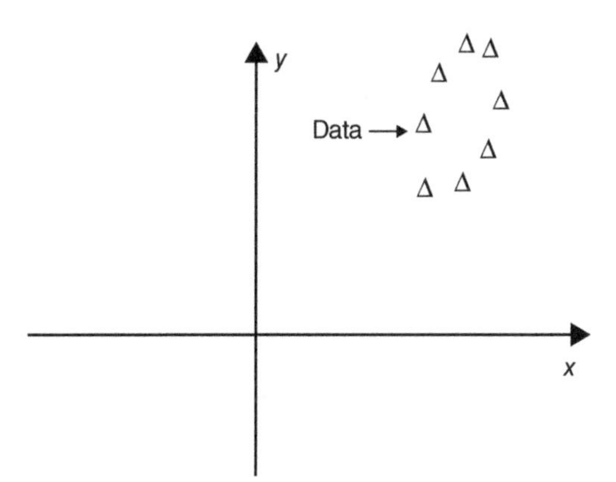

Figure 11.3 2D data set.

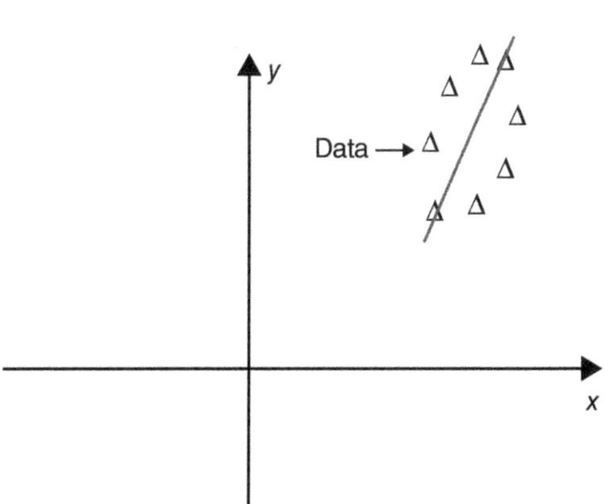

Figure 11.4 First PCA axis.

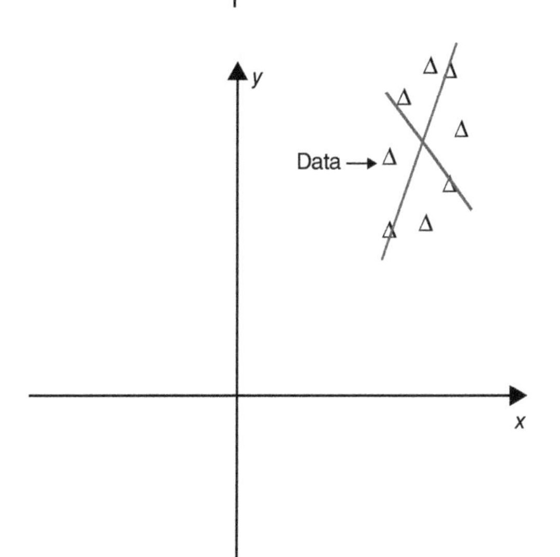

Figure 11.5 Second PCA axis.

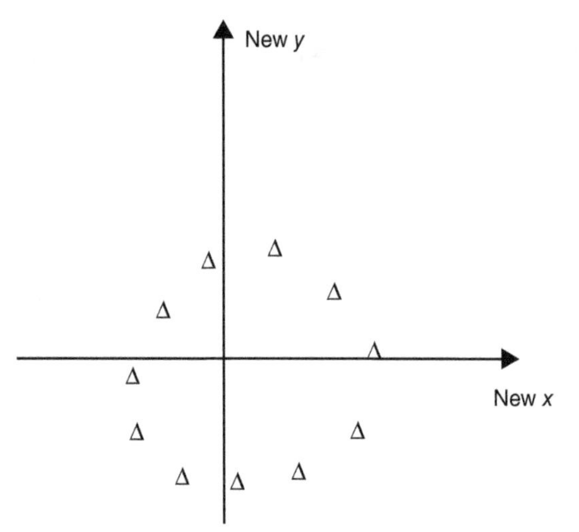

Figure 11.6 New axis.

The eigenvectors have given us a much more useful axis to frame the data. We can now reframe the data in these new dimensions as shown in Figure 11.6.

We observe that nothing has been done to the data itself. We are just viewing the data from a different angle. So obtaining the eigenvectors gets you from one set of axes to another. These axes are much more intuitive to the shape of the data now. These directions are where there is most variation, and that is where there is more information. If there were no variation in the data (e.g. everything was equal to 1), there would be no information. In this instance, the eigenvalue for that dimension would equal zero because there is no variation. The old axes were well defined (age (x) and hours spent on Internet (y)), whereas the new ones are not.

Example 11.1 (Calculating the population principal components) Suppose the random variables X_1 and X_2 have the covariance matrix

$$\Sigma = \begin{bmatrix} 5 & 2 \\ 2 & 2 \end{bmatrix}.$$

It may be verified that the eigenvalue–eigenvector pairs are

$$\lambda_1 = 6, \quad e_1' = [0.894, 0.447],$$
$$\lambda_1 = 1, \quad e_1' = [0.447, -0.894].$$

Therefore, the principal components become

$$Y_1 = e_1'\mathbf{X} = 0.894X_1 + 0.447X_2,$$
$$Y_2 = e_1'\mathbf{X} = 0.447X_1 - 0.894X_2.$$

Remarks 11.1

- PCA is not scale invariant, i.e. a feature changes if length scales are multiplied by a common factor. Also changing the units of each column will change the results of PCA.
- Parameters with high variance will influence the principal components more.
- A good practice is to center and scale your variables (prcomp in R does not scale automatically).
- Because PCA is based off on orthogonal vectors, the signs for each vector are arbitrary, i.e. one could run PCA on the same data set and get all the same numbers except the sign would be flipped.

As the dimensionality of data increases, the manageability and effectiveness of the data tend to decrease. On a high level, increase in dimensionality is related to the fact that as variables or features are added to a data set, the average and minimum distance between points or observations increase. In general, creating good predictions becomes more difficult as the distance between the known points and the unknown points increases. Additionally, features in the data set may not add much value or predictive power in the context of the target (dependent) variable. These features do not improve the model, rather they increase noise in the dataset, as well as the overall computational load of the model.

When collecting data or modeling a data set, it is not always straightfoward to know which variables are important. There is no guarantee that the variables you picked or were provided are the right variables. PCA looks at the overall structure of the continuous variables in a data set to extract meaningful signals from the noise in the data set. It aims to eliminate redundancy in variables while preserving important information.

11.5 How PCA Works

PCA is a transformation method that creates (weighted linear) combinations of the original variables in a data set, with the intent that the new combinations will capture as much variance (i.e. the separation between points) in the dataset as possible while eliminating correlations (i.e. redundancy). PCA creates the new variables by transforming the original observations in a dataset to a new set of variables using the eigenvectors and eigenvalues calculated from a covariance matrix of your original variables. The formal steps in PCA are as follows:

Step 1: The first step of PCA is centering the values of all of the input variables (e.g. subtracting the mean of each variable from the values), making the mean of each variable equal to zero. Centering is an important preprocessing step because it ensures that the resulting components are only looking at the variance within the dataset, and not capturing the overall mean of the dataset as an important variable (dimension). Without mean-centering, the first principal component found by PCA might correspond with the mean of the data instead of the direction of maximum variance.

Step 2: Next we scale the variables. This step is needed if the variables have different units.

Step 3: The covariance matrix of the data needs to be calculated. Covariance is measured between two variables (dimensions) at a time and describes how related the values of the variables are to one another. A large covariance value (positive or negative) indicates that the variables have a strong linear relationship with one another. Covariance values close to 0 indicate a weak or nonexistent linear relationship.

We simplify the above steps into the following algorithm.

11.5.1 Algorithm

Let x_i denote a p-dimensional vectors (data points), $i = 1, \dots, n$

1. Standardize the data (this step is needed if the variables have different units).
2. Calculate the $p \times p$ covariance matrix Σ.

$$\Sigma = \frac{1}{n}\sum_{i=1}^{n}(x_i - \overline{x})(x_i - \overline{x})^T, \quad \text{where } \overline{x} = \frac{1}{m}\sum_{i=1}^{n}x_i.$$

3. Calculate the eigenvectors of the covariance matrix (orthonormal basis).
4. PCA basis vectors = the eigenvectors of Σ.
5. Select p eigenvectors that correspond to the largest p eigenvalues to be the new basis.

Normalizing the data prior to performing PCA can be important, particularly when the variables have different units or scales. PCA assumes that the data can be approximated by a linear structure and that the data can be described with fewer features. It assumes that a linear transformation can and will capture the most important aspects of the data. It also assumes that high variance in the data means that there is a high signal-to-noise ratio. Dimensionality reduction does result in a loss of some information. By not keeping all the eigenvectors, there is some information that is lost. However, if the eigenvalues of the eigenvectors that are not included are small, you are not losing too much information. Another consideration to make with PCA is that the variables become less interpretable after being transformed.

We illustrate an application of PCA to real data sets.

11.6 Application

The datasets used in this application corresponds to the weekly rates of return for five stocks (JP Morgan, Citibank, Wells Fargo, Royal Dutch Shell, and ExxonMobil) listed on the New York Stock Exchange were determined for the period January 2004 through December 2005. The weekly rates of return are defined as (current week closing price–previous week closing price)/(previous week closing price), adjusted for stock splits and dividends. All the analysis were performed using the R statistical software.

Figure 11.7 is a scatterplot of Royal Dutch Shell stock versus Exxon Mobil stock. From the plot, we observe that there exists some linear relationship between the two stocks.

The principal components of the five stocks are given as:

```
                         PC1        PC2         PC3        PC4
JP_Morgan         -0.4690832  0.3680070 -0.60431522  0.3630228
Citibank          -0.5324055  0.2364624 -0.13610618 -0.6292079
Wells_Fargo       -0.4651633  0.3151795  0.77182810  0.2889658
Royal_Dutch_Shell -0.3873459 -0.5850373  0.09336192 -0.3812515
Exxon_Mobil       -0.3606821 -0.6058463 -0.10882629  0.4934145
                         PC5
JP_Morgan          0.38412160
Citibank          -0.49618794
Wells_Fargo        0.07116948
Royal_Dutch_Shell  0.59466408
Exxon_Mobil       -0.49755167

Importance of components:
                          PC1    PC2    PC3     PC4     PC5
Standard deviation     1.5612 1.1862 0.7075 0.63248 0.50514
Proportion of Variance 0.4874 0.2814 0.1001 0.08001 0.05103
Cumulative Proportion  0.4874 0.7689 0.8690 0.94897 1.00000
```

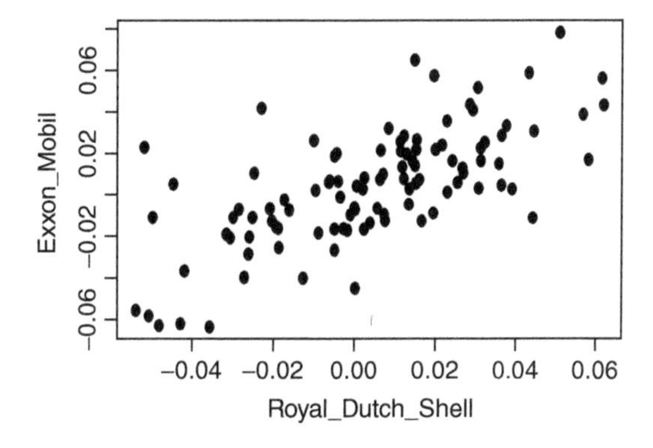

Figure 11.7 Scatterplot of Royal Dutch Shell stock versus Exxon Mobil stock.

The first component (PC1) is a roughly equally weighted sum, or "index" of the five stocks. The second component (PC2) represents a contrast between the banking stocks (JP Morgan, Citibank, Wells Fargo) and the oil stocks (Royal Dutch Shell, Exxon-Mobil). Thus, we can conclude that based on (PC1) and (PC2) most of the variation in these stock returns are due to market activity and uncorrelated industry activity.

Also the results from the proportion of variance explains the percentage of the total variation in the dataset. For example, (PC1) explains about 49% of the total variance, which means that nearly half of the information in the dataset (five variables) can be explained by just that one Principal Component. (PC2) explains 28% of the total variance. Therefore, by knowing the position of a sample in relation to just (PC1) and (PC2), we can get a very accurate view on where it stands in relation to other samples, as just (PC1) and (PC2) can explain 77% of the variance.

11.7 Problems

1 Discuss two advantages of using principal components analysis.

2 Discuss why we sometimes need to standardize a datasets before we apply PCA.

3 What is the difference between a variance–covariance matrix and a correlation matrix?

4 Given the variance–covariance matrix,

$$\Sigma = \begin{bmatrix} 10 & 5 \\ 5 & 6 \end{bmatrix}.$$

Compute the eigenvalues λ_1 and λ_2 and the corresponding eigenvectors. Please ensure that the eigenvectors are orthogonal.

5 Compute the population principal components Y_1 and Y_2 for the following variance–covariance matrix.

$$\Sigma = \begin{bmatrix} 1 & 3 \\ 3 & 50 \end{bmatrix}$$

6 Explain the sample variability with two sample principal components and one sample principal component.

7 Perform a principal components analysis using the sample covariance matrix of the data given in Example 3.2.1.

8 For the following statements state whether True or False and explain your reasoning.
(a) Subsequent principal components are always orthogonal to each other.
(b) The output of PCA is a new representation of the data that is always a higher dimensionality than the original feature representation.

9 Select any complete data sets that contains two variables.
(a) Visualize the data sets to identify any pattern or trend in the data.
(b) Draw all the principal components.

12

Discriminant and Cluster Analysis

12.1 Introduction

This chapter is divided into two parts: Discriminant analysis and cluster analysis. In the first part, we will present the discriminant analysis technique. Discriminant analysis finds a set of prediction equations based on explanatory (independent) variables that are used to classify individuals into groups. In practice, there are two possible goals in a discriminant analysis: finding a predictive equation for classifying new individuals and interpreting the predictive equation to better understand the relationships that may exist among the variables. Discriminant analysis is very similar to multiple regression analysis; however, the main difference between these two techniques is that regression analysis deals with a continuous dependent variable, while discriminant analysis deals with a discrete dependent variable. We will conclude the first part of this chapter by presenting some real applications to seismic and financial time series.

In the second part of this chapter, we will discuss cluster analysis. Cluster analysis or clustering is the task of grouping a set of objects in such a way that objects in the same group (called a cluster) are more similar (in some sense) to each other than to those in other groups (clusters). Cluster analysis is different from discriminant analysis in the sense that there are no predefined classes. Typical applications of cluster analysis include to get insight into data distribution, and it is also a preprocessing step for other algorithms.

Before we proceed with the discussion of discriminant and cluster analysis, we introduce the concept of distance.

12.2 Distance

Most multivariate techniques are based upon the concept of distance. If we consider the point $A = (x_1, x_2)$ in the plane, the **Euclidean or straight line distance**,

Data Science in Theory and Practice: Techniques for Big Data Analytics and Complex Data Sets, First Edition. Maria Cristina Mariani, Osei Kofi Tweneboah, and Maria Pia Beccar-Varela.

$d(O,A)$, from A to the origin $O = (0,0)$ is defined as

$$d(O,A) = \sqrt{x_1^2 + x_2^2}. \tag{12.1}$$

In general if $A = (x_1, x_2, \ldots, x_p)$, the Euclidean distance from A to the origin $O = (0, 0, \ldots, 0)$ is

$$d(O,A) = \sqrt{x_1^2 + x_2^2 + \cdots + x_p^2}. \tag{12.2}$$

The Euclidean distance between two arbitrary points A and B with coordinates $A = (x_1, x_2, \ldots, x_p)$ and $B = (y_1, y_2, \ldots, y_p)$ is given by

$$d(A,B) = \sqrt{(x_1 - y_1)^2 + (x_2 - y_2)^2 + \cdots + (x_p - y_p)^2}. \tag{12.3}$$

Another distance measure is the Minkowski metric. The Minkowski metric between the points A and B with coordinates $A = (x_1, x_2, \ldots, x_p)$ and $B = (y_1, y_2, \ldots, y_p)$ is given by

$$d(A,B) = \left[\sum_{i=1}^{p} |x_i - y_i|^m\right]^{\frac{1}{m}}. \tag{12.4}$$

If $m = 1$, $d(A,B)$ measures "city-block" (Manhattan) distance between two points in p dimensions. For $m = 2$, $d(A,B)$ becomes the Euclidean distance. In general, varying m changes the weight given to larger and smaller differences.

Two additional popular measures of distance or dissimilarity are given by the Canberra metric and the Czekanowski coefficient. Both of these measures are defined for nonnegative variables only. We have

$$\text{Canberra metric} \quad d(A,B) = \sum_{i=1}^{p} \frac{|x_i - y_i|}{(x_i + y_i)}. \tag{12.5}$$

$$\text{Czekanowski coefficient} \quad d(A,B) = 1 - \frac{2\sum_{i=1}^{p} \min(x_i, y_i)}{\sum_{i=1}^{p} (x_i + y_i)}. \tag{12.6}$$

12.3 Discriminant Analysis

Discriminant analysis is a tool that characterizes some classes or groups based on their similar behaviors. The goal of discriminant analysis is to develop discriminant functions (linear combination of explanatory variables) that will discriminate between the categories of the dependent variable. It helps the researchers and practitioners to examine whether significant differences exist among the groups, in terms of the independent variables. It also evaluates the accuracy of the classification.

For example, a researcher could perform a discriminant analysis to determine patients at high or low risk for stroke. The analysis might classify patients into high- or low-risk groups or categories, based on personal attributes such as cholesterol level, body mass, minutes of exercise per week, packs of cigarettes per day, etc. The Internal Revenue Service (IRS) also uses discriminant analysis to identify people they may want to audit.

There are several different ways to conduct two-group discriminant analysis. Two-group discriminant analysis deals with classifying observations into one of two groups, based on two or more quantitative, predictor variables.

In this book, we will discuss the discriminant analysis based on Kullback–Leibler and Chernoff distances.

12.3.1 Kullback–Leibler Divergence

The Kullback–Leibler (K–L) divergence is used to measure the distance between two probability distributions. It helps to measure how one probability distribution is different from other. To define the K–L divergence (or K–L distance), we consider a statistical decision problem of classifying a random observation x as one of two possible classes (C_1 and C_2). We define the probabilities $w_1 = \Pr(C_1) > 0$, $w_2 = \Pr(C_2) = 1 - w_1 > 0$ as " a priori" class probabilities, and $p_1(x) = \Pr(x|C_1)$ and $p_2(x) = \Pr(x|C_2)$ as class conditional probabilities satisfying $p(x) = w_1 p_1(x) + w_2 p_2(x)$. The K–L distance is then defined as:

$$\text{K–L}(p_1 : p_2) = \int p_1(x) log \frac{p_1(x)}{p_2(x)}\, dx, \tag{12.7}$$

where $\text{K–L}(p_1 : p_2) \neq \text{KL}(p_2 : p_1)$.

The K–L divergence measures the expected number of extra bits required to code samples from the class conditional probability $p_1(x)$ when using a code based on $p_2(x)$, rather than using a code based on $p_1(x)$. Even though the K–L divergence measures the "distance" between two distributions, it is not a distance measure.

12.3.2 Chernoff Distance

Chernoff distance or Chernoff-α coefficient is also useful in discriminating the statistical samples or populations. It measures the similarity of two probability distributions, which is defined as:

$$C_\alpha(p_1 : p_2) = \min_{0 \le \alpha \le 1} \int p_1(x)^{1-\alpha} p_2(x)^\alpha \, dx. \tag{12.8}$$

The optimal Chernoff α-coefficient is obtained using the regularization technique for α. Since it is a measure of similarity ($0 < C_\alpha(p_1, p_2) \leq 1$) relating to the overlapping of the densities p_1 and p_2, equivalently, the Chernoff information is defined as:

$$C(p_1 : p_2) = \max_{0\leq\alpha\leq 1} -\log \int p_1(x)^{1-\alpha} p_2(x)^\alpha \, dx. \tag{12.9}$$

Before we proceed, we discuss the concept of discrete Fourier transform (DFT).

The DFT is the equivalent of the continuous Fourier transform for signals known only at N instants separated by sample times T (i.e. a finite sequence of data). It is usually used in practical applications since we generally cannot observe signals continuously. The DFT vector $X(\omega_k)$ is normally distributed with means $M_j(\omega_k)$ and spectral matrices $f_j(\omega_k)$ at frequencies $\omega_k = k/n$, where $k = 0, 1, \ldots, [n/2]$. At this point, we obtain the means and spectral matrices for two populations ($\pi_j, j = 1, 2$) for example earthquakes and explosions groups. The DFT vector $X(\omega_k)$ is uncorrelated at different frequencies ω_k and ω_l for $k \neq l$. Since $X(\omega_k)$ is a time series vector, we assume that the means are equal and the covariance matrices are different. Now, from Whittle approximation, the log likelihood of probability density $p_j(X)$ (Shumway and Stoffer 2010) is defined as follows:

$$\ln p_j(X) = \sum_{0<\omega_k<1/2} [-\ln |f_j(\omega_k)| - X^*(\omega_k) f_j^{-1}(\omega_k) X(\omega_k)] \tag{12.10}$$

which can be written as:

$$\ln p_j(X) = \sum_{0<\omega_k<1/2} [-\ln |f_j(\omega_k)| - \text{tr}\{I(\omega_k) f_j^{-1}(\omega_k)\}]. \tag{12.11}$$

The periodogram matrix is $I(\omega_k) = X(\omega_k)X^*(\omega_k)$, where $X^*(\omega_k)$ is the complex conjugate of DFT vector. For equal "prior" probabilities, we may assign an observation X into population π_i whenever $\ln p_i(X) < \ln p_j(X)$ for $j \neq i, j = 1, 2$.

Using the definition of K–L information, the expectation of periodogram matrix can be expressed as the approximation of the spectral matrix (Kakizawa et al. 1998):

$$E_j I(\omega_k) = f_j(\omega_k).$$

Now, the ratio of densities is approximated as:

$$\ln \frac{p_1(X)}{p_2(X)} = \sum_{0<\omega_k<\frac{1}{2}} \left[-\ln \frac{|f_1(\omega_k)|}{|f_2(\omega_k)|} - \text{tr}\{(f_2^{-1}(\omega_k) - f_1^{-1}(\omega_k)) I(\omega_k)\} \right]. \tag{12.12}$$

Equivalently,

$$I(f_1 : f_2) = \frac{1}{n} E_1 \ln \frac{p_1(X)}{p_2(X)}$$

$$= \frac{1}{n} \sum_{0<\omega_k<\frac{1}{2}} \left[\text{tr}\{f_1(\omega_k) f_2^{-1}(\omega_k)\} - \ln \frac{|f_1(\omega_k)|}{|f_2(\omega_k)|} - p \right]. \tag{12.13}$$

The Chernoff information as a measure of disparity between the densities:

$$B_\alpha(p_1; p_2) = -\ln E_2 \left\{ \left(\frac{p_2(x)}{p_1(x)} \right)^\alpha \right\}.$$

The measurement is optimized by regularizing the parameter α, where $0 < \alpha \leq 1$. We will optimize the measurement as follows:

$$B_\alpha(f_1 : f_2) = \frac{1}{2n} \sum_{0<\omega_k<1/2} \left[\ln \frac{|\alpha f_1(\omega_k) + (1-\alpha) f_2(\omega_k)|}{|f_2(\omega_k)|} - \alpha \ln \frac{|f_1(\omega_k)|}{|f_2(\omega_k)|} \right]. \tag{12.14}$$

Note that $B_\alpha(p_1 : p_2) = B_{1-\alpha}(p_2 : p_1)$. When we scale $B_\alpha(p_1; p_2)$ by $\alpha(1-\alpha)$, it converges to $I(p_1 : p_2)$ for $\alpha \to 0$, and $I(p_2 : p_1)$ for α 1. So the Chernoff distance tends to behave as K–L distance when α approaches 0 and 1.

We will discuss two applications of the discriminant analysis technique. The first and second application deals with applying the discriminant technique to sets of seismic time series and financial time series, respectively.

12.3.3 Application – Seismic Time Series

We begin by briefly presenting some background infomation of the seismic time series. The earthquake series used in this example consist of a set of magnitude 3.0–3.3 aftershocks of a recent $M = 5.2$ intraplate earthquake in June, 2014. These earthquakes were located near Clifton, Arizona. A set of mining explosions cataloged with similar magnitudes as the earthquakes ($M = 3.0$–3.5) were selected from the same region (within a radius of 6.2 miles). We then choose an unknown seismic signal that occurred in March, 2001, located in the same region, Arizona. The aim of selecting the new unknown signal is to identify whether it was an earthquake or an explosion series. The data contains information about the longitude, latitude, the average distance to seismic events, average azimuth, and the magnitude of each seismic event in the region (see Table 12.1).

The discriminant analysis technique is used to classify an unknown seismic event in Arizona.

Table 12.1 Events information.

Events	Station	Magnitude	Latitude	Longitude
EA-1	TUC	3.3	32.59°	−109.12°
EA-2	TUC	3.2	32.63°	−109.01°
EA-3	TUC	3.2	32.64°	−109.11°
EA-4	TUC	3	32.58°	−109.08°
EA-5	TUC	3	32.53°	−109.06°
EX-1	TUC	3.2	32.65°	−109.08°
EX-2	TUC	3.1	32.59°	−109.05°
EX-3	TUC	3	32.67°	−109.08°
EX-4	TUC	3.1	32.65°	−109.09°
EX-5	ANMO	3.2	32.65°	−109.08°
UN	ANMO	3.2	32.64°	−109.15°

We applied the K–L and Chernoff distance techniques on earthquake and explosion groups. These techniques measure the similarity of two statistical samples or populations. The statistics in Eq. (12.13) are $n = 1024$, window size $k = 256$, frequency $\omega_k = 0.25$. From Eq. (12.13), we obtain the K–L divergence using the diagonal elements of the spectral matrices. We then optimized the Chernoff coefficient, α (0.58), to estimate the maximum value of Chernoff disparity $B_\alpha(\hat{f}_1, \hat{f}_2)$ in Eq. (12.14). The K–L and Chernoff distances of the unknown seismic event are 0.219 and 0.061, respectively. From Table 12.2, we see that all explosions have positive K–L and Chernoff distances, where the earthquakes have negative distances. Figure 12.2 shows the classification of earthquakes and explosions by using the Chernoff differences along with the K–L differences. It is clear that the points in the first quadrant are classified as mining explosions and the points in the third

Table 12.2 Discriminant scores for earthquakes and explosions groups.

Events	K–L	Chernoff	Events	K–L	Chernoff
EA1	−3.313	−0.082	EX1	0.330	0.012
EA2	−2.273	−0.064	EX2	1.922	0.060
EA3	−1.854	−0.056	EX3	1.125	0.068
EA4	−0.452	−0.029	EX4	0.725	0.085
EA5	−0.151	−0.013	EX5	0.222	0.060

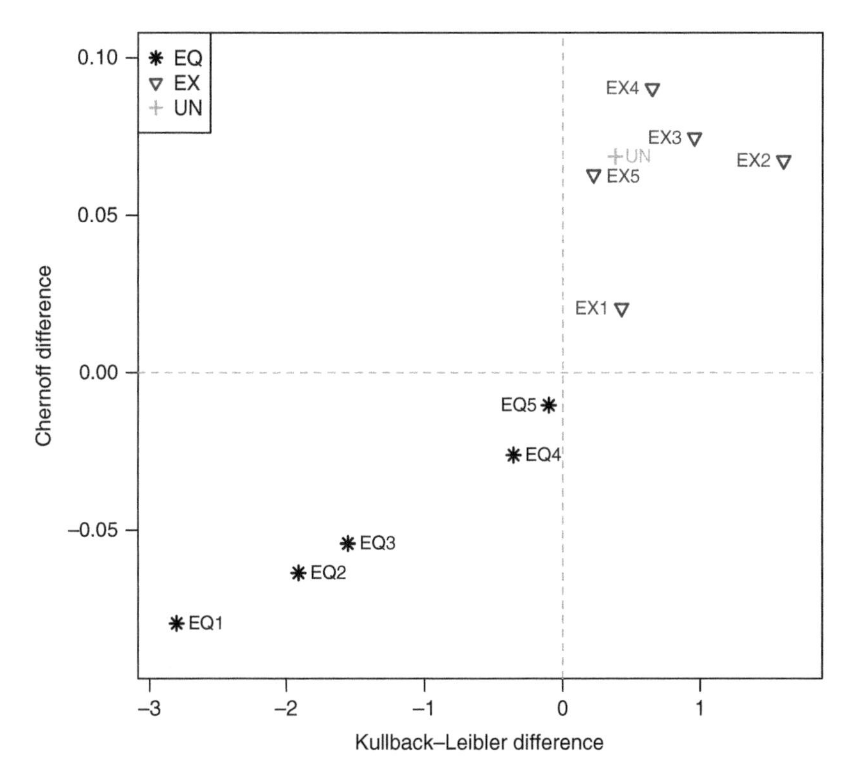

Figure 12.1 Classification (by quadrant) of earthquakes and explosions using the Chernoff and Kullback–Leibler differences.

quadrant are classified as earthquakes. So we conclude that the unknown event is classified as an explosion.

12.3.4 Application – Financial Time Series

The data used in this example corresponds to a sampled minute by minute time series recorded on 15 September 2008 for the Lehman Brothers collapse and 6 May 2010 for the flash crash event. These time series were made up of 1024 data points. The time series used contained the following companies: ExxonMobil Corporation (XOM), Walmart Retail Company (WMT), Verizon Communications, Inc. (VZ), United Technologies Corporation (UTX), and McDonald's Corporation (MCD). We also selected two stock market data (Citigroup, Inc. from 2009 and Iamgold Corporation [IAG] from 2011) to determine whether the Lehman Brothers had an influence on each of them and if the flash crash had an influence in the IAG from 2011.

Table 12.3 Discriminant scores for Lehman Brothers collapse and Flash crash event.

Stocks	K–L scores	Chernoff scores	Stocks	K–L scores	Chernoff scores
XOM-Lehman	−0.425	−0.022	XOM-Flash	0.085	0.017
WMT-Lehman	−0.159	−0.009	WMT-Flash	0.068	0.016
VZ-Lehman	−0.156	−0.009	VZ-Flash	0.102	0.018
UTX-Lehman	−0.436	−0.022	UTX-Flash	0.163	0.034
MCD-Lehman	−0.101	−0.006	MCD-Flash	0.124	0.025

As we did in the first application in the previous section, the K–L and Chernoff distance techniques are applied to the Lehman Brothers collapse and Flash crash event based on their frequency domain. An important feature of these techniques is that they measure the similarity of two statistical samples or populations. Equation (12.13) helps us to obtain the K–L divergence using the diagonal elements of these spectral matrices. We then optimized the Chernoff coefficient, α (0.58), to estimate the maximum value of Chernoff disparity $B_\alpha(\hat{f}_1,\hat{f}_2)$ in Eq. (12.14). The K–L and Chernoff distances of Citigroup (2009) stock market are obtained as −0.106 and −0.006, respectively. Similarly, K–L and Chernoff distances of IAG (2011) stock market are obtained as −1.283 and −0.033, respectively. From Table 12.3, we see that the Lehman Brothers collapse have negative K–L and Chernoff distances and the flash crash have positive K–L and Chernoff distances. Thus, we have correctly discriminated between the two events. Similarly, in Table 12.4, we observe that the Citigroup (2009) and IAG (2011) have negative K–L and Chernoff distances. Figure 12.2 shows the classification of Lehman Brothers collapse and Flash crash event using the Chernoff differences along with the K–L differences. It is clear that the points in the first quadrant are classified as Flash-Crash events, and the points in the third quadrant are classified as Lehman Brothers collapse. Therefore, we conclude that the Lehman Brothers collapse had an influence on these two events.

In the second part of this chapter, we discuss the concept of cluster analysis.

Table 12.4 Discriminant scores for Citigroup in 2009 and IAG stock in 2011.

Stocks	K–L scores	Chernoff scores
CITI (2009)	−0.106	−0.006
IAG (2011)	−1.283	−0.033

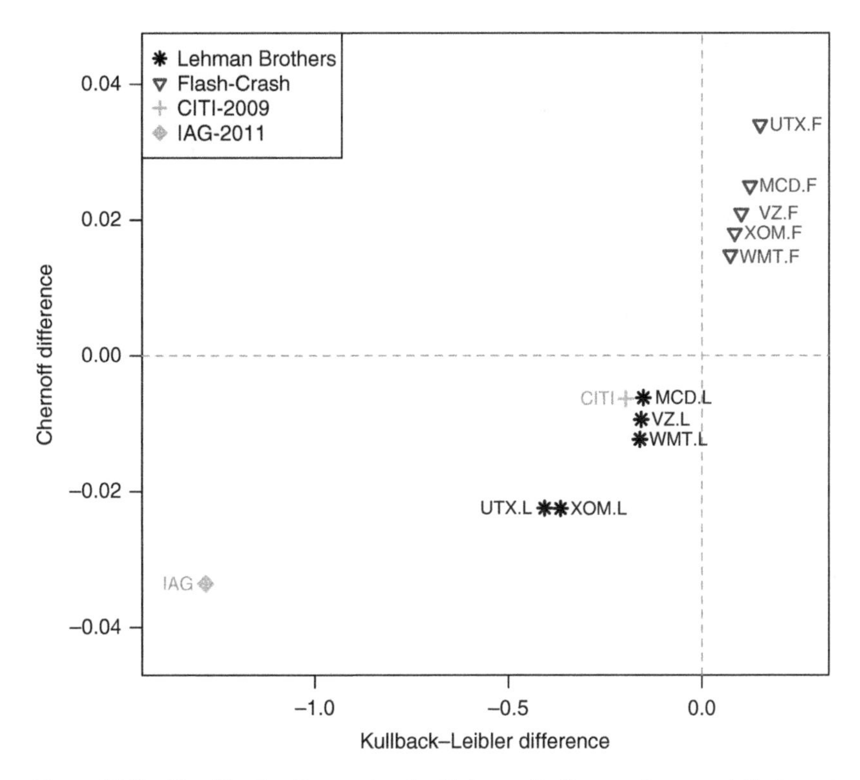

Figure 12.2 Classification (by quadrant) of Lehman Brothers collapse and Flash crash event using the Chernoff and Kullback–Leibler differences.

12.4 Cluster Analysis

Clustering is a technique of grouping datasets in such a way that the data in the same group have more similar properties than other groups. Examples of clustering are the following:

- Help businesses discover distinct groups in their customer bases, and then use this knowledge to develop targeted marketing programs.
- Identifying categories of life insurance policy holders with a high average claim cost.
- Grouping earth quake epicenters along a country or continent faults, etc.

In general, a good clustering method will produce high-quality clusters with high intra-class similarity and low inter-class similarity. The quality of a clustering result depends on both the similarity measure used by the method and its implementation. Similarity is expressed in terms of a distance function, which is typically metric, i.e. $d(i, j)$. In practice, the definitions of distance functions are usually very different for interval-scaled, boolean, categorical, and ordinal variables.

There are several approaches to clustering and they are:

- *Partitioning algorithms*: Construct various partitions and then evaluate them by some criterion.
- *Hierarchy algorithms*: Create a hierarchical decomposition of the set of data (or objects) using some criterion.
- *Density-based*: Based on connectivity and density functions.
- *Grid-based*: Based on a multiple-level granularity structure.
- *Model-based*: A model is formulated for each of the clusters and the idea is to find the best fit of that model to each other.

In this chapter, we will focus on the first approach, i.e. partitioning algorithms: Construct various partitions and then evaluate them by some criterion.

12.4.1 Partitioning Algorithms

For the partitioning method, we construct a partition of a database D of n objects into a set of k clusters.

Given a k, find a partition of k clusters that optimizes the chosen partitioning criterion. The two ways of selecting k is by using the k-means and the k-medoids.

We explain each type in more details.

12.4.2 *k*-Means Algorithm

The k-means algorithm is an iterative algorithm that divides the dataset into k pre-defined unique nonoverlapping clusters were each data point belongs to only one group. It tries to make the inter-cluster data points as similar as possible while also keeping the clusters as different as possible. It works by assigning data points to subgroups such that the sum of the squared distance between the data points and the cluster's centroid (mean) is at the minimum. The centroid is defined as the arithmetic mean of all the data points that belong to that cluster. The less variation we have within clusters, the more homogeneous (similar) the data points are within the same cluster.

The process is composed of these three steps:

Step 1: Partition the dataset into k initial clusters.

Step 2: Proceed through the list of items in the datasets, assigning an item to the cluster whose centroid is nearest. Distance is usually computed using Euclidean distance with either standardized or unstandardized observations. Recalculate the centroid for the cluster receiving the new item and for the cluster losing the item.

Step 3: Repeat Step 2 until no more reassignments take place.

In practice, instead of starting with a partition of all items into k preliminary groups in Step 1, we could specify k initial centroids and then proceed to Step 2. The final assignment of items to clusters will be dependent upon the initial partition or the initial selection of seed points. Experience suggests that most major changes in assignment occur with the first reallocation step.

12.4.3 *k*-Medoids Algorithm

The k-medoids or partitioning around medoids (PAM) algorithm is a clustering algorithm which is similar to the k-means algorithm. In contrast to the k-means algorithm, k-medoids chooses data points as centers (medoids) and can be used with arbitrary distances, while in k-means the center of a cluster is not necessarily one of the input data points (it is the average between the points in the cluster). The algorithm for the k-medoids is as follows:

Step 1. Initialize: Randomly select k of the n data points as the medoids.
Step 2. Assignment step: Associate each data point to the closest medoid.
Step 3. Update step: For each medoid m and each data point o associated to m swap m and o and compute the total cost of the configuration (that is, the average dissimilarity of o to all the data points associated to m). Select the medoid o with the lowest cost of the configuration.
Step 4. Repeat alternating Steps 2 and 3 until there is no change in the assignments.

Next, we briefly define the J-Divergence. The J-Divergence is defined as a symmetric disparity measure:

$$J(f_1 : f_2) = I(f_1 : f_2) + I(f_2 : f_1). \tag{12.15}$$

The definition of the first and second term follows from Eq. (12.13). We use the disparity as a quasi-distance between the sample spectral matrix of a single vector $\mathbf{x}$, and the population π_j ($j = 1, 2$):

$$J(\hat{f} : f_j) = I(\hat{f} : f_j) + I(f_j : \hat{f}). \tag{12.16}$$

Eq. (12.16) implies that the quasi-distance between sample series with estimated spectral matrices $\hat{f}_i$ and f_j, follows the approximation:

$$J(\hat{f}_i;\hat{f}_j) = \frac{1}{n} \sum_{0<\omega_k<1/2} \left[\mathrm{tr}\{\hat{f}_i(\omega_k)\hat{f}_j^{-1}(\omega_k)\} + \mathrm{tr}\{\hat{f}_j(\omega_k)\hat{f}_i^{-1}(\omega_k)\} - 2p \right] \tag{12.17}$$

for $i \neq j$. The PAM algorithm is intended to obtain a sequence of objects called medoids that are located in the clusters. So it partitions the datasets of n objects into k clusters, where both the dataset and the number k are inputs of the algorithm. This approach minimizes the total dissimilarities between the objects and their closest selected object. For details of the PAM algorithm, see Kaufman and Rousseeuw (1990).

Next we present two applications of the cluster analysis technique. We remark that the data sets used in this example are the same data used in Section 12.3.3.

12.4.4 Application – Seismic Time Series

We performed the clustering technique on earthquakes and explosions data and on the unknown seismic event. We used the PAM algorithm with symmetric divergence to obtain two clusters using the five earthquakes, five explosions, and one unknown seismic event. In Figure 12.3, the circles indicate earthquake events and triangles present the explosion events. Since the unknown event belongs to the same ellipse as explosions, we classified it as an explosion event.

12.4.5 Application – Financial Time Series

We used the PAM algorithm to cluster the Lehman Brothers collapse and Flash crash event and their effects. We also selected the data of Citigroup (2009) and IAG stock (2011) to see whether the market data of these years were influenced by the crashes of Lehman Brothers collapse (2008) and also verify if the stock IAG stock (2011) was affected by the Flash crash event (2010). In Figure 12.4, the symbols

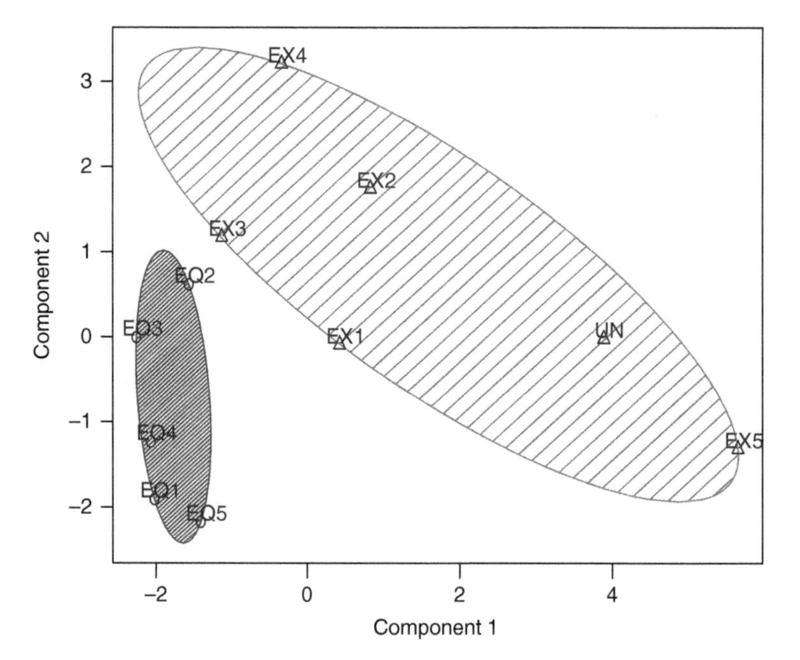

Figure 12.3 Clustering results for the earthquake and explosion series based on symmetric divergence using PAM algorithm.

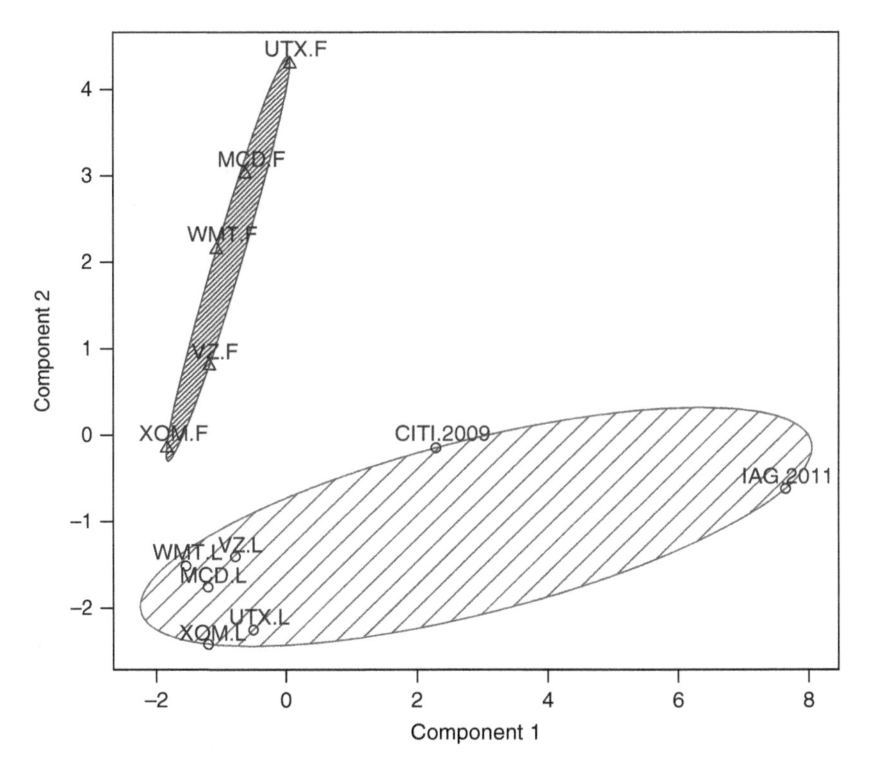

Figure 12.4 Clustering results for the Lehman Brothers collapse, Flash crash event, Citigroup (2009), and IAG (2011) stock data based on symmetric divergence using the PAM algorithm.

of triangle indicate the flash crash event classification and the symbols of circles indicate the Lehman Brothers collapse classification. From the figure, we see that the stocks from Citigroup (2009) and IAG (2011) were influenced by the Lehman Brothers collapse since they are in the same ellipse. On the other hand, the stocks from IAG (2011) had no influence from the flash crash event.

12.5 Problems

1 Discuss the difference between discriminant and cluster analysis.

2 Compute the Euclidean distance between the following points:
   (a) $P(20, 80)$ and $Q(30, 44)$
   (b) $P(20, 80)$ and $Q(90, 40)$
   (c) $P(30, 44)$ and $Q(90, 40)$

3 Compute the Manhattan distance between the following points:
(a) $A(8, -1)$ and $B(-1, 2)$
(b) $A(2, 4)$ and $B(0, 3)$
(c) $A(20, 34)$ and $B(10, 40)$

4 Discuss the difference between k-medoids algorithm and k-means algorithm.

5 Discuss the difference between k-means and hierarchical clustering algorithms.

6 Explain a way to improve supervised learning algorithms with clustering.

7 Draw a flowchart diagram for the k-medoids algorithm and k-means algorithm described in this chapter.

8 Discuss some advantages and disadvantages of using the k-medoids algorithm and k-means algorithm.

9 In the following statements, state whether True or False and explain your reasoning.
(a) In discriminant analysis, there is no multicollinearity in the predictor variables. Therefore, there is no ambiguous measure of the relative importance of the predictors in discriminating between groups.
(b) Discriminant functions are linear combinations of the predictor or independent variables, which will best discriminate between the categories of the dependent variable.
(c) In two-group discriminant analysis it is possible to derive only one discriminant function.

10 What are some real-life applications of the discriminant and cluster analysis?

13

Multidimensional Scaling

13.1 Introduction

In this chapter, we will discuss other techniques for visualizing high-dimensional data namely, multidimensional scaling (MDS). In Chapter 11, we discussed the principal components analysis (PCA) technique for visualizing high-dimensional data sets. MDS is a visualization technique which is used to display the information contained in a distance matrix. A distance matrix is a square matrix (two-dimensional array) containing the distances, taken pairwise, between the elements of a set.

MDS is the art of visualizing the level of similarity of individual cases of a dataset. They translate information about the pairwise "distances" among a set of m objects into a configuration of m points mapped into an abstract Cartesian space. It is a form of nonlinear dimensionality reduction, and MDS can be also used to reveal hidden trends in a correlation matrix.

In this chapter, we will discuss two general methods for solving the MDS problem. The first is called metric multidimensional scaling (MMDS) which tries to reproduce the original metric or distances. The second method, called non-metric multidimensional scaling (NMMDS), assumes that only the ranks of the distances are known. Similar to PCA, MDS can be used with supplementary or illustrative elements, which are projected onto the dimensions after they have been computed. The MMDS solution may result in projections of data objects that conflict with the ranking of the original observations. The NMMDS solves this problem by iterating between a monotizing algorithmic step and a least-squares projection step.

Before we discuss the MDS technique in more detail, we present an example to motivate the reader.

Data Science in Theory and Practice: Techniques for Big Data Analytics and Complex Data Sets, First Edition. Maria Cristina Mariani, Osei Kofi Tweneboah, and Maria Pia Beccar-Varela.

13.2 Motivation

The following example will help explain what MDS does. Consider the following set of data in Table 13.1.

A scatter plot of these data points are shown on Figure 13.1.

Figure 13.1 helps us to visually assess the distance between each pair of points. From the figure, the point A is near B, but far from C and D. In fact the points C and D each seem to be by themselves. The actual distance between two points i

Table 13.1 Data matrix.

Label	X	Y
A	1	6
B	1	5
C	1	1
D	2	2

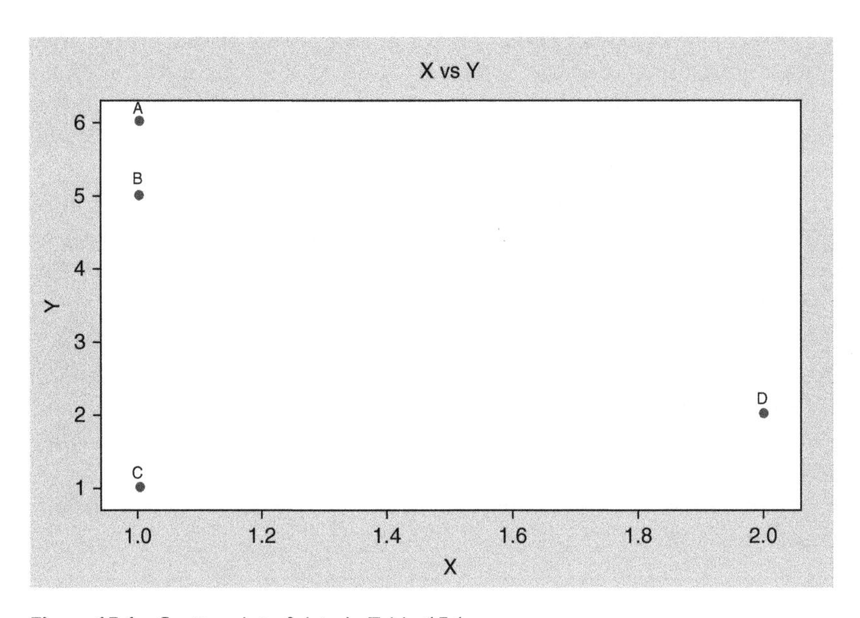

Figure 13.1 Scatter plot of data in Table 13.1

and j may be computed numerically using the Euclidean distance formula:

$$d_{ij} = \sqrt{\sum_{k=1}^{p}(x_{ik} - x_{jk})^2}, \tag{13.1}$$

where p is the number of dimensions (which is 2 in our example), d_{ij} is the distance, and x_{ik} is the data value of the ith row and kth column.

In this case, the computed distance matrix for our data matrix is given as:

For example, the distance from A to C is calculated as follows:

$$d_{AC} = \sqrt{(1-1)^2 + (6-1)^2} = 5.0 \tag{13.2}$$

The reader can verify that the distance from C to A is also 5.0. Since the distance from A to C is the same as the distance from C to A, the distance matrix is symmetric.

The task attempted by MDS is that given only a distance matrix as shown in Table 13.2, we find the original data so that a visual can be generated by using for example a scatter plot.

Next, we mention some of the challanges facing MDS. To begin with, as the number of objects increases, the possible number of dimensions increases as well. For example, if we have three objects, we will have to define at most a two-dimensional plane. With four objects, we will have to find a three-dimensional space. And so on, with each new object adding one more possible dimension.

Secondly, if the data are shifted in such a way that their positions relative to each other are maintained (i.e. rotated, translated, or transposed), the computed distance matrix will be the same. Hence, the distance matrix could have come from numerous sets of data.

A third challenge comes when the distances themselves are not actually known. In this instance one might only be given knowledge of their relative size.

MDS techniques have proved useful because scenarios often occur where the actual coordinates of the objects are not known, but some type of distance matrix is available. This is usually the case in social sciences where people cannot draw an overall picture of a group of objects, but they can express how different individual

Table 13.2 Distance matrix.

	A	*B*	*C*	*D*
A	0.0000	1.0000	5.0000	4.1231
B	1.0000	0.0000	4.0000	3.1623
C	5.0000	4.0000	0.0000	1.4142
D	4.1231	3.1623	1.4142	0.0000

pairs of objects are. From these pair-wise differences, MDS often can provide a useful picture.

13.3 Number of Dimensions and Goodness of Fit

The main tasks researchers and practitioners have is determining the number of dimensions in the MDS model. Each dimension denotes a different underlying factor. The objective of the MDS analysis is to keep the number of dimensions as small as possible for better visualization. Usually, the practitioner will select two or at most, three dimensions. If more dimensions are required, we may decide that MDS is not appropriate for your data.

The usual technique is to solve the MDS problem for a number of dimension values and select the smallest number of dimensions that achieves a reasonably small value of stress. The program displays a simple bar chart of the stress values to aid in the selection of the number of dimensions.

As in any data analysis and numerical problems, an expression is needed to express how well a particular set of data are represented by the model that the analysis imposes. In the case of MDS, we are trying to model distances. Hence, the most obvious choice for a goodness-of-fit statistic is one based on the differences between the original distances and their predicted values. Such a measure is called stress and is calculated as follows:

$$\text{Stress} = \sqrt{\frac{\sum (d_{ij} - \hat{d}_{ij})^2}{\sum d_{ij}^2}}, \tag{13.3}$$

where d_{ij} is the original distance and $\hat{d}_{ij}$ is the predicted distance based on the MDS model. Note that this predicted value depends on the number of dimensions kept and the algorithm that you used (metric versus nonmetric).

From (13.3), we observe that MDS fits with stress values near zero are the best. The author in Kruskal (1964) gave the following stress values to describe the goodness of fit based on his experience (Table 13.3):

Table 13.3 Stress and goodness of fit.

Stress	Goodness of fit
0.200	Poor
0.100	Fair
0.050	Good
0.025	Excellent
0.000	Perfect

However, empirical evidence shows that acceptable values of stress depends on the quality of the distance matrix and the number of objects in that matrix.

Before we discuss the two main types of MMDS and NMMDS, we will present a brief discussion of proximity measures.

13.4 Proximity Measures

Proximity measures characterize the similarity or dissimilarity that exists between objects. Dissimilarities, i.e. d_{ij} represent the distance between two objects and similarities, i.e. s_{ij} represent how close (in some sense) two objects are. In contrast to cases in which we distinguish only between "similar" and "dissimilar" objects i and j, proximities measure the degree of similarity by a real number s_{ij}, typically between 0 and 1. The larger the value of s_{ij}, the larger the similarity between i and j, and $s_{ij} = 1$ implies maximum similarity. A dual approach measures the dissimilarity between i and j by a numerical value $d_{ij} \geq 0$ (with 0 being the minimum dissimilarity). In practice, there are many ways to find appropriate values for s_{ij} or d_{ij}.

Similarity measures can be are converted to dissimilarities using the formula:

$$d_{ij} = \sqrt{s_{ii} + s_{jj} - 2s_{ij}}, \tag{13.4}$$

where d_{ij} represents a dissimilarity and s_{ij} represents a similarity.

We recall that dissimilarity measures are described in Section 12.2.

Quite generally, a similarity measure s can be obtained from a dissimilarity measure d by a decreasing function h such as, for example

$$s = h(d) = 1 - e^{-d}$$

or

$$s = \frac{(d_0 - d)}{d_0},$$

where d_0 is the maximum observed dissimilarity value.

However, a range of special definitions has been proposed for binary data where each component takes only two alternatives 1 and 0. Please refer to Johnson and Wichern (2014) for more details on similarity measure for binary data.

13.5 Metric Multidimensional Scaling

Metric Multidimensional Scaling (MMDS) also known as classical MDS transforms a distance matrix into a set of coordinates such that the Euclidean distances derived from these coordinates approximate the original distances.

MMDS begins with a $(n \times n)$ distance matrix D with elements d_{ij} where $i,j = 1, \ldots, n$. The objective of MMDS is to find a configuration of points in

p-dimensional space from the distances between the points such that the coordinates of the n points along the p dimensions yield a Euclidean distance matrix whose elements are as close as possible to the elements of the given distance matrix $\mathcal{D}$.

13.5.1 The Classical Solution

The classical solution is based on a distance matrix that is computed from a Euclidean geometry.

Definition 13.1 An $(n \times n)$ distance matrix $\mathcal{D} = (d_{ij})$ is Euclidean if for some points $x_1, \dots, x_n \in \mathbb{R}^p$; $d_{ij}^2 = (x_i - x_j)^\top (x_i - x_j)$.

The following result tells us whether a distance matrix is Euclidean or not.

Theorem 13.1 Define $\mathcal{A} = (a_{ij}), a_{ij} = -\frac{1}{2} d_{ij}^2$ and $\mathcal{B} = \mathcal{HAH}$ where $\mathcal{H}$ is the centering matrix. $\mathcal{D}$ is Euclidean if and only if $\mathcal{B}$ is positive semidefinite. If $\mathcal{D}$ is the distance matrix of a data matrix $\mathcal{X}$, then $\mathcal{B} = \mathcal{H}\mathcal{X}\mathcal{X}^T\mathcal{H}$, where $\mathcal{B}$ is called the inner product matrix.

The task of MMDS is to find the original Euclidean coordinates from a given distance matrix. Let the coordinates of n points in a p dimensional Euclidean space be given by x_i $(i = 1, \dots, n)$ where $x_i = (x_{i1}, \dots, x_{ip})^\top$.

Let $\mathcal{X} = (x_1, \dots, x_n)^\top$ be the coordinate matrix and assume $\bar{x} = 0$. The Euclidean distance between the ith and jth points is defined by:

$$d_{ij}^2 = \sum_{k=1}^{p} (x_{ik} - x_{jk})^2.$$

The general b_{ij} term of $\mathcal{B}$ is defined by:

$$b_{ij} = \sum_{k=1}^{p} x_{ik} x_{jk} = x_i^\top x_j.$$

It is possible to derive $\mathcal{B}$ from the known squared distances d_{ij}, and then from $\mathcal{B}$ the unknown coordinates. We describe as follows. We begin by letting

$$\begin{aligned} d_{ij}^2 &= x_i^\top x_i + x_j^\top x_j - 2x_i^\top x_j \\ &= b_{ii} + b_{jj} - 2b_{ij}. \end{aligned} \tag{13.5}$$

Centering of the coordinate matrix $\mathcal{X}$ implies that $\sum_{i=1}^{n} b_{ij} = 0$. Summing 13.5 over i, over j, and over i and j, we obtain the following three equations:

$$\frac{1}{n} \sum_{i=1}^{n} d_{ij}^2 = \frac{1}{n} \sum_{i=1}^{n} b_{ii} + b_{jj},$$

$$\frac{1}{n}\sum_{j=1}^{n} d_{ij}^2 = b_{ii} + \frac{1}{n}\sum_{j=1}^{n} b_{jj},$$
$$\frac{1}{n^2}\sum_{i=1}^{n}\sum_{j=1}^{n} d_{ij}^2 = \frac{2}{n}\sum_{i=1}^{n} b_{ii}. \tag{13.6}$$

From (13.5) and (13.6), solving for b_{ij}, we obtain:

$$b_{ij} = -\frac{1}{2}(d_{ij}^2 - d_{i\bullet}^2 - d_{\bullet j}^2 + d_{\bullet\bullet}^2).$$

With $a_{ij} = -\frac{1}{2}d_{ij}^2$, and

$$a_{i\bullet} = \frac{1}{n}\sum_{j=1}^{n} a_{ij},\ a_{\bullet j} = \frac{1}{n}\sum_{i=1}^{n} a_{ij} \quad \text{and} \quad a_{\bullet\bullet} = \frac{1}{n^2}\sum_{i=1}^{n}\sum_{j=1}^{n} a_{ij}$$

Thus we get:

$$b_{ij} = a_{ij} - a_{i\bullet} - a_{\bullet j} + a_{\bullet\bullet}.$$

Next define the matrix $\mathcal{A}$ as (a_{ij}), and observe that:

$$\mathcal{B} = \mathcal{H}\mathcal{A}\mathcal{H}.$$

The inner product matrix $\mathcal{B}$ can be expressed as:

$$\mathcal{B} = \mathcal{X}\mathcal{X}^\top,$$

where $\mathcal{X} = (x_1, \dots, x_n)^\top$ is the $(n \times p)$ matrix of coordinates.

As required, in Theorem 13.1 the matrix $\mathcal{B}$ is symmetric, positive semi-definite and of rank p, and hence, it has p nonnegative eigenvalues and $n - p$ zero eigenvalues. $\mathcal{B}$ can now be written as:

$$\mathcal{B} = \Gamma\Lambda\Gamma^\top,$$

where $\Lambda = \text{diag}(\lambda_1, \dots, \lambda_p)$, the diagonal matrix of the eigenvalues of $\mathcal{B}$, and $\Gamma = (\gamma_1, \dots, \gamma_p)$, the matrix of corresponding eigenvectors. Hence, the coordinate matrix $\mathcal{X}$ containing the point configuration in $\mathbb{R}^p$ is given by:

$$\mathcal{X} = \Gamma\Lambda^{\frac{1}{2}}.$$

Remarks 13.1 From above we see that MMDS is identical to principal components analysis. However, PCA is more focused on the dimensions themselves, and seek to maximize explained variance, whereas MDS is more focused on relations among the scaled objects.

In practice, the algorithm for MMDS is given as follows:

Step 1: Begin with distances d_{ij}.
Step 2: Define $\mathcal{A} = -\frac{1}{2}d_{ij}^2$.

Step 3: Set $\mathcal{B} = (a_{ij} - a_{i\bullet} - a_{\bullet j} + a_{\bullet\bullet})$.
Step 4: Find the eigenvalues $\lambda_1, \dots, \lambda_p$ and the associated eigenvectors $\gamma_1, \dots, \gamma_p$ where the eigenvectors are normalized so that $\gamma_i^\top \gamma_i = \lambda_i$.
Step 5: Select an appropriate number of dimensions p (ideally $p = 2$).
Step 6: The coordinates of the n points in the Euclidean space are given by $x_{ij} = \gamma_{ij}\lambda_j^{1/2}$ for $i = 1, \dots, n$ and $j = 1, \dots, p$.

13.6 Nonmetric Multidimensional Scaling

We recall from Section 13.5 that the goal of MMDS is to transform a distance matrix into a set of coordinates of the points in p-dimensional space, so that there is a good agreement between the observed proximities and the inter-point distances.

The development of Nonmetric Multidimensional Scaling (NMMDS) was motivated by two main weaknesses in the MMDS:

1. the definition of an explicit functional connection between dissimilarities and distances in order to derive distances out of given dissimilarities, and
2. the restriction to Euclidean geometry in order to determine the object configurations.

The idea of a NMMDS is to demand a less-rigid relationship between the final configuration of the points and the distances. In NMMDS, it is assumed only this relationship can be described by some monotone function. Assume that an unknown monotonic increasing function f,

$$d_{ij} = f(\delta_{ij}), \tag{13.7}$$

is used to generate a set of distances d_{ij} as a function of given dissimilarities δ_{ij}. Where f has the property that if $\delta_{ij} < \delta_{rs}$, then $f(\delta_{ij}) < f(\delta_{rs})$. The scaling is based on the rank order of the dissimilarities. The nonmetric multidimensional scaling is therefore ordinal in character. The most common approach used to determine the elements d_{ij} and to obtain the coordinates of the objects $x_1, x_2, \dots, x_n$ given only rank order information is an iterative process commonly referred to as the Shepard–Kruskal algorithm (Kruskal 1964, Shepard 1962). We present the algorithm as follows.

13.6.1 Shepard–Kruskal Algorithm

Step 1: Compute the Euclidean distances $d_{ij}^{(0)}$ from an arbitrarily chosen initial configuration $\mathcal{X}_0$ in dimension p^*, provided that all objects have different coordinates. One might use MMDS to obtain these initial coordinates.

Step 2: Identify the disparities $\hat{d}_{ij}^{(0)}$ from the distances $d_{ij}^{(0)}$ by constructing a monotone regression relationship between the $d_{ij}^{(0)}$s and δ_{ij}s, under the requirement that if $\delta_{ij} < \delta_{rs}$, then $\hat{d}_{ij}^{(0)} \le \hat{d}_{rs}^{(0)}$. This is called the weak monotonicity requirement. To obtain the disparities $\hat{d}_{ij}^{(0)}$, a useful approximation method is the pool-adjacent violators (PAV) algorithm (Ayer et al. 1955).

Step 3: The spatial configuration of $\mathcal{X}_0$ is altered to obtain $\mathcal{X}_1$. From $\mathcal{X}_1$ the new distances $d_{ij}^{(1)}$ can be obtained which are more closely related to the disparities $\hat{d}_{ij}^{(0)}$ from Step 2.

In order to assess how well the derived configuration fits the given dissimilarities, Kruskal suggests a measure called STRESS1 that is defined by

$$\text{STRESS1} = \left(\frac{\sum_{i<j} (d_{ij} - \hat{d}_{ij})^2}{\sum_{i<j} d_{ij}^2} \right)^{\frac{1}{2}}.$$

An alternative stress measure is given by

$$\text{STRESS2} = \left(\frac{\sum_{i<j} (d_{ij} - \hat{d}_{ij})^2}{\sum_{i<j} (d_{ij} - \bar{d})^2} \right)^{\frac{1}{2}},$$

where $\bar{d}$ denotes the average distance.

We remark that NMMDS is only based on the rank order of dissimilarities. The object of NMMDS is to create a spatial representation of the objects with low dimensionality.

In practice, the algorithm for NMMDS is given as follows:

Step 1: Select an initial configuration.
Step 2: Find d_{ij} from the configuration.
Step 3: Fit $\hat{d}_{ij}$, the disparities, by the PAV algorithm.
Step 4: Find a new configuration $\mathcal{X}_{n+1}$ by using the steepest descent.
Step 5: Go to Step 2.

13.7 Problems

1 Show that

(a) $b_{ii} = a_{\bullet\bullet} - 2a_{i\bullet};\; b_{ij} = a_{ij} - a_{i\bullet} - a_{\bullet j} + a_{\bullet\bullet};\; i \neq j$

(b) $\mathcal{B} = \sum_{i=1}^{p} x_i x_i^\top$

(c) $\sum_{i=1}^{n} \lambda_i = \sum_{i=1}^{n} b_{ii} = \frac{1}{2n \sum_{i,j=1}^{n} d_{ij}^2}.$

2 Explain why nonmetric multidimensional scaling is ordinal in character.

3 In the following statements, state whether True or False and explain your reasoning.
(a) NMMDS procedures assume that the input data are ordinal, but they result in metric output.
(b) The MMDS procedures assume that input data are metric and the output is also metric.
(c) The objective in MDS is to obtain a spatial map that best fits the input data in the smallest number of dimensions.
(d) Stress values indicate the proportion of variance of the optimally scaled data that is not accounted for by the MDS model.

4 Discuss the main difference between MDS and PCA.

5 Discuss some applications of MDS.

6 Use the values in Table 13.4, perform the following task.
(a) Find the scatter plot.
(b) Generate a distance (dissimilarity) matrix from the multivariate data. Please use the Euclidean distance.
(c) Measure how well this configuration (arrangement of sites on the ordination) match the original distance (dissimilarity) matrix.

7 Let S be the centered inner product matrix with elements $< xi - \bar{x}, x_j - \bar{x} >$. Let $\lambda_1 > \lambda_2 > \cdots > \lambda_k$ be the k-largest eigenvalues of S. Let D be a diagonal matrix with diagonal entries $\sqrt{\lambda_1}, \sqrt{\lambda_2, \dots, \lambda_k}$. Find the solutions z_t to the following scaling problem:

$$S_C(z_1, z_2, \dots, z_N) = \sum_{i,i'} (s_{ii'} - \langle z_i - \bar{z}, z_{i'} - \bar{z} \rangle)^2.$$

Table 13.4 Data matrix.

Label	*X*	*Y*
Site 1	2	6
Site 2	4	5
Site 3	1	1
Site 4	2	2
Site 5	2	3

8 Generate 100 observations of three variates X_1, X_2, X_3 according to:

$$X_1 \sim Z_1,$$
$$X_2 = x_1 + 0.005.Z_2,$$
$$X_3 = 5.Z_3,$$

where Z_1, Z_2, Z_3 are independent standard normal variates. Compute the leading principal component direction.

9 From Problem 9, show that the leading principal component aligns itself in the maximal variance direction X_3, while the leading factor essentially ignores the uncorrelated component X_3, and picks up the correlated component $X_2 + X_1$.

14

Classification and Tree-Based Methods

14.1 Introduction

In many instances, the response variable Y to be predicted is qualitative or categorical (consists of attributes, labels, or nonnumerical entries) instead of quantitative (numerical measurements or counts). For example a student's major, place of birth, race, eye color, etc. In this chapter, we study techniques for predicting qualitative responses, a process that is known as classification. Classification is the process of finding a model or function which helps in separating the data into a category or class. In classification, data is categorized under different labels according to some input parameters and then the labels are predicted for the data. Normally, methods used for classification first predict the probability of each of the categories of a categorical variable, as the basis for making the classification. There are several classification techniques that are used to predict a qualitative response. In this chapter, we will discuss some widely used classifiers: logistic regression, linear discriminant analysis, and tree based-models (one single decision tree and random forest). We will discuss more advanced techniques in later chapters, such as support vector machines (Chapter 16) and neural networks (Chapter 17). Please refer to Breiman (1996), Breiman (2001), Bishop (2006), and Hastie et al. (2008b) for a comprehensive study and more details of the algorithms discussed in this chapter.

14.2 An Overview of Classification

Classification problems occur often in real life and some examples include:

1. A financial institution must be able to determine whether a customer will default payments of loans, on the basis of income, educational level, past payment history, and several other factors.

Data Science in Theory and Practice: Techniques for Big Data Analytics and Complex Data Sets,
First Edition. Maria Cristina Mariani, Osei Kofi Tweneboah, and Maria Pia Beccar-Varela.

2. A system administrator wants to determine if an email is a spam and should be delivered to the Junk folder.
3. A researcher wants to identify images of single digits 0–9 correctly.
4. A scientist wants to identify disease based on the gene expression levels. Gene expression is the process by which the genetic code, that is, the nucleotide sequence of a gene is used to direct protein synthesis and produce the structures of the cell.

In the classification setting, we have a set of training observations $(x_1, y_1), \dots, (x_n, y_n)$ that can be used to build a model. The goal is to let the model to perform well not only on the training data, but also on test observations that were not used to train the model. Usually in a data sample, the training observations are implemented to build up a model, while the test observations is used to validate the model built.

14.2.1 The Classification Problem

Classification is supervised learning for which the true class labels for the data points are given in the training data. The set up for a classification problem is as follows:

- Training data: $(x_1, y_1), \dots, (x_n, y_n)$.
- The feature vector $X = (X_1, X_2, \dots, X_p)$, where each variable X_j is quantitative for $j = 1, \dots, p$
- The response variable Y is categorical where $Y \in Y = 1, 2, \dots, K$ form a predictor $Y(x)$ to predict Y based on X, where K denotes the number of categories.

$Y(x)$ divides the input space into a collection of regions, each labeled by one class.
We begin our discussion with the logistic regression model.

14.2.2 Logistic Regression Model

Logistic regression estimates the posterior probabilities of classes given X directly without assuming the marginal distribution on X. Logistic Regression is a powerful classification algorithm that is used to predict the probability of a categorical variable. We assume that the predictors (x_k) are independent of each other, so the model has no multicollinearity. We express the model as:

$$\text{logit}(p(x)) = \beta_0 + \beta_1 x_1 + \beta_2 x_2 + \beta_3 x_3 + \cdots + \beta_k x_k, \tag{14.1}$$

where p is the probability of presence of the characteristic of interest and $\beta_0, \beta_1, \dots, \beta_k$ are the coefficient parameters. The logit transformation is defined as the log odds that is $\ln \frac{p}{1-p}$. Therefore, the logistic regression model is similar to a linear regression, but it is constructed using the natural logarithm of the "odds" of the target variable. Thus:

$$\ln \frac{p(x)}{1 - p(x)} = \beta_0 + \beta_1 x_1 + \beta_2 x_2 + \beta_3 x_3 + \cdots + \beta_k x_k, \tag{14.2}$$

where $x = (x_1, \ldots, x_p)$ are k predictors. Equation (14.2) can be rewritten as

$$p(x) = \frac{e^{\beta_0 + \beta_1 x_1 + \cdots + \beta_k x_k}}{1 + e^{\beta_0 + \beta_1 x_1 + \cdots + \beta_k x_k}}. \tag{14.3}$$

Since logistic regression predicts probabilities, rather than classes, we can fit it into the data using the likelihood technique. Maximum likelihood is a very general approach that is used to fit many of the nonlinear models. The mathematical details of maximum likelihood are beyond the scope of this book. Please refer to Casella and Berger (2002) for details of the maximum likelihood method. The likelihood helps to find the best model that explains the datasets well. The data set used in this study contains a vector of features (x_i) and an observed class (y_i). We assume the probability of that class is either p, when $y_i = 1$, or $1 - p$, when $y_i = 0$. So the likelihood function of Eq. (14.2) is as follows:

$$L(\beta_0 + \beta_1 x_1 + \cdots + \beta_k x_k) = \prod_{i=1}^{N} p(x_i)^{y_i} (1 - p(x_i))^{1 - y_i}. \tag{14.4}$$

The coefficients $\beta_0, \beta_1, \ldots, \beta_k$ are unknown and must be estimated based in the training data.

We now seek to estimate the parameters $\beta_0, \beta_1, \ldots, \beta_k$ that maximize the likelihood function $L(\beta_0 + \beta_1 x_1 + \cdots + \beta_k x_k)$ in Eq. (14.4). We maximize the logarithm of the likelihood function as follows:

$$l(\beta) = \sum_{i=1}^{N} (y_i \beta^T x_i - \log(1 + e^{\beta^T x_i})). \tag{14.5}$$

Next, we use some regularization techniques to obtain a parsimonious model with important features from the original model. A parsimonious model is a model that achieves a desired level of goodness of fit or prediction with a few explanatory variables. The regularization technique penalizes the magnitude of coefficients of features thereby minimizing the error between predicted and actual observations. In this book, the regularization techniques used in the logistic regression are the l_1 and l_2 regularization. In l_1 regularization, we shrink the weights using the absolute values of the weight coefficients. In the l_2 regularization, the weights are shrunk by computing the Euclidean norm of the weight coefficients.

14.2.2.1 l_1 Regularization

We use the least absolute shrinkage and selection operator (lasso) regularization technique by adding an l_1 penalty term in Eq. (14.5). This forces the absolute value sum of the regression coefficients to be less than a fixed value. This is due to the fact that the tuning parameter makes certain coefficients to be set to zero, effectively by choosing a simpler model that does not include those coefficients. So we maximize the penalized versions as follows:

$$l_\lambda(\beta) = \sum_{i=1}^{N} (y_i \beta^T x_i - \log(1 + e^{\beta^T x_i})) - \lambda \sum_{j=1}^{p} |\beta_j|, \tag{14.6}$$

where λ is a tuning parameter that controls the strength of penalty term. The parameter λ is selected in a way that the resulting model minimizes the out of sample error. The out of sample error is a measure of how the accurately a model is able to correctly predict outcome for observations that was not part of the data sample.

14.2.2.2 l_2 Regularization

We also use Ridge-regression by adding an l_2 penalty term in Eq. (14.5). This regularization technique overcomes the multicollinearity problem in our data. When we have multicollinearity in the data, the variance of estimation is large. So the parameter estimation may be far from the true value. To overcome this issue, we add a degree of bias to the regression estimates and shrink the estimators to the true parameters. We maximize the l_2 penalized versions as follows:

$$l_\lambda(\beta) = \sum_{i=1}^{N} (y_i \beta^T x_i - \log(1 + e^{\beta^T x_i})) - \lambda \sum_{j=1}^{p} \beta_j^2, \tag{14.7}$$

where λ controls the amount of regularization.

We can access the performance of the logistic regression by obtaining the classifier and applying the classifier to the training data set and determine what percentage of data is misclassified since we know the true labels in the training data.

14.3 Linear Discriminant Analysis

Discriminant analysis is another classification method that estimates the within-class density of X given the class label. Combined with the prior probability of classes, the posterior probability of Y can be obtained by the Bayes rule (Casella and Berger 2002). We state Bayes' formula as follows:

Theorem 14.1 (Bayes' rule) Let $A_1, A_2, \ldots$ be a partition of a sample space, and let B be any set. Then for each $i = 1, 2, \ldots,$

$$P(A_i|B) = \frac{P(B|A_i)P(A_i)}{\sum_{j=1}^{\infty} P(B|A_j)P(A_j)}.$$

The objective of linear discriminant analysis (LDA) is to project a data set onto a lower dimensional space with good class-separability in order to avoid overfitting by minimizing the error in parameter estimation and reduce computational costs for a given classification task.

Under LDA, we assume that the density for X, given every class k is following a Gaussian distribution. Here is the probability density formula for a multivariate Gaussian distribution:

$$f_k(x) = \frac{1}{(2\pi)^{p/2}|\Sigma_k|^{1/2}} e^{-\frac{1}{2}(x-\mu_k)^T \Sigma_k^{-1}(x-\mu_k)} \tag{14.8}$$

where p is the dimension and Σ_k is the covariance matrix. This involves the square root of the determinant of this matrix. In this case, we are doing matrix multiplication. The vector x and the mean vector μ_k are both column vectors as discussed in Chapter 3.

For the LDA method, we assume that for different k the covariance matrix is identical, i.e. $\Sigma_k = \Sigma, \forall k$. By making this assumption, the classifier becomes linear. The main difference between LDA and quadratic discriminant analysis (QDA) is that we do not assume that the covariance matrix is identical for different classes.

Due to the fact that the covariance matrix determines the shape of the Gaussian density, in LDA, the Gaussian densities for different classes have the same shape but are shifted versions of each other (that is, different mean vectors).

We now present the optimal classification of the LDA.

14.3.1 Optimal Classification and Estimation of Gaussian Distribution

We discuss the optimal classification based on the Bayes rule. Bayes rule assumes that a class has the maximum posterior probability given the feature vector X. Since the log function is an increasing function, the maximization is equivalent because whatever gives you the maximum should also give you a maximum under a log function. Next, we plug in the density of the Gaussian distribution assuming identical covariance and then multiplying the prior probabilities we have:

$$\begin{aligned}
\hat{Y}(x) &= \arg\max_k P(Y = k|X = x) \\
&= \arg\max_k f_k(x)\pi_k \\
&= \arg\max_k \log(f_k(x)\pi_k) \\
&= \arg\max_k \left[-\log((2\pi)^{p/2}|\Sigma|^{1/2}) - \frac{1}{2}(x-\mu_k)^T\Sigma^{-1}(x-\mu_k) + \log(\pi_k)\right] \\
&= \arg\max_k \left[-\frac{1}{2}(x-\mu_k)^T\Sigma^{-1}(x-\mu_k) + \log(\pi_k)\right].
\end{aligned}$$

We note that

$$-\frac{1}{2}(x-\mu_k)^T\Sigma^{-1}(x-\mu_k) = x^T\Sigma^{-1}\mu_k - \frac{1}{2}\mu_k^T\Sigma^{-1}\mu_k - \frac{1}{2}x^T\Sigma^{-1}x.$$

The above relation is given as exercise in the problems section of this chapter. Therefore, we have

$$\hat{Y}(x) = \arg\max_k \left[x^T\Sigma^{-1}\mu_k - \frac{1}{2}\mu_k^T\Sigma^{-1}\mu_k + \log(\pi_k)\right]. \tag{14.9}$$

which is the final classifier.

The LDA gives a linear boundary because the quadratic term is dropped. Thus from (14.9), have

$$\hat{Y}(x) = \arg\max_k \left[x^T\Sigma^{-1}\mu_k - \frac{1}{2}\mu_k^T\Sigma^{-1}\mu_k + \log(\pi_k)\right]. \tag{14.10}$$

From (14.10), we can define the linear discriminant function as:

$$\hat{Y}(x) = \arg\max_k \delta_k(x), \tag{14.11}$$

where $\delta_k(x) = x^T\Sigma^{-1}\mu_k - \frac{1}{2}\mu_k^T\Sigma^{-1}\mu_k + \log(\pi_k)$.

The decision boundary between class k and l is:

$$\{x : \delta_k(x) = \delta_l(x)\}$$

or equivalently,

$$\log\frac{\pi_k}{\pi_l} - \frac{1}{2}(\mu_k+\mu_l)^T\Sigma^{-1}(\mu_k-\mu_l) + x^T\Sigma^{-1}(\mu_k-\mu_l) = 0. \tag{14.12}$$

In a binary classification problem, i.e. $k = 1$ and $l = 2$, we would define a constant $a_0 = \log\frac{\pi_1}{\pi_2} - \frac{1}{2}(\mu_1+\mu_2)^T\Sigma^{-1}(\mu_1-\mu_2)$, where π_1 and π_2 are prior probabilities for the two classes and μ_1 and μ_2 are mean vectors. Next we define $(a_1, a_2, \ldots, a_p)^T = \Sigma^{-1}(\mu_1-\mu_2)$ and classify to class 1 if $a_0 + \sum_{j=1}^{p} a_j x_j > 0$ and to class 2 if otherwise.

In the above discussion, we assume that we have the prior probabilities for the classes and within-class densities. In practice, we have only a set of training observations. Therefore, we will have to find the π_ks and the $f_k(x)$. We can estimate the π_ks and the parameters in the Gaussian distributions by using the maximum likelihood estimator:

$$\hat{\pi}_k = N_k/N,$$

where N_k is the number of class-k samples and N is the sample size in the training observations. The prior probabilities for class k can be obtained by counting the frequency of data points in the class.

The mean vector for every class is obtained using the following formula:

$$\hat{\mu}_k = \sum_{g_i=k} x^{(i)}/N_k$$

The covariance matrix formula is given as

$$\hat{\Sigma} = \sum_{k=1}^{K} \sum_{g_i=k} (x^{(i)} - \hat{\mu}_k)(x^{(i)} - \hat{\mu}_k)^T / (N - K),$$

where $x^{(i)}$ denotes the ith sample vector.

Thus to use LDA to obtain a classification rule, the first step involves estimating the parameters using the formulas above, then go back and find the linear discriminant function and choose a class according to the discriminant functions.

14.4 Tree-Based Methods

In this section, we discuss tree-based methods for classification. These involve dividing the predictor space into a number of simple regions. In order to make a prediction for a given observation, we use the mean or the mode of the training observations in the region to which it belongs. Since the set of splitting rules used to divide the predictor space can be summarized in a tree, these types of approach are known as decision tree methods.

Generally, tree-based methods are simple and useful for interpretation. In this chapter, we also introduce one single decision tree and random forests.

We begin with the one single decision tree technique.

14.4.1 One Single Decision Tree

A decision tree is a flowchart-like structure made of nodes and branches. At each node, a split on the data is performed based on one of the input features, generating two or more branches as output. Several splits are made in the upcoming nodes and increasing numbers of branches are generated to partition the original data. This process is recursive and continues until a node is generated where all or almost all of the data belong to the same class and further splits are no longer possible.

A criterion used for making the splits is the classification error rate. Since we plan classification to assign an observation in a given region to the most commonly occurring error rate class of training observations in that region, the classification error rate is simply the fraction of the training observations in that region that do not belong to the most common class:

$$E = 1 - \max_k \hat{p}_{mk},$$

where $\hat{p}_{mk}$ is the proportion of training observations in the mth region that are from the kth class. However, it turns out that classification error is not sufficiently sensitive for tree-growing, and in practice, other measures are preferable. The Gini index is defined by:

$$G = \sum_{k=1}^{K} \hat{p}_{mk}(1 - \hat{p}_{mk}).$$

This index measures the total variance across the K classes. We observe that the Gini index takes on a small value if all of the $\hat{p}_{mk}$s are close to zero or one.

In practice when building a classification tree, the Gini index is typically used to evaluate the quality of a particular split, since it is more sensitive to node purity than the classification error rate.

Tree-structured classifiers are constructed by repeated splits of the space X into smaller and smaller subsets, beginning with X itself.

In general, the construction of a tree involves the following three elements:

- How do we decide which node (region) to split and how to split it?
- How do we decide when to declare a node terminal (leaf node) and stop splitting?
- How do we assign class labels?

The decision tree classifies new data points as follows. We let a data point pass down the tree and identify which leaf node it lands in. The class of the leaf node is assigned to the new data point. Usually, all the points that land in the same leaf node will be given the same class.

A class assignment rule assigns a class $j = 1, \dots, K$ to every terminal (leaf) node $t \in \tilde{T}$. The class is assigned to node t . $\tilde{T}$ is denoted by $\kappa(t)$. If $\kappa(t) = 2$, all the points in node t would be assigned to class 2.

If we use 0–1 loss, the class assignment rule is defined as

$$\kappa(t) = \arg\max_{j} \; p(j|t)$$

where we pick the majority class or the class with the maximum posterior probability. The 0–1 loss function is an indicator function that returns one when the target and output are different and zero otherwise. This means that when we misclassify, we lose 1 point and we do not lose anything if we make the correct classification.

14.4.2 Random Forest

The random forest technique is a type of additive model that predicts the data by combining decisions from a sequence of base models. It reduces the variance by avoiding overfitting of the model.

The class of base models can be expressed as follows:

$$Y(x) = f_0(x) + f_1(x) + f_2(x) + \cdots , \tag{14.13}$$

where the final model Y is the sum of simple base models f_i. We define each base classifier as a simple decision tree. In fact it is an ensemble technique that considers multiple learning algorithms to obtain the best predictive model. At this point, all the base models or trees are made independently using a different subsample of the data. Once we have a new generated training set, we divide it randomly into two parts. The two-third samples are used to build a tree and the one-third samples are used to obtain the predictions of trees. We take the majority vote of these one-third predictions as the predicted value for the data point and then we estimate the error. For a full detail study of Random Forest, the reader is referred to the reference in Hastie et al. (2008b). We now present the algorithm of random forest that is used in this study:

1. We first select a random sample of size N with replacement from the data.
2. Select a random sample without replacement of the predictors.
3. Construct a split by using predictors selected in Step 2.
4. Repeat Steps 2 and 3 for each subsequent split until the tree is as large as desired.
5. Drop the out-of-bag data down the tree. We then store the class assigned to each observation along with each observation's predictor values.
6. Repeat Steps 1–5 for large number of times.
7. For each observation in the dataset, we count the number of trees that is classified in one category over the number of trees.
8. Assign each observation to a final category by a majority vote over the set of trees. Thus, if 51% of the time over a large number of trees a given observation is classified as a "1," that becomes its classification.

The random forest includes three main tuning parameters such as node size, number of trees (ntree), and number of predictors sampled (mtry) for splitting. To build a best predictive model, we estimate the best tuning parameters and important variables using mean decrease accuracy (MDA) and mean decrease Gini (MDG) indices. The MDA determines the importance of a variable by measuring the change in prediction accuracy, when the values of the variable are randomly permuted compared to the original observations. However, the MDG index is a measure of how each variable contributes to the homogeneity of the nodes and leaves in the resulting random forest. For the details of these methodologies, consult Breiman (2001) and references therein.

When used for classification, a random forest obtains a class vote from each tree, and then classifies using majority vote, i.e. random forest makes a prediction for each test instance and the final output prediction is the one that receives more than half of the votes. An important feature of random forests is its use of out-of-bag samples. An out-of-bag error estimate is almost identical to that obtained by N-fold cross validation. Cross validation is one of the techniques used to test the effectiveness of a model, it is also a resampling procedure used to evaluate a model whenever we have a limited data. To perform cross validation, one needs to keep aside a sample of the data which is not used to train the model and later use this sample for testing and validation.

Unlike many other nonlinear estimators, random forests can be fit in one sequence, with cross validation being performed along the way. Once the out-of-bag error stabilizes, the training can be terminated.

14.5 Applications

In this section, we briefly discuss some real-life applications of logistic regression methodology, LDA, one single decision tree, and random forest described above.

In this application, our focus is on credit risk scoring where we examined the impact of the choice of different machine learning models to identifying credit card defaults of a bank in Taiwan. This study took payment data in October, 2005, from an important bank (a cash and credit card issuer) in Taiwan, and the targets were credit card holders of the bank. The dataset contains 30 000 observations, characterized by the same 23 labeled variables, each of them representing either a default or not a default (binary value) of credit card clients. Among the total 30 000 observations, 6636 observations (22.12%) are the cardholders with default payment. The source of the data sets is UCI Machine Learning Respository.

We randomly divided our data sets into the training sample and the test sample with a ratio of 2 : 1. We used the training sample to train four models and then used the test sample to compare them. Using the 23 features, we will present the results when the four models were applied to our credit default test dataset.

There are several performance measures to compare the performance of the models. However, we will present results based on the C index (area under the receiver operating characteristic [ROC] curve [i.e. AUC]) and mean square error (MSE) criteria. The area under the ROC curve (AUC) is a performance metrics for binary classifiers. By comparing the ROC curves with AUC, it captures the extent to which the curve is up in the Northwest corner. A higher AUC is good. A score of 0.5 is no better than random guessing. A score of 0.9 would be a very good model but a score of 0.9999 would be too good to be true and will indicate overfitting. Overfitting is an error that results when a function is too closely fit to a limited set of data points.

Table 14.1 reports the four models performances on the test dataset. From Table 14.1, the model with the highest C index is Random forest followed by the one single decision tree.

Next, we investigate the performance of four algorithms using only the top 10 variables (see Table 14.2) selected by the top Random Forest algorithm. The results for the C index and MSE are provided in Table 14.3.

Table 14.1 Models' performances on the test dataset with 23 variables using AUC and mean square error (MSE) values for the five models.

Model	AUC	MSE
LDA	0.709	0.1633
l_2 penalized logistic model	0.715	0.2401
One single decision tree	0.738	0.1467
Random forests	0.760	0.1381

Table 14.2 Top 10 variables selected by the Random forest algorithm.

Variable	Random forest
$X1$	PAY_0
$X2$	PAY_2
$X3$	PAY_AMT1
$X4$	BILL_AMT3
$X5$	PAY_3
$X6$	BILL_AMT5
$X7$	BILL_AMT4
$X8$	BILL_AMT6
$X9$	BILL_AMT2
$X10$	PAY_AMT5

Table 14.3 Performance for the four models using the top 10 features from model Random forest on the test dataset.

Model	AUC	MSE
LDA	0.704	0.1633
l_2 penalized logistic model	0.709	0.2298
One single decision tree	0.723	0.1473
Random forests	0.701	0.1381

From Table 14.3, the one single decision tree performed significantly better compared to the other three models in terms of the C index and MSE. Using only the top 10 features, Random forest model poorly performed on these new datasets. We remark that all the values in Tables 14.1 and 14.3 were based on the test set performance.

In summary, we built binary classifiers based on machine learning models on real data in predicting customers credit card default probability payments in Taiwan. The tree-based models, turn out to be the best and stable binary classifiers as they properly create split directions, thus keeping only the efficient information. From the financial perspective, the profile of the selected top 10 variables from the Random Forest model will be essential in deciding whether a customer will default payments. In practice, requesting a few lists of variables from clients can help speed up the time to deliver a decision on a credit card request.

Other applications of the machine learning techniques discussed in this chapter includes the diagnosis and prognosis of cancer and heart disease (Mariani et al. 2019). In this study, the authors compared the predictive ability of several machine learning algorithms to breast cancer and heart disease. The authors predicted the test data based on the important variables and computed the prediction accuracy using the ROC curve. The Random Forest algorithm produced the best performance in analyzing the breast cancer and heart disease data.

14.6 Problems

1 Describe the difference between a regression problem and a classification problem. Give two examples each.

2 A Pew Internet Poll asked cell phone owners about how they used their cell phones. One question asked whether or not, during the past 30 days, they had used their phone while in a store to call a friend or family member for advice about a purchase they were considering. The poll surveyed 1003 adults living in the United States by telephone. Of these, 462 responded that they had used their cell phone while in a store within the last 30 days to call a friend or family member for advice about a purchase they were considering. Suppose that you want to investigate differences in cell phone use among customers of different ages. You create an indicator explanatory variable x that has the value 1 if the customer is 20 years of age or less and is 0 if the customer over 20 years of age.

(a) What proportion of those surveyed reported that they used their cell phone while in a store within the last 30 days to call a friend or family member for advice about a purchase they were considering?

(b) Describe the statistical model for logistic regression in this setting.
(c) Interpret the regression slope in terms of an effect based on a difference in age of one year.

3 Explain what is wrong and why: If $b_1 = 5$ in a logistic regression analysis with one explanatory variable, we estimate that the probability of an event is multiplied by 5 when the value of the explanatory variable increases by one unit.

4 Distinguish between overfitting and underfitting? How can we avoid these errors?

5 Explain the bias-variance trade-off of a predictive model.

6 Using any datasets of your choice, apply any of the techniques described in this chapter to solve a classification problem. Datasets can be accessed from the Internet from the University of California Irvine (UCI) Machine Learning Repository.

7 Describe the process of constructing tree-structured classifiers.

8 Given a covariance matrix Σ, a vector x and a mean vector μ_k show that:

$$-\frac{1}{2}(x-\mu_k)^T\Sigma^{-1}(x-\mu_k) = x^T\Sigma^{-1}\mu_k - \frac{1}{2}\mu_k^T\Sigma^{-1}\mu_k - \frac{1}{2}x^T\Sigma^{-1}x.$$

9 Show that for $K = 2$ class classification, only one tree needs to be grown at each gradient-boosting iteration.

10 Assume that $\hat{\beta}_j/||\hat{\beta}_j||^2$ is a the normalized version for a coefficient estimate $\hat{\beta}_j$. Show that as a tuning parameter $\lambda \to \infty$, the normalized ridge-regression estimates converge to the renormalized partial-least-squares one-component estimates.

15

Association Rules

15.1 Introduction

In this chapter, we will discuss association rule which is an important tool for mining commercial data bases. Association rule finds joint values of the variables $X = (X_1, X_2, \ldots, X_p)$ that appear most frequently in the database. It is a machine learning technique for finding interesting hidden associations and relationships among large sets of data items. A typical example is market basket analysis. Market basket analysis is one of the key techniques used by retailers to show associations and relations between items. The technique is based upon the theory that if you buy a certain group of items, you are more (or less) likely to buy another group of items. This helps retailers to identify patterns between the items that people buy together frequently.

In Section 15.2, we will discuss the market basket analysis in detail and present some examples.

15.2 Market Basket Analysis

In the wholesale and retail businesses, market basket analysis helps managers better understand and ultimately serve their customers by highlighting purchasing patterns. As mentioned earlier, this modeling approach is based on the theory that customers who buy a certain item (or group of items) are more (or less) likely to buy another specific item (or group of items). Other applications of market basket analysis includes, analysis of debit and credit card purchases, analysis of telephone calling patterns, identification of fraudulent car insurance claims, and several others. A predictive market basket analysis can be used to identify sets of item purchases (or events) that generally occur in sequence.

We illustrate the idea of market basket analysis by using a transaction data. Many companies have access to the transaction database of their customers which

Data Science in Theory and Practice: Techniques for Big Data Analytics and Complex Data Sets,
First Edition. Maria Cristina Mariani, Osei Kofi Tweneboah, and Maria Pia Beccar-Varela.

includes how much, how often, and what they spend their money on. This is true for credit and debit card issuers; however, any retailer that captures information on its customers' purchases possesses potentially invaluable transaction data. A transaction database consists of a set of transactions where each transaction is a set of items. Typically, a transaction is a single customer purchase and the items are the things bought.

Given a set of transactions, we can find rules that will predict the occurrence of an item based on the occurrences of other items in the transaction. Table 15.1 is an example of market basket transaction data. In the table each row represents a transaction with dates, transaction ID and items that were purchased. For example in row one, a customer with transaction ID 1 on 05/02/2020 purchased the items milk, diaper, and coke.

Market basket transaction data can be represented in a binary format as shown in Table 15.2, where each row corresponds to a transaction and each column corresponds to an item. The items are treated as a binary variable whose value is one if the item is present in a transaction and zero if otherwise.

Table 15.1 Market basket transaction data.

Date	Transaction ID	Items
05/02/2020	1	Milk, Diaper, Coke
05/02/2020	2	Milk, Beer
05/06/2020	3	Book, Toy
05/07/2020	4	Diaper
05/07/2020	5	Diaper, Toy
05/08/2020	6	Coke, Milk, Toy

Table 15.2 A binary 0/1 representation of market basket transaction data.

Date	Transaction ID	Milk	Diaper	Beer	Coke	Toy	Book
05/02/2020	1	1	1	0	1	0	0
05/02/2020	2	1	0	1	0	0	0
05/06/2020	3	0	0	0	0	1	1
05/07/2020	4	0	1	0	0	0	0
05/07/2020	5	0	1	0	0	1	0
05/08/2020	6	1	0	0	1	1	0

The representation of the transactional data in Table 15.2 is very simple since it ignores certain aspects of the data such as the quantity of items sold.

15.3 Terminologies

In this section, we briefly define some terminologies in association rule analysis that will be useful throughout this chapter.

15.3.1 Itemset and Support Count

Let $S = \{s_1, s_2, \dots, s_K\}$ be the set of all items in a market basket transactional data and $T = \{t_1, t_2, \dots, t_N\}$ be the set of all transactions. The collection of zero or more items is termed an itemset. For instance, the items $\{Milk, Beer\}$ in the second row of Table 15.1 is an example of a 2-itemset. An important property of an itemset is its support count which is defined as the frequency of occurrence of a particular itemset. Mathematically, the support count, $\sigma(A)$, for an itemset A can be defined as:

$$\sigma(A) = \text{card}(\{t_i | A \subseteq t_i,\ t_i \in T\}),$$

where card(.) denotes the number of elements in the set.

In Table 15.1, we see that the support count of {Milk, coke } is 2 because only two transactions contain all two items.

15.3.2 Frequent Itemset

A frequent itemset is defined as an itemset whose support is greater than or equal to the minimum support threshold. One major challenge in mining frequent itemsets from a large data set is the fact that such mining often generates a huge number of itemsets satisfying the minimum support (min sup) threshold, especially when min sup is set low. This is because if an itemset is frequent, each of its subsets is frequent as well. In fact a long itemset will contain a combinatorial number of shorter frequent sub-itemsets. Therefore to overcome this difficulty, we introduce the concepts of closed frequent itemset and maximal frequent items.

15.3.3 Closed Frequent Itemset

An itemset A is closed in a data set D if there exists no proper super-itemset B such that B has the same support count as A in D. In other words, a closed itemset is an itemset in which all immediate supersets have different support count. An itemset A is a closed frequent itemset in set D if A is both closed and frequent in D. In other words, a closed frequent itemset is a closed itemset that satisfies the minimun support threshold.

15.3.4 Maximal Frequent Itemset

An itemset A is a maximal frequent itemset in a data set D if A is frequent, and there exists no super-itemset B such that $A \subset B$ and B is frequent in D. In other words, maximal frequent itemset are frequent itemset that cannot be extended with any item without making them infrequent.

15.3.5 Association Rule

An association rule is a process of determining the likelihood that two pieces of information appear together. It can be expressed as a probability or a percentage. Let us consider an instance where $S = s_1, s_2, \dots, s_K$ denotes the set of all distinct items in the market basket transactional data, where the total number of all items can be expressed as its cardinality, i.e. $K = \text{card}(S)$. The association rule is an implication expression of the form

$$A \Rightarrow B,$$

where $A, B \subset S$ are any two disjoint itemsets. We say that A is the antecedent, B is the consequent, and $\text{card}(A) = |A|$, is the size or length of set A.

For example in Table 15.2, the first person purchased milk, diaper, and coke. We see the association rule as:

$$\{milk, diaper\} \Rightarrow \{coke\}.$$

This rule is interpreted as the event that a customer who purchased milk and diaper also grabs a coke as well. The strength of this association can be measured by the support, confidence, and lift of the rule. In the following definitions, A and B are events rather than sets, e.g. $\Pr(A \cap B)$ denotes the proportions of transactions that include items in event A and items in event B.

Association rule is very useful in analyzing datasets. The manager of a grocery store for example could utilize association rule to know if certain groups of items are consistently purchased together and use this data for adjusting store layouts, cross-selling, and promotions based on statistics.

15.3.6 Rule Evaluation Metrics

We now discuss on some metrics that are used to evaluate the association rules. We begin with support.

- *Support*: It is a measure of how frequently the collection of items occur together as a percentage of all transactions. This metric is estimated by the proportion of transactions that contain both item sets A and B out of all transactions. We define the support of $A \Rightarrow B$ as:

$$\text{support}\ (A \Rightarrow B) = \Pr(A \cap B).$$

In other words, $\Pr(A)$ is the support of itemset A, and $\Pr(B)$ is the support of itemset B.

- *Confidence*: It is defined as the ratio of the number of transactions that includes all items in $\{B\}$ as well as the number of transactions that includes all items in $\{A\}$ to the number of transactions that includes all items in $\{A\}$. So it is the conditional probability of event B given event A. The confidence of $A \Rightarrow B$ is defined mathematically as:

$$\text{confidence}\ (A \Rightarrow B) = \Pr(B|A) = \frac{\Pr(A \cap B)}{\Pr(A)}.$$

  Another name for confidence is strength.
- *Lift*: The lift can be interpreted as a general measure of association between two item sets. Lift (A, B) denotes the ratio of the probability of containing event B under the condition of containing event A to the probability of containing event B without event A. The lift of $A \Rightarrow B$ is defined mathematically as:

$$\text{lift}\ (A \Rightarrow B) = \frac{\Pr(A \cap B)}{\Pr(A) \cdot \Pr(B)},$$

  is the ratio of the observed support to that expected if the two rules were independent. The lift can be interpreted as a general measure of association between two itemsets. Lift $(A \Rightarrow B) > 1$ indicates positive correlation and Lift $(A \Rightarrow B) < 1$ indicates negative correlation, while Lift $(A \Rightarrow B) = 1$ indicates that zero correlation. Since support and lift cannot capture implication rule (i.e. $A \Rightarrow B$ and $B \Rightarrow A$), we define a new rule called conviction.
- *Conviction*: Conviction measures the implication strength of the rule from statistical independence has value 1 if the items are unrelated. Mathematically, the conviction of a rule can be defined as:

$$\text{conviction}(A \Rightarrow B) = \frac{1 - \Pr(B)}{1 - \Pr(B|A)} = \frac{\Pr(B')}{\Pr(B'|A)} = \frac{\Pr(B')\Pr(A)}{\Pr(B' \cap A)},$$

  where B' denotes the complement of B. Conviction is sensitive to rule direction, conviction $(A \Rightarrow B) \neq \text{conviction}(B \Rightarrow A)$.

There are other evaluation measures such as the **leverage (Piatetsky-Shapiro)** of a rule and **coverage** of a rule. The leverage (Piatetsky-Shapiro) measures the difference of actual and expected occurrences of A and B together in the data set and what is expected if A and B are statistically dependent. The coverage of a rule is a simple measure of how often an itemset appears in the data set.

In Section 15.4, we present the Apriori algorithm which is one of the most popular algorithm for mining association rules. The algorithm finds the most frequent combinations in a database and identifies association rules between the items.

15.4 The Apriori Algorithm

Apriori algorithm is an iterative process for mining frequent itemsets in transaction databases. In 1994, R. Agarwal and R. Srikant improved this algorithm by exploiting one nice property of itemsets of different sizes and named it as Apriori Hastie et al. (2008a). The Apriori algorithm finds a solution by using the property that, if an itemset $A \subset A'$, then $\Pr(A) = \text{support}(A) \geq \text{support}\,(A') = \Pr(A')$. In other words, *if an itemset is frequent, then all of its subsets must also be frequent.* The algorithm is usually followed by two steps; "join" and "prune" in order to find the most frequent itemsets. The "join" step generates $(K+1)$ itemset from K-itemsets by joining each item with itself. The "Prune" step reduces the size of the candidate itemsets if the candidate item does not meet minimum support. The iteration steps of the Apriori algorithm are presented below:

1. In the first step, a threshold or minimum support (min sup) is defined in order to find the frequent items. Each item is taken as a 1-itemsets candidate. Then the algorithm will count the occurrences of each item.
2. The set of 1-itemsets whose occurrence is satisfying the min sup are determined. Only those candidates which count more than or equal to minimum support are taken ahead to the next iteration and the others are pruned.
3. We follow a "join" step by generating a group of 2-itemset by combining items with itself.
4. The 2-itemset candidates are pruned using the minimum support threshold value and we will have 2-itemsets with minimum support only.
5. The next iteration generates 3-itemsets using "join" and "prune" steps. This iteration will follow the anti-monotone property of the support measure (i.e. the support for an itemset never exceeds the support for its subsets). In this instance, the subsets of 3-itemsets, that is the 2-itemset subsets of each group fall in the minimum support. If all 2-itemset subsets are frequent then the superset will be frequent, otherwise it is pruned.
6. The next step will follow making 4-itemset by joining 3-itemset with itself and pruning if its subset does not meet the minimum support criteria. The algorithm is stopped when the most frequent itemset is obtained.

We now present a Pseudo code for the Apriori algorithm:

```
C_k: Candidate item set of size k
    L_k: Frequent item set of size k
    L_1: {frequent items};
For (k = 1; L_k! = Ø; k + +) do begin
C_{k+1} = candidates generated from L_k;
For each transaction t in database do
            Increment the count of all candidates in C_{k+1}
            Those are contained in t
L_{k+1} = candidates in C_{k+1} with minimum support
    End
    Return ∪_k L_k
```

15.4.1 An example of the Apriori Algorithm

We now present an example with transaction data shown in Table 15.3.

In Table 15.3, there are four distinct transaction ID (TID), and each transaction involves between 2 and 4 grocery items. Using the Apriori algorithm to the dataset of transactions, we identify all frequent $k = 1, 2, 3$ itemsets with 50% minimum support.

Step 1: We first create a table containing support count of each item presented in the dataset that is called C_1 (i.e. candidate set).

Itemset	Sup
{*Coke*}	2
{*Beer*}	3
{*Diaper*}	3
{*Bread*}	1
{*Milk*}	3

Table 15.3 Grocery transactional data.

TID	Items
10	Coke, Diaper, Bread
20	Beer, Diaper, Milk
30	Coke, Beer, Diaper, Milk
40	Beer, Milk

The minimum support of this problem is $50\% \times 4 = 2$. In this example, the number of transaction is 4. Thus 50% of this value is 2. We compare the candidate set or (C_1) items support count with the minimum support count. If the support count of C_1 is less than the minimum support, we remove those items. Thus, we obtain the itemset L_1 as follows:.

Itemset	Sup
$\{Coke\}$	2
$\{Beer\}$	3
$\{Diaper\}$	3
$\{Milk\}$	3

Step 2: We generate a candidate set C_2 using all possible combination of two items from L_1 (join step ($K = 2$).

Itemset
$\{Coke, Beer\}$
$\{Coke, Diaper\}$
$\{Coke, Milk\}$
$\{Beer, Diaper\}$
$\{Beer, Milk\}$
$\{Diaper, Milk\}$

we find the support count of these itemsets by searching in each dataset.

Itemset	Sup
$\{Coke, Beer\}$	1
$\{Coke, Diaper\}$	2
$\{Coke, Milk\}$	1
$\{Beer, Diaper\}$	2
$\{Beer, Milk\}$	3
$\{Diaper, Milk\}$	2

After counting their supports, the algorithm eliminates all candidate itemsets of C_2 whose support counts are less than minimum support.

Itemset	Sup
$\{Coke, Diaper\}$	2
$\{Beer, Diaper\}$	2
$\{Beer, Milk\}$	3
$\{Diaper, Milk\}$	2

Step 3: We generate candidate set C_3 using L_2 (join step).

Itemset
$\{Beer, Diaper, Milk\}$

Check if all subsets of these itemsets are frequent or not and if not, we remove that itemset.

Itemset	Sup
$\{Beer, Diaper, Milk\}$	2

Finally, we top since no frequent itemsets can be found. Thus, we have discovered all the frequent itemsets that is, $\{Coke\}$, $\{Beer\}$, $\{Diaper\}$, $\{Milk\}$, $\{Coke, Diaper\}$, $\{Beer, Diaper\}$, $\{Beer, Milk\}$, $\{Diaper, Milk\}$, $\{Beer, Diaper, Milk\}$.

There are two important characteristics of the Apriori algorithm. Firstly, it is a sequential process which takes into account the itemset lattice one level at a time, i.e. from frequent 1-itemsets to the maximum size of frequent itemsets. Secondly, at each iteration (level), it generates new candidate itemsets from the frequent itemsets of the previous iteration. The support for each candidate is counted and compared to the minimum support. The total number of iterations of this algorithm is $K_{\max} + 1$, where $K_{\max}$ is the maximum size of the frequent itemsets.

The Apriori algorithm reduces the number of candidate itemsets and finds the frequent itemset. But the algorithm is computationally expensive if the candidate itemset is very large or the minimum support is low. However, we may use an alternative solution to reduce the number of comparisons by using advanced data structures, such as hash tables and sorting the candidate itemsets.

15.5 Applications

We present an application of association rules analysis to biblical data (Luhby 2009). The biblical dataset consist of a parsed-version (with punctuation and stop words removed) of the King James Bible. The goal is to find the words that commonly occur together in sentences using association rules. We will find top five rules in decreasing order of confidence, lift and conviction for itemsets of size 2 and 3 which satisfy the 1% support threshold.

The data was loaded into `Python` and transformed into a transaction data frame using the *mlxtend library*, where each sentence was read as a transaction. In order to count the occurrences of every word, a sparse matrix was created using *sMat* command in python.

This matrix was used to create a list of frequent itemsets using the Apriori algorithm from the mlxtend library, and a relatively small support of 1%, which was appropriate for this use case, as we have a big number of transactions and a lot of words to choose from.

```
freqItemsets.head()
SPINoREMSPI
     support      itemsets
0   0.019099       (about)
1   0.022990   (according)
2   0.035176       (after)
3   0.020932       (again)
4   0.044693     (against)
```

The association rules were created based on the frequent itemsets list:

```
rules.head()
SPINoREMSPI
  antecedents consequents     ...     leverage  conviction
0    (against)      (lord)    ...     0.003538    1.112048
1       (lord)   (against)    ...     0.003538    1.017580
2         (am)      (lord)    ...     0.004989    1.321886
3       (lord)        (am)    ...     0.004989    1.024479
4       (said)  (answered)    ...     0.008850    1.084174
```

There are several metrics to measure the strength of an association rule, so the top five rules for itemsets of size 2 and 3 based on three different metrics are shown in Sections 15.5.1–15.5.3.

15.5.1 Confidence

Confidence is the conditional probability of the consequent given the antecedent. The larger this factor, the more creditable a rule is. The top five rules in order of confidence are the following:

```
        antecedents consequents  confidence
304    (shalt, thy)       (thou)    1.000000
298   (thee, shalt)       (thou)    1.000000
275   (shalt, lord)       (thou)    0.997382
258     (thy, hast)       (thou)    0.974432
270   (thus, saith)       (lord)    0.949451
```

The confidence of 1 or 100% means that 100% of the words that has `"shalt"` and `"thy"` also had `"thou"` appearing. The same also applies to the words `"thee"` and `"shalt"`. We can conclude that the words shalt and thy, as well as thee and shalt, imply the word thou with the highest confidence.

15.5.2 Lift

The lift is a general measure of association between the two item sets. Values greater than one indicate positive correlation, while values less than one indicate negative correlation. The top five rules in order of lift are the following:

```
        antecedents consequents       lift
269    (thus, lord)      (saith)  22.182650
268   (saith, lord)       (thus)  21.669011
276    (thou, lord)      (shalt)   9.122143
256     (thou, thy)       (hast)   8.898690
306     (thou, thy)      (shalt)   8.545587
```

We can observe that these rules have a high degree of positive since all the lift values are greater than one.

15.5.3 Conviction

Finally, conviction is a metric that captures the notion of implication rules, and it solves the problem of support and lift in distinguishing the rule direction. The top five rules in order of conviction are the following:

```
        antecedents consequents  conviction
304    (shalt, thy)       (thou)         inf
298   (thee, shalt)       (thou)         inf
275   (shalt, lord)       (thou)  334.331372
258     (thy, hast)       (thou)   34.230554
270   (thus, saith)       (lord)   15.541888
```

We notice that the first two rules have an infinite conviction because the corresponding confidence is equal to one as we saw in the confidence example.

15.6 Problems

1 Use the Apriori algorithm to complete the association rule generation process from following transactions tables. Experiment with different values of support and confidence to observe how they control the number of association rules generated.

2 (a) What is the support of $\{5\}$, $\{2,4\}$, and $\{2,4,5\}$ from Table 15.4?

Table 15.4 Transaction data.

Date	Number
101	1, 3,4, 5, 6, 7, 9, 10
102	2, 4, 5, 7, 8, 9, 10
103	1, 2, 3, 4, 10
104	2, 4 5, 6, 7, 10
105	1, 3 4, 5, 7, 8

(b) What is the confidence of $\{2,4\} \rightarrow \{5\}$ and $\{5\} \rightarrow \{2,4\}$ in Table 15.4? Is the confidence symmetric?

3 Show two rules that have a confidence of 0.7 or greater for an itemset containing three items.

4 Suppose that a frequent itemset consists of $k = 4$ items $\{x, y, z, w\}$. How many association rules (of any size) can be formed from data with $K = 4$ items and what are they? How many association rules can be formed from this itemset (exactly $K = 4$ items)?

5 Suppose the confidence for rule $\{y, z, w\} \rightarrow \{x\}$ is low (i.e. $C(\{y, z, w\} \rightarrow \{x\}) < z$).
(a) Do we need to calculate $C(\{w\} \rightarrow \{x, y, z\})$?
(b) Do we need to calculate $C(\{x, y\} \rightarrow \{z, w\})$?

6 Explain the efficiency of Apriori approach for high frequency data mining.

7 Explain a method to optimize the Apriori algorithm for Frequent Items Mining.

8 Consider the following beverage preferences from 1000 people:

Beverages	Coffee	Non-coffee
Tea	150	50
Non-tea	650	150

(a) What is the support of $\{Tea\} \rightarrow \{Coffee\}$?
(b) What is the confidence of $\{Tea\} \rightarrow \{Coffee\}$?
(c) What is the lift of $\{Tea\} \rightarrow \{Coffee\}$?

9 Explain whether it is possible to use $\text{conv}(A \Rightarrow B) = \frac{1-S(B)}{1-\text{conf}(A \rightarrow B)}$ instead of association rules, where $S(B) = P(B)$ is the support of B and $\text{conf}(A \rightarrow B) = P(B|A)$ is the confidence of rule $A \rightarrow B$.

10 Explain how the above measure avoids the problems associated with both the confidence and the lift measures.

16

Support Vector Machines

16.1 Introduction

In this chapter, we discuss the support vector machine (SVM), a methodology for classification tasks. SVM was first developed in the 1963 by Vladimir Vapnik and Alexey Chervonenkis. In subsequent years, the model has evolved considerably into one of the most flexible and effective machine learning algorithm.

The SVM algorithm works by looking for a linearly separable hyperplane, or a decision boundary separating members of one class from the other. If such a hyperplane exists, the algorithm ends. However, if such a hyperplane does not exist, SVM uses a nonlinear mapping to transform the training data into a higher dimension. Then it searches for the linear best separating hyperplane. The SVM algorithm finds this hyperplane using support vectors and margins. As a training algorithm, SVM may not be very fast compared to some other classification methods such as the logistic regression model but owing to its ability to model complex nonlinear boundaries, SVM is highly preferred since it produces significant accuracy with less computation power. In addition, it is comparatively less prone to overfitting, that is the model performs well on the training data, but it does not generalize well. SVM has been applied to several classification problem such as image classification, classification of satellite data, handwritten digit recognition, etc.

16.2 The Maximal Margin Classifier

Before we begin our discussion of maximal margin classifier, we would present some definitions.

Definition 16.1 (Linear equation) A linear equation in three variables x, y, and z is any equation of the form

$$ax + by + cz = d,$$

Data Science in Theory and Practice: Techniques for Big Data Analytics and Complex Data Sets, First Edition. Maria Cristina Mariani, Osei Kofi Tweneboah, and Maria Pia Beccar-Varela.

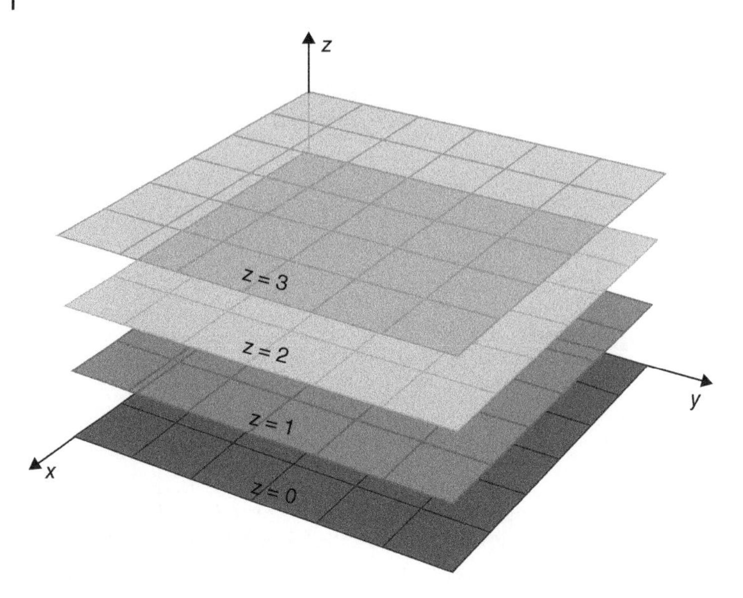

Figure 16.1 The *xy*-plane and several other horizontal planes.

where a, b, c, d are constants and the coefficients a, b, c are not all zero. Any such equation defines a plane in $\mathbb{R}^3$.

Below are some examples of linear equations and their corresponding planes:

- The equation $z = 0$ defines the *xy*-plane in $\mathbb{R}^3$, since the points on the *xy*-plane are precisely those points whose *z*-coordinate is zero. If d is any constant, the equation $z = d$ defines a horizontal plane $\mathbb{R}^3$ which is parallel to the *xy*-plane. Figure 16.1 shows several such planes.
- The equation $y = 0$ define the *xz*-plane, and equations of the form $y = d$ define planes parallel to these. Figure 16.2 shows several such planes of the form $y = d$.
- The equation $x + y + z = 1$ defines a slanted plane in $\mathbb{R}^3$, which goes through the points $(1, 0, 0)$, $(0, 1, 0)$, and $(0, 0, 1)$. This plane is shown in Figure 16.3.

Definition 16.2 (Hyperplanes) A linear equation of the form

$$a_n x_n + a_2 x_2 + \cdots + a_n x_n = d,$$

where $a_1, a_2, \ldots, a_n$ and d are constants. Any such equation defines a hyperplane in $\mathbb{R}^n$.

A hyperplane in $\mathbb{R}^3$ is a plane, and a hyperplane in $\mathbb{R}^4$ is similar to a plane, except that it is three-dimensional. In general, the word hyperplane refers to an $(n - 1)$-dimensional flat in $\mathbb{R}^n$. For example, hyperplanes in $\mathbb{R}^7$ are six-dimensional flats.

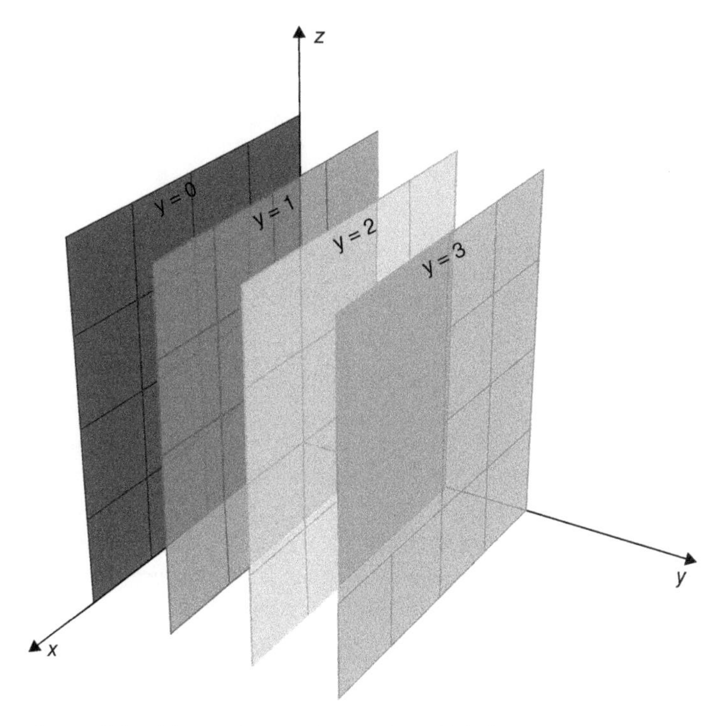

Figure 16.2 The *xz*-plane and several parallel planes.

Figure 16.3 The plane $x + y + z = 1$.

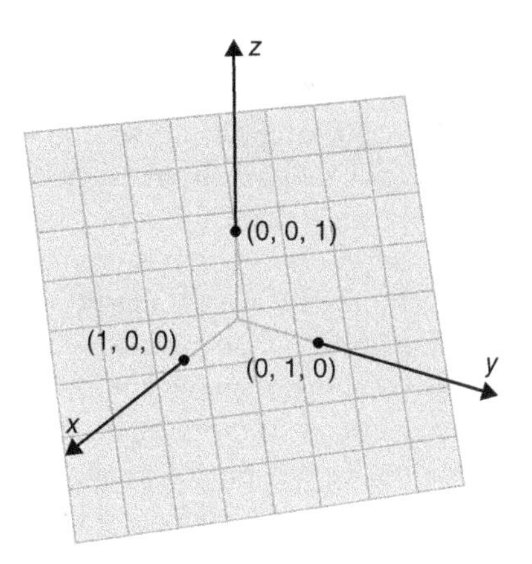

Example 16.1 Find an equation for the plane that is parallel to the plane $x + y + 4z = 6$ and goes through the point (4, 2, 3). Solution: The required plane must have an equation of the form:

$$x + y + 4z = d$$

for some constant d. Substituting the point (4, 2, 3) gives

$$d = 4 + 2 + 4(3) = 18.$$

Thus, the desired plane is defined by the equation:

$$x + y + 4z = 18.$$

Remarks 16.1 The equation for a plane is not unique. For example, the planes defined by the equations

$$x + y + 4z = 17 \text{ and } 2x + 2y + 8z = 34$$

are the same, since the second equation is just twice the first equation.

In general, if our data can be perfectly separated using a hyperplane, then there will in fact exist an infinite number of such hyperplanes. To be able to construct a classifier based upon a separating hyperplane, we must have a way to decide which of the infinite possible separating hyperplanes to use. One way to go about this is by using the maximal margin hyperplane, i.e. separating hyperplane that best separates the hyperplane farthest minimum distance from the training observations. We can then classify a test observation based on which side of the maximal margin hyperplane it lies. This is known as the *maximal margin classifier*. The objective is that a classifier that has a large margin on the training data will also have a large margin on the test data, thereby classifying the test observations correctly.

Even though the maximal margin classifier is a very simple way to perform classification. In practice, the existence of such a hyperplane may not always exists especially for complex data sets. In that case, the maximal margin classifier provides a poor solution. In instances like this, the concept can be extended where a hyperplane exists which almost separates the classes, using what is known as a *soft margin*. Soft margin is when we extend maximal margin classifier to cases in which the data are not linearly separable. In this case, some observations are permitted to be on the incorrect side of the margins. The margin is soft as a small number of observations violate the margin. The softness is controlled by slack variables which control the place values of the observations relative to the margins and separating hyperplane. In an optimization problem, a slack variable is a variable that is added to an inequality constraint to transform it into an equality. The support vector classifier maximizes a soft margin. The optimization problem can be modified as

$$y_i(\theta_0 + \theta_1 x_{1i} + \theta_2 x_{2i} + \cdots + \theta_n x_{ni}) \geq 1 - \epsilon_i \quad \text{for every observation,}$$

where $\epsilon_i \geq 0$ and $\sum_{i=1}^{n} \epsilon_i \leq C$. The ϵ_i is the slack variables corresponding to the ith observation and C is regularization parameter.

16.3 Classification Using a Separating Hyperplane

Suppose we have a $n \times p$ data matrix $\mathbf{X}$ that consists of n training observations in p-dimensional space,

$$\mathbf{X} = \begin{bmatrix} x_{1,1} & x_{1,2} & \cdots & x_{1,k} & \cdots & x_{1,p} \\ x_{2,1} & x_{2,2} & \cdots & x_{2,k} & \cdots & x_{2,p} \\ \vdots & \vdots & & \vdots & \vdots & \vdots \\ x_{j,1} & x_{j,2} & \cdots & x_{j,k} & \cdots & x_{j,p} \\ \vdots & \vdots & & \vdots & \vdots & \vdots \\ x_{n,1} & x_{n,2} & \cdots & x_{n,k} & \cdots & x_{n,p} \end{bmatrix}$$

and these observations fall into two classes, i.e. $y_1, \ldots, y_n \in \{-1, 1\}$ where -1 denotes one class and 1 the other class.

We also have a test observation, a p-vector of observed features

$$x^* = \begin{bmatrix} x_1^* \\ x_2^* \\ \vdots \\ x_p^* \end{bmatrix}.$$

Our objective is to develop a classifier based on the training data that will correctly classify the test observation using its feature characteristics.

We begin by first assuming that it is possible to construct a hyperplane that separates the hyperplane training observations perfectly according to their class labels.

We label the observations from the first class (i.e. lower left) as $y_i = 1$ and those from the second class (i.e. upper right) as $y_i = -1$ as shown in Figure 16.4. In Figure 16.4, there are two classes of observations, each of which has measurements on two variables. Three separating hyperplanes, out of many possible, are shown in the figure.

Then, a separating hyperplane has the property that

$$H_1 : \theta_0 + \theta_1 x_{1i} + \theta_2 x_{2i} + \cdots + \theta_p x_{pi} \geq 1, \quad \text{for } y_i = 1 \tag{16.1}$$

and

$$H_2 : \theta_0 + \theta_1 x_{1i} + \theta_2 x_{2i} + \cdots + \theta_p x_{pi} \leq -1, \quad \text{for } y_i = -1. \tag{16.2}$$

Alternatively, one can write

$$y_i(\theta_0 + \theta_1 x_{1i} + \theta_2 x_{2i} + \cdots + \theta_p x_{pi}) \geq 1, \quad \text{for every observation} \tag{16.3}$$

for all $i = 1, \ldots, n$.

If a separating hyperplane exists, we can use it to construct a classifier: a test observation is assigned a class depending on which side of the hyperplane it is

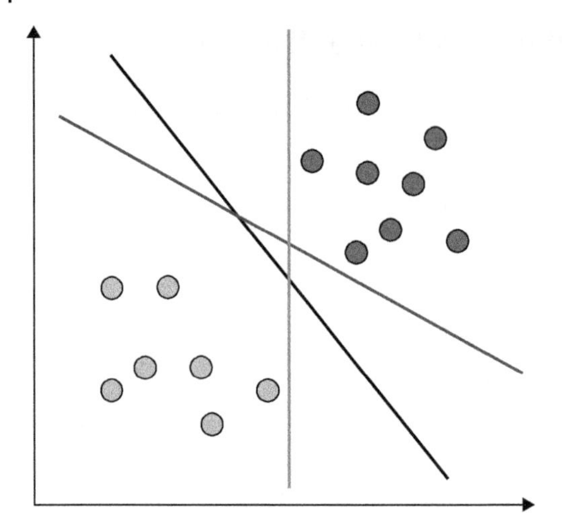

Figure 16.4 Two class problem when data is linearly separable.

located. Thus, we classify the test observation x^* based on the sign of

$$f(x^*) = \theta_0 + \theta_1 x_1^* + \theta_2 x_2^* + \cdots + \theta_p x_p^*.$$

If $f(x^*)$ is positive, then we assign the test observation to class 1 and if $f(x^*)$ is negative, we assign the test observation to class −1.

If the data is not linearly separable as shown in Figure 16.5, then SVM can be extended to give a good result. There are two main steps for nonlinear generalization of SVM. In the first step, we transform the training data into a higher dimensional data using a nonlinear mapping. Once the data is transformed into

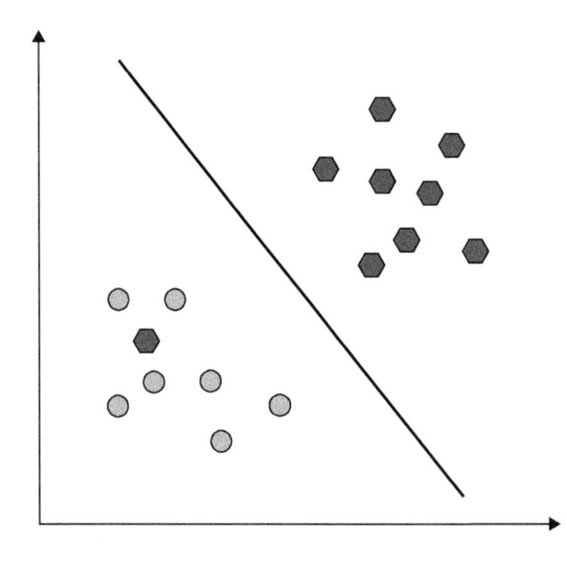

Figure 16.5 Two class problem when data is not linearly separable.

the new higher dimension, we proceed to the second step which involves finding a linear separating hyperplane in the new space. The maximal marginal hyperplane found in the new space corresponds to a nonlinear separating hypersurface in the original space.

16.4 Kernel Functions

Handling the nonlinear transformation of training data into higher dimension is a difficult task. In fact the procedure may be computationally intensive. Therefore to avoid some of this problem, the concept of Kernel functions is introduced. A kernel is a function that quantifies the similarity of two observations. It is a generalization of the inner product of nonlinear transformation and is denoted by

$$K(X_1, X_2) = \sum_{j=1}^{p} x_{ij} x_{i'j} \tag{16.4}$$

for $i = 1, \dots, n$. Where $X_1 = (x_{11}, x_{12}, \dots, x_{1n})$ and $X_2 = (x_{21}, x_{22}, \dots, x_{2n})$.

Equation (16.5) is a linear kernel since the support vector classifier is linear in the features. If we replace every instance of $\sum_{j=1}^{p} x_{ij} x_{i'j}$ in (16.4) the quantity

$$K(X_1, X_2) = \left(1 + \sum_{j=1}^{p} x_{ij} x_{i'j}\right)^d \tag{16.5}$$

we obtain a polynomial kernel of degree d, where d is a natural number.

Some of the other common kernels are sigmoid kernel and Gaussian radial basis function. In practice, there is no rule to determine which kernel will provide the most accurate result in a given situation and also the accuracy of SVM does not depend on the choice of the kernel. Please refer to Hastie et al. (2008b) and James et al. (2013) for a comprehensive study and more details of the algorithms discussed in this chapter.

16.5 Applications

In this section, we present an application of the SVM technique to the credit card defaults payment data discussed in Chapter 14. Please refer to Section 14.5 for a description of the data sets.

The dataset studied contains 5000 observations, characterized by the same 23 labeled variables, each of them representing either a default or not a default (binary value) of credit card clients. Among the total 5000 observations, 1107 observations (22.14%) are the cardholders with default payment.

We randomly divided our data sets into the training sample and the test sample with a ratio of 2 : 1. We used the training sample to train the SVM model and then used the test sample to compare them. Using the 23 features, we present the results when the SVM was applied to our credit default test dataset.

In this example, we used both the linear SVM and nonlinear SVM using the radial kernel. The results obtained are based on the C index and mean square error (MSE) criteria. Table 16.1 displays the models performances on the test dataset with 23 variables using the C index and MSE.

Table 16.1 Models performances on the test dataset.

Model	C index	MSE
Linear SVM	0.595	0.208
Nonlinear SVM	0.598	0.217

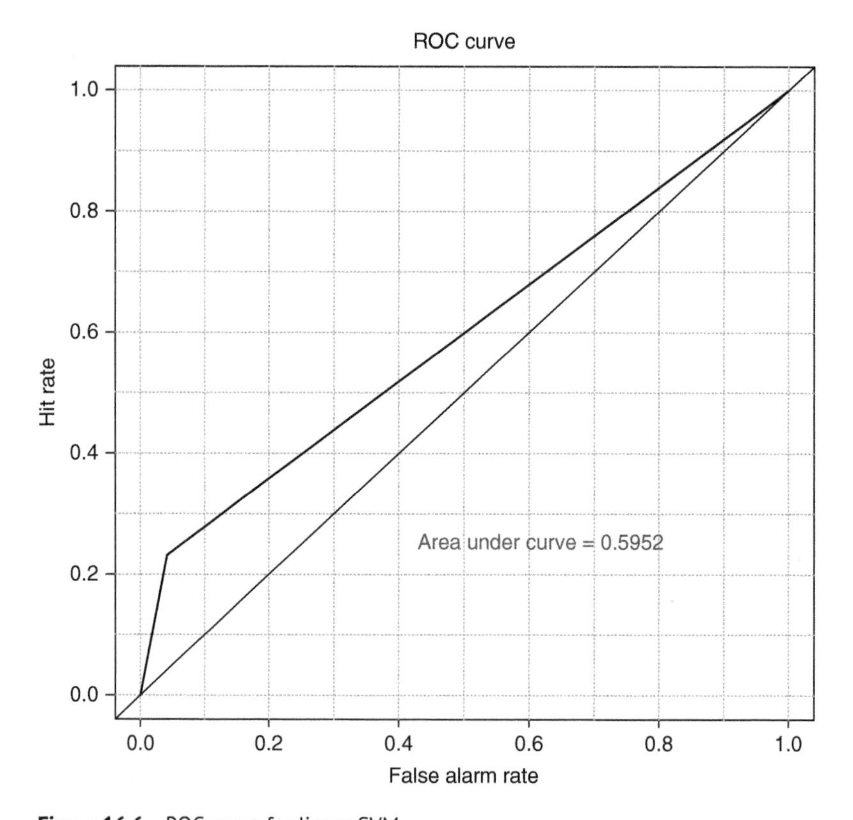

Figure 16.6 ROC curve for linear SVM.

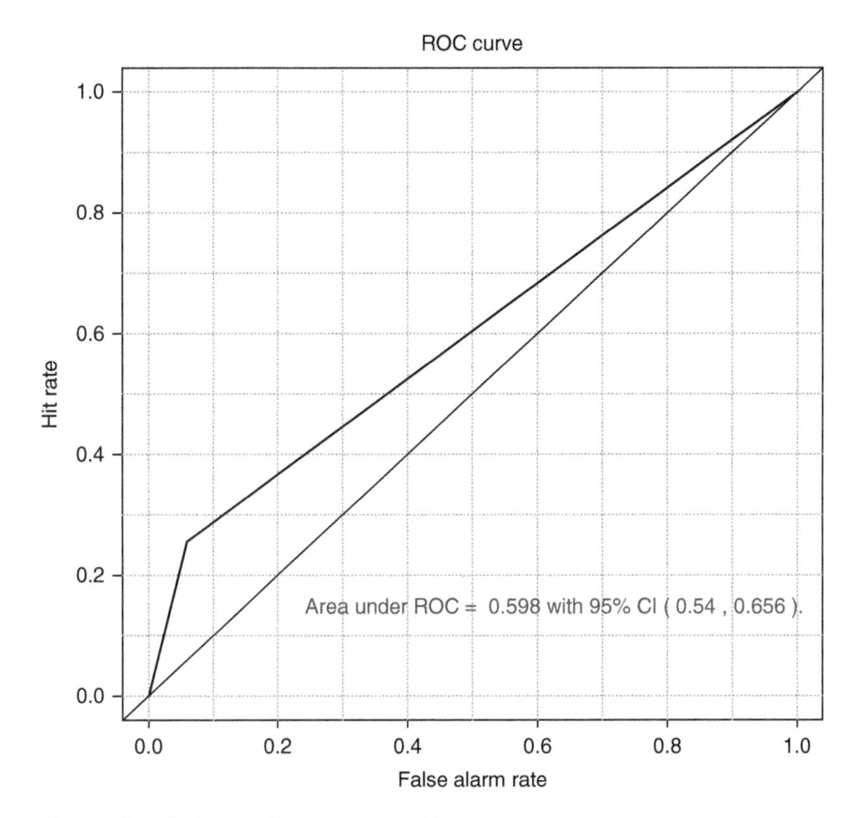

Figure 16.7 ROC curve for nonlinear SVM.

From the values in the Table 16.1, we observe that according to the C index metric, the difference between the linear SVM and nonlinear SVM are not very significant. This is because the data is almost linearly separable. The receiver operating characteristic curve, or ROC curve corresponding to the linear SVM and the nonlinear SVM are given in Figures 16.6 and 16.7, respectively. The ROC curve shows the performance of a classification model at all classification thresholds. This curve plots two parameters, namely, True Positive Rate (Hit Rate) and False Positive Rate (False Alarm Rate). The true positive rate is used to measure the percentage of actual positives which are correctly identified. The false positive rate of a test is defined as the probability of falsely rejecting the null hypothesis.

16.6 Problems

1 Find the equation of the plane that is parallel to $2x - 4y + 2z = 5$ and goes through the point $(1, 4, 5)$.

2 Sketch the hyperplane $2x - 4y = 5$. Indicate the set of points for which $2x - 4y - 5 > 0$, as well as the set of points for which $2x - 4y - 5 < 0$.

3 Find an equation for the hyperplane in $\mathbb{R}^4$ that goes through the point $(3, 1, 5, 3)$ and has normal vector $(-1, 1, -1, 1)$.

4 Let H be the hyperplane in $\mathbb{R}^4$ defined by the equation $x_1 + 4x_2 + 4x_3 + 6x_4 = 6$. Find the distance from the point $(3, 4, 5, 6)$ to H.

5 Generate a simulated two-class data set with 200 observations and 2 features in which there is a visible but non-linear separation between the two classes.
(a) Show that in this setting, as SVM with a polynomial kernel (with degree $d > 1$) will outperform a support vector classifier on the training data.
(b) Make plots and report the training and test error rates.

6 We are given $n = 5$ observations in $p = 2$ dimensions. Each observation has an associated class label as shown in Table 1.

Observation	X_1	X_2	Y
1	5	5	Blue
2	2	5	Blue
3	3	2	Red
4	5	4	Red
5	5	2	Red

(a) Sketch the observations.
(b) Sketch the optimal separating hyperplane, and provide the equation for this hyperplane.
(c) Describe the classification rule for the maximal margin classifier.
(d) On your sketch, indicate the margin for the maximal margin hyperplane.

7 Discuss the difference between soft margin SVM and hard margin SVM algorithms.

8 Suppose that the points $(+1, -1)$ are linearly separable. Find a separating hyperplane $\beta^T x + \beta_0 = 0$ that completely separates $+1$ from -1's, with minimum margin.

9 Suppose a optimization problem is defined as:

$$\min_{\beta,\beta_0} \frac{1}{2}||\beta||^2 + C.\sum_{i=1}^{n} \xi,$$
$$\text{s.t.} \xi \geq 0 \quad \text{and} \quad y_i(\beta^T x_i + \beta_0) \geq 1 - \xi_i$$

for $i = 1, 2, \ldots, n$. Solve the problem using Lagrangian and Karush–Kuhn–Tucker (KKT) conditions.

17

Neural Networks

17.1 Introduction

In this chapter, we present another class of learning methods for classification problems. We begin our discussion by first introducing perceptrons.

17.2 Perceptrons

Perceptrons were developed in the mid 1990s by Frank Rosenblatt. It is an algorithm for learning a binary classifier called a threshold function: a function that maps its input $\mathbf{x}$ to an output value $f(\mathbf{x})$:

$$f(\mathbf{x}) = \begin{cases} 1, & \text{if } \mathbf{w} \cdot \mathbf{x} + b > 0, \\ 0, & \text{if otherwise,} \end{cases}$$

where $\mathbf{w}$ is a vector of real-valued weights, $\mathbf{w} \cdot \mathbf{x}$ denotes the dot product, i.e. $\mathbf{w} \cdot \mathbf{x} = \sum_{i=1}^{m} w_i x_i$ where m is the input units to the perceptron and b is the bias. The bias shifts the decision boundary away from the origin and does not depend on any input units.

A perceptron takes several binary input units, $x_1, x_2, \dots$ and produces a single binary output unit. The algorithm works by introducing weights, $w_1, w_2, \dots,$ which expresses the importance of each inputs $\mathbf{x}$ to the output units $f(\mathbf{x})$.

17.3 Feed Forward Neural Network

A feed forward neural network is an artificial neural network where connection between units do not form a cycle. A neural network is a two-stage regression or classification model, typically represented by a network diagram as shown in Figure 17.1.

Data Science in Theory and Practice: Techniques for Big Data Analytics and Complex Data Sets, First Edition. Maria Cristina Mariani, Osei Kofi Tweneboah, and Maria Pia Beccar-Varela.

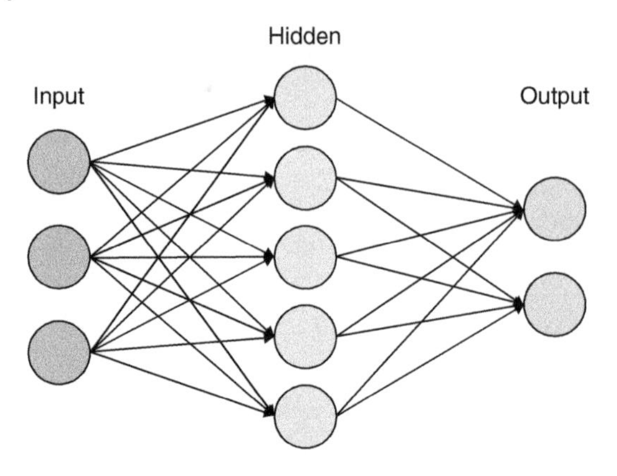

Figure 17.1 Single hidden layer feed-forward neural networks.

For the regression case, typically we only have one output variable or unit. Neural network has many known applications in data science, statistics, and related disciplines.

Feed forward neural networks as shown in Figure 17.1 have used to solve several classification tasks such as handwriting recognition, speech recognition, stock market prediction, and several others. In this figure, the feed forward neural networks is thought of as a network of "neurons" organized in three layers. The inputs layer, hidden layer, and the outputs layer. There are simplest networks which contain no hidden layers, and this is equivalent to the classical linear regression model. Each perceptron in one layer is connected to every perceptron on the next layer. Thus information is constantly moved from one layer to the next.

The units in the middle of the network in Figure 17.1, are called hidden units. This is because these units are not directly observed. In this particular figure, we have only one hidden unit but in general, we can have more than one hidden layer.

The input layers describe the number of variables or features the neural network uses to make its prediction. The choice of the number of hidden layers depends on the classification problem you want to solve. Generally, one to five hidden layers works well for most problems. The output layer represents the number of predictions one wants to make.

For regression problems, $K = 1$ and in this case there is only one output unit Y_1. Since the network can handle multiple numeric responses, we will discuss the general case in more details. For K-class classification, there are K units at the top, with the kth unit modeling the probability of class k. There are K output measurements $Y_k, k = 1, \dots, K$, each being coded as a 0–1 variable for the kth class.

The hidden units Z_m are created from linear combinations of the inputs variables. In fact can be thought of as the basis expansion of the input units X. The target Y_k is modeled as a function of linear combinations of the Z_m which is defined as follows:

$$Z_m = \sigma(\alpha_{0m} + \alpha_m^T X), \ m = 1, \dots, M, \tag{17.1}$$

$$T_k = \beta_{0k} + \beta_k^T Z, \ k = 1, \dots, K, \tag{17.2}$$

$$f_k(X) = g_k(T), \ k = 1, \dots, K, \tag{17.3}$$

where $Z = (Z_1, \dots, Z_M)$ and $T = (T_1, \dots, T_K)$. The activation function $\sigma(v)$ is normally chosen to be the sigmoid or logistic function $\sigma(v) = \frac{1}{1+e^{-v}}$. Other types of activation function includes, hyperbolic tangent function, rectified linear unit function etc. The output function $g_k(T)$ allows a final transformation of the vector of outputs T. For regression problems, we usually choose the identity function $g_k(T) = T_k$. We remark that if the activation function σ is the identity function, then the neural network model reduces to a linear model in the inputs.

When fitting the neural network model, we seek values of the weights that make the model fit the training data well. Let θ denote the complete set of weights which consists of $\alpha_{0m}, \alpha_m; m = 1, \dots, M$ for $M(p+1)$ weights and $\beta_{0k}, \beta_k; k = 1, \dots, K$ for $K(M+1)$ weights. For regression problems, we use the sum-of-square errors as our error function which is defined as:

$$R(\theta) = \sum_{k=1}^{K} \sum_{i=1}^{N} (y_{ik} - f_k(x_i))^2. \tag{17.4}$$

However for classification problems, we either use the cross entropy or squared error function which is defined as:

$$R(\theta) = -\sum_{k=1}^{N} \sum_{i=1}^{K} y_{ik} \log f_k(x_i), \tag{17.5}$$

and the corresponding classifier is $\arg\max_k f_k(x)$. The general approach for minimizing the error function $R(\theta)$ is by the gradient descent a called back-propagation algorithm. Please refer to Hastie et al. (2008b) for details of the algorithm. Advantages of the back-propagation algorithm are its simplicity and local nature. In the back propagation algorithm, each hidden unit passes and receives information only to and from units that share a connection. Therefore, it can be implemented efficiently on a parallel architecture computer.

However, a disadvantage of the feed forward neural networks discussed above is the fact that it does not take its history into account. Therefore, in the following sections, we will discuss two modifications of the feed forward neural network which works in dynamical environments.

We begin our discussion with the recurrent neural networks.

17.4 Recurrent Neural Networks

Consider a sequence of inputs $x_1, \dots, x_T$ with $x_i \in \mathbb{R}^D$. At time-step t, the recurrent neural network takes the current input x_t together with the previous hidden state $h_{t-1} \in \mathbb{R}^p$ to calculate the output. In particular, at each time-step, we apply the recurrence formula:

$$h_t = f_W(x_t, h_{t-1}), \tag{17.6}$$

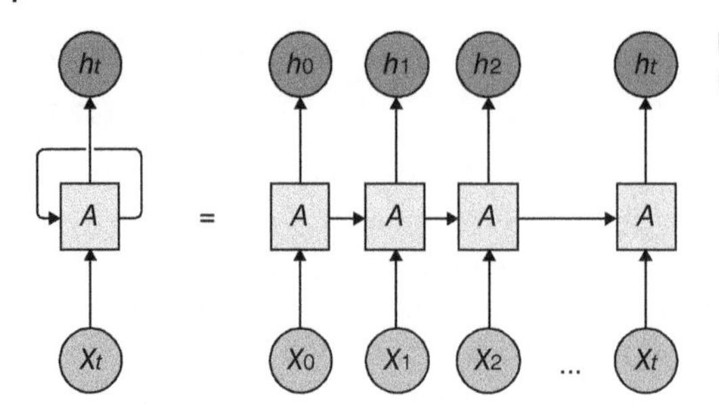

Figure 17.2 Simple recurrent neural network.

where f_W is a function with parameters W. These parameters are the same for every time-step. The output can then be computed as:

$$y_t = gW_2(h_t), \tag{17.7}$$

where gW_2 can be modeled as a simple feed-forward network.

The recurrent neural networks is a class of artificial neural networks where connections between nodes form a directed graph along a temporal sequence. This allows it to exhibit temporal dynamic behavior. Figure 17.2 represents the simple recurrent neural network architecture. In simple recurrent neural networks, the output can be used as inputs while having hidden states. Recurrent neural networks are usually used in natural language processing and speech recognition.

17.5 Long Short-Term Memory

The long short-term memory is a modification of the recurrent neural networks. The model introduces the memory cell, a unit of computation that replaces traditional artificial neurons in the hidden layer of the network. With these memory cells, the networks are able to effectively associate memories and input remote in time, hence well suited to grasp the structure of data dynamically over time with high prediction accuracy.

The gates produce an activation function, usually created through a sigmoid or exponential linear function, which controls the extent of which new information flows into the cells. The equations of the forward pass of a long short-term memory unit with a forget gate can be expressed mathematically as:

$$f_t = \sigma_g(W_f x_t + U_f h_{t-1} + b_f), \tag{17.8}$$

$$i_t = \sigma_g(W_i x_i + U_i h_{t-1} + b_i), \tag{17.9}$$

$$o_t = \sigma_g(W_o x_t + U_o h_{t-1} + b_o), \tag{17.10}$$

$$c_t = f_t \circ c_{t-1} + i_t \circ \sigma_c(W_c x_t + U_c h_{t-1} + b_c), \tag{17.11}$$

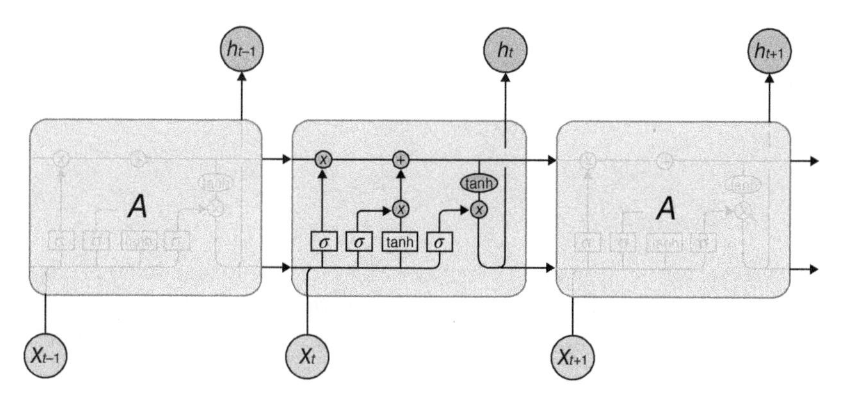

Figure 17.3 Long short-term memory unit.

where x_t is the input feature, h_t is the hidden state, c_t is the context vector, o_t is the output gate, f_t is the forget gates, and i_t is the input gates. $W \in \mathbb{R}^{p \times D}$, $U \in \mathbb{R}^{p \times p}$ are weight matrices and $b \in \mathbb{R}^p$ is a bias vector parameters which need to be learned during training.

In Figure 17.3, the merging arrows describe concatenation. The upper flow is the flow of the context vector, the lower part the flow of h_t. The merging arrows as seen in the lower left part denote concatenation. The concatenation follows from the fact that $W_f x_t + U_f h_{t-1} = W_f^*[x_t, h_{t-1}]$ for matrix $W_f^* = [W_f, U_f]$.

The function of the long short-term memory unit can be described as a series of pipes which control the flow of information.

The hidden state h_t is created through the context vector c_t which passes through the output gate o_t. The context vector c_t is a description of the history of the time series weighted with new input. The forget and input gates f_t, i_t control how much previous versus new information should be included in the context. The output gate o_t then decides how much of this context should be taken into consideration when constructing the current hidden state h_t.

17.5.1 Residual Connections

For deep layers with many hidden layers, it is common to encounter the problem of vanishing or exploding gradients. During training, gradients are propagated throughout the network by the chain rule. Thus, if the gradients of the loss function are small, they will eventually vanish given a large deep network. Exploding gradients on the other hand results when large error gradients accumulate and result in very large updates to the model weights during training. Residual connections provides a great boost to the networks ability to overcome vanishing or exploding gradients. A residual connection connects the output of one earlier layer to the input of another future layer. The input of the model are combined using a simple sum. Adding residual connections between the layers can be expressed as:

$$\hat{h}_l = F(W, h_{l-1}),$$

$$h_l = \hat{h}_l + h_{l-1}.$$

17.5.2 Loss Functions

We can evaluate the performance of our network by using the mean squared error as described in Section 17.3. We are now faced with the task of evaluating the performance of our network and finding a way to tune its parameters to accurately model its given task.

17.5.3 Stochastic Gradient Descent

The process of optimizing the weight w is to find the minimum value of the error function $E(w)$. Given that $E(w)$ is smooth and a continuous function of w, we can find the gradient $\nabla E(w)$ which points in the direction of the greatest increase of the error function.

The weight w is defined such that:

$$\nabla E(w) = 0 \tag{17.12}$$

is called a stationary point. The error function could be highly non-convex (i.e. the matrix of all second partial derivatives (the Hessian) is neither positive semidefinite, nor negative semidefinite.), giving us several points where the gradient vanishes. To find an optimal solution would thus require us to compare several local minima to find a sufficiently good solution. In this book, we evaluate the gradient for each training example and update the weights as follows:

$$w^{(t+1)} = w^{(t)} - \eta \nabla E_n(w^t). \tag{17.13}$$

This version of gradient descent is called stochastic gradient descent. For the stochastic gradient descent technique, we evaluate the error function $E(w)$ as the sum of individual errors for each training example.

The learning rate fills a crucial role in the stochastic gradient descent algorithm. A small learning rate yields an update scheme which could be very slow to converge. On the other hand, a large learning rate could overshoot and fail to converge completely. It is common to adjust the learning rate at various epochs of the algorithm to avoid these problems. There are several techniques for adjusting the gradient descent algorithm to achieve convergence. One technique is by using the Adam optimizer (Kingma and Ba 2015). The Adam optimizer updates the scheme which computes adaptive learning rates. Adam keeps track of both past gradients and past squared gradients and computes an estimate of the mean and variance of the gradients. The algorithms leverage the power of adaptive learning rates methods to find individual learning rates for each parameter.

17.5.4 Regularization – Ensemble Learning

Given the large number of parameters in a neural network, it is very prone to overfit the data. We recall that overfitting our model means our model has learned the training set too well and cannot generalize well to unseen data. We can overcome this problem using ensemble learning to increase the predictive performance of our classifiers by decreasing generalization error. Ensemble learning is the idea that a large group of people are collectively better at decision-making than even individual experts. Thus, given a set of k trained models $\mathcal{M} = \{M_1, \ldots, M_k\}$, we can combine these into a single classifier by considering the average of their outputs, i.e.

$$M(x) = \frac{1}{k}\sum_{i=1}^{k} M_i(x).$$

17.6 Application

In this section, we apply the recurrent neural network and long short-term model to financial stock market data.

17.6.1 Emergent and Developed Market

We studied the emergent market daily indices corresponding to two countries: Philippines (PSI), from 2 July 1997 to 25 October 2001 and Thailand (SETI), from 2 July 1997 to 25 October 2001. We also studied the developed market daily index corresponding to United States (Nasdaq), from 2 January 1997 to 31 December 2001. The financial time series contains information on the date of stock, stock open price, stock close price, stock low price, and stock high price.

17.6.2 The Lehman Brothers Collapse

The data used in this study corresponds to a sampled minute by minute time series recorded on 15 September 2008 for the Lehman Brothers collapse. The time series used contained the following companies: JPMorgan Chase & Co. (JPM) and Walmart retail company (WMT). The financial time series contains information on the date of stock, stock open price, stock close price, stock low price, and stock high price.

17.6.3 Methodology

The following are the steps used to perform our analysis.

S:1. *Raw data*: Historical stock data is collected for the prediction of future stock prices.

S:2. *Data preprocessing*:
 - *Data cleaning*: Dealing with missing values.
 - *Data transformation*: Normalization.
 - *Data splitting*: 60% for training, 20% for validation, and 20% for testing.

S:3. *Feature extraction*: We choose the feature from date, open, high, low and close prices.

S:4. *Training neural network*: The data is fed to the neural network and trained for prediction assigning random biases and weights.

S:5. *Validating neural network*: Sample of data is used to provide an unbiased evaluation of a model fit on the training dataset while tuning model parameters.

S:6. *Output generation*: The output value generated by the output layer of the recurrent neural network and long short-term memory are compared with the target value.

17.6.4 Analyses of Data

Below are the setup used in the network architectures. Number of inputs = 4, number of neurons = 200, number of outputs = 4, number of layers = 2, learning rate = 0.001, batch size = 50, and number of epochs = 100.

The batch size is the number of samples processed before the model is updated. The number of epochs is the number of complete passes through the training dataset. All the analyses were performed using Python 3.6 with Tensorflow (an open source machine learning library).

17.6.4.1 Results of the Emergent Market Index

The results for the emergent market index are shown in Figures 17.4–17.5.

17.6.4.2 Results of the Developed Market Index

The results for the developed market index are shown in Figures 17.6–17.8.

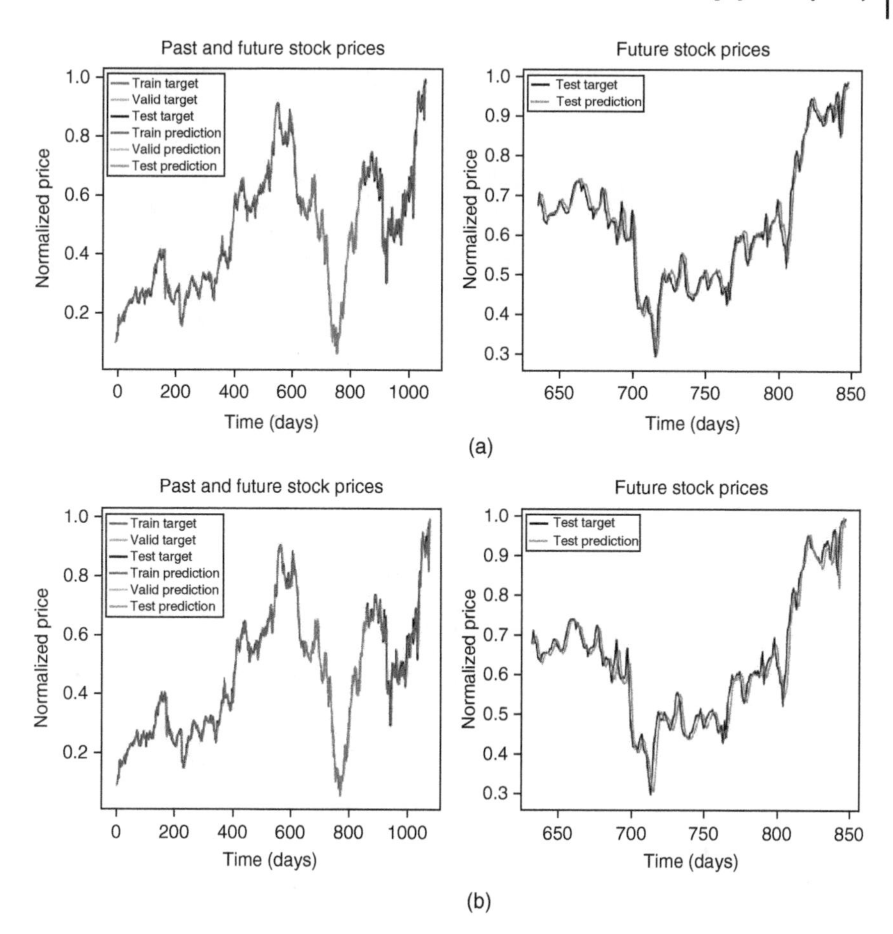

Figure 17.4 Philippines (PSI). (a) Basic RNN. (b) LTSM.

17.7 Significance of Study

Due to the popularity of stock market trading, researchers are constantly finding out new methods for prediction. In this application, we used the recurrent neural network and long short-term memory unit to help investors, analysts, or stakeholders interested in investing in the stock market by providing them a good

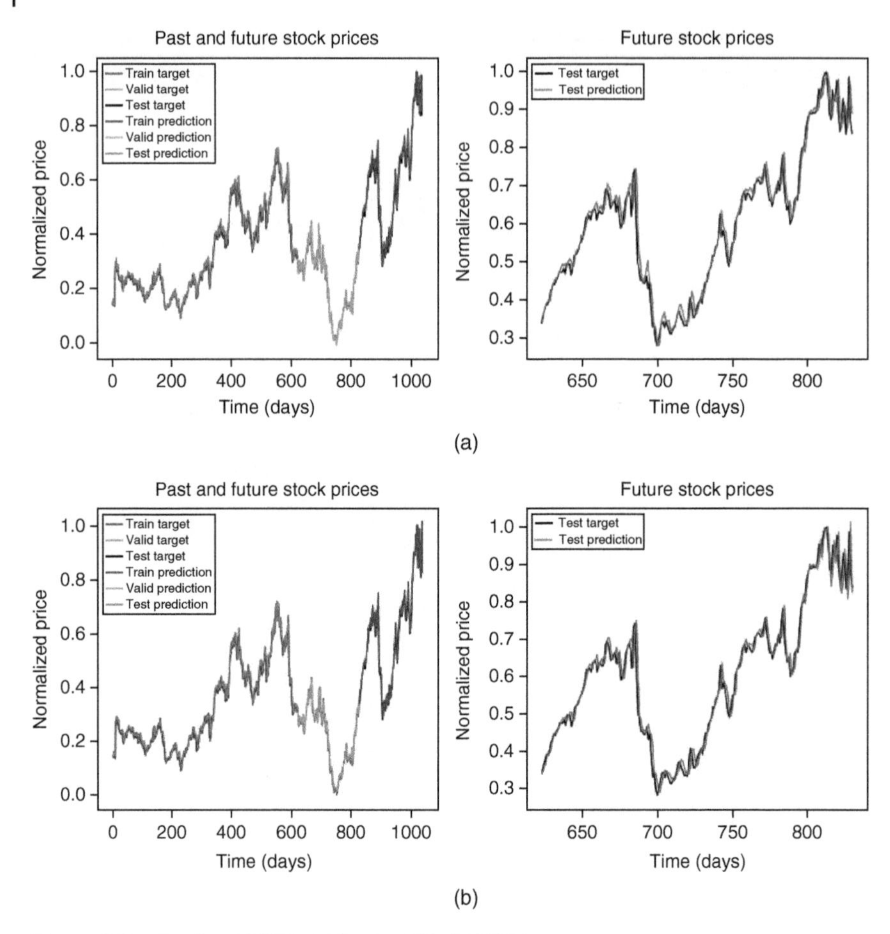

Figure 17.5 Thailand (SETI). (a) Basic RNN. (b) LTSM.

knowledge of the future situation of the stock market. From the results obtained, we observe that the recurrent neural network and the long short-term memory behaved similarly on the stock market data.

17.8 Problems

1 Determine whether the following statements are true or false and explain your reasoning.

(a) For effective training of a neural network, the network should have at least 5–10 times as many weights as there are training samples.

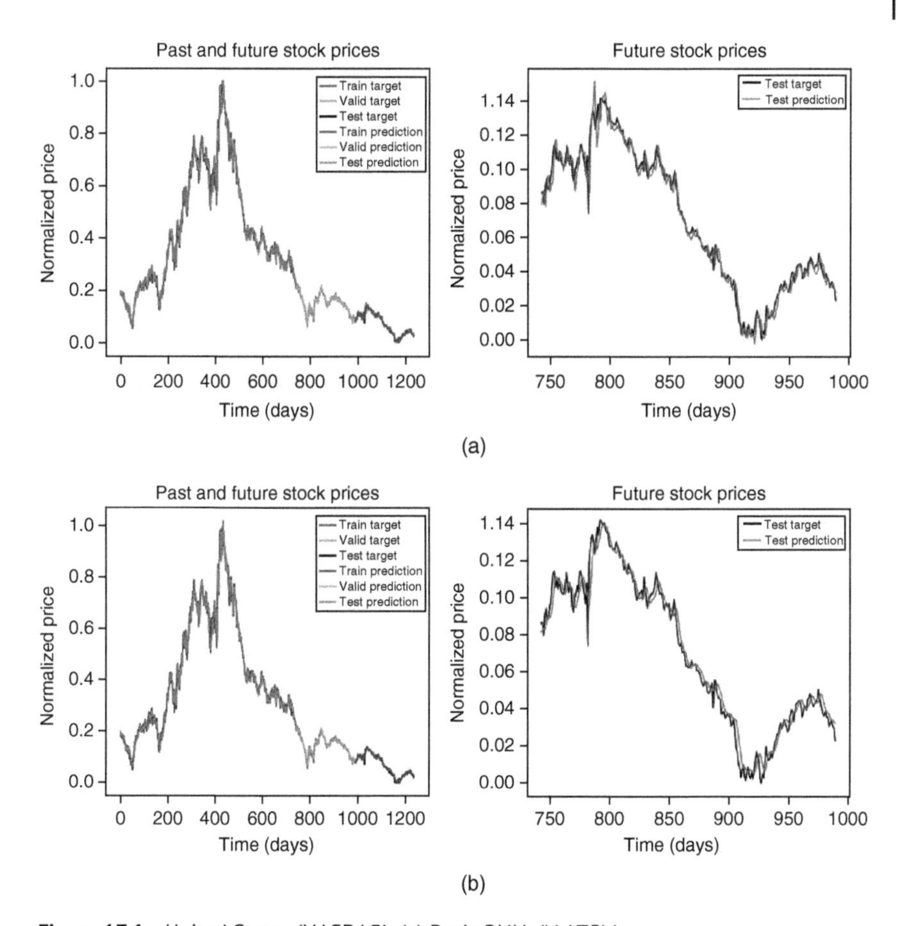

Figure 17.6 United States (NASDAQ). (a) Basic RNN. (b) LTSM.

(b) A perceptron is guaranteed to perfectly learn a given linearly separable function within a finite number of training steps.
(c) The more hidden-layer units a multilayered perceptron neural network has, the better it can predict desired outputs for new inputs that it was not trained with.

2 The perceptron algorithm cycles through the training examples, applying the following rule:

$$
\begin{aligned}
& z^{(i)} \leftarrow \mathbf{w}^T \mathbf{x}^{(i)}, \\
& \text{If } z^{(i)} t^{(i)} \leq 0, \\
& \qquad \mathbf{w} \leftarrow \mathbf{w} + t^{(i)} \mathbf{x}^{(i)},
\end{aligned}
$$

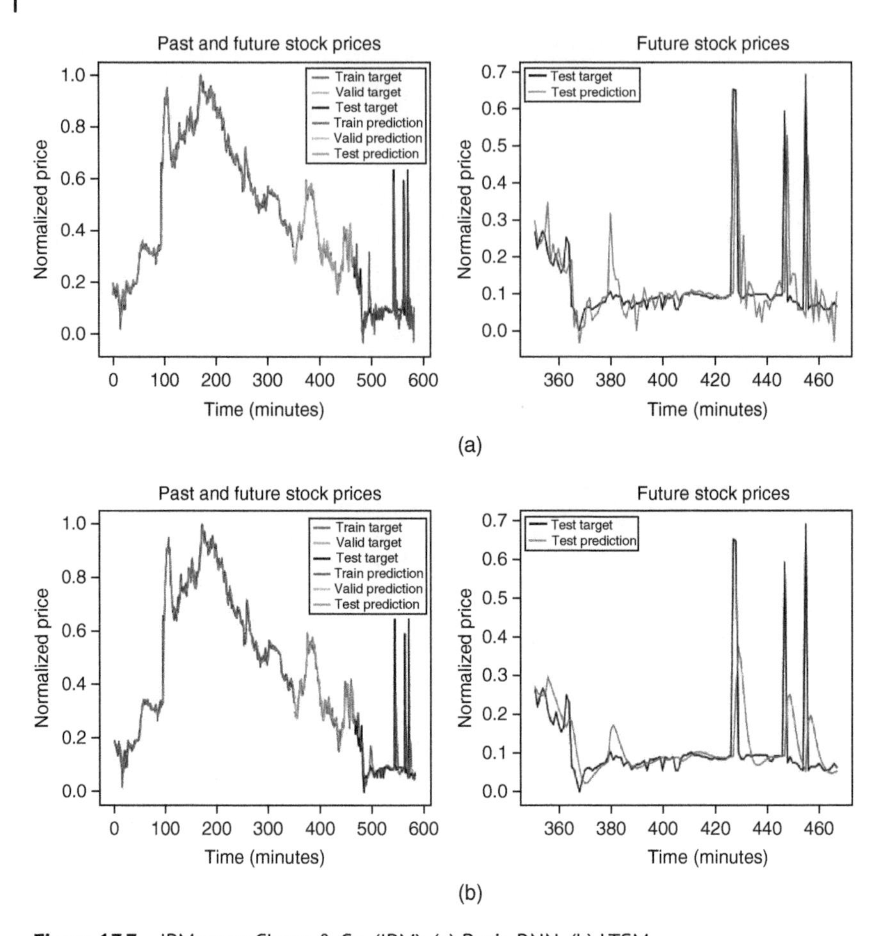

Figure 17.7 JPMorgan Chase & Co. (JPM). (a) Basic RNN. (b) LTSM.

where the targets take values in $\{0, 1\}$. Suppose we make the inequality strict in the conditional, i.e. we update the weights only if $z^{(i)}t^{(i)} < 0$. What would go wrong? You may assume the weights are initialized to 0.

3 Given single artificial neuron (unit) with three inputs $\mathbf{x} = (x_1, x_2, x_3)$ that receive only binary signals (either 0 or 1). Suppose that the weights corresponding to the three inputs have the following values, $w_1 = 3, w_2 = -5$, $w_3 = 2$ and the activation of the unit is given by the step-function:

$$\psi(v) = \begin{cases} 1, & \text{if } v \geq 0, \\ 0, & \text{if otherwise.} \end{cases}$$

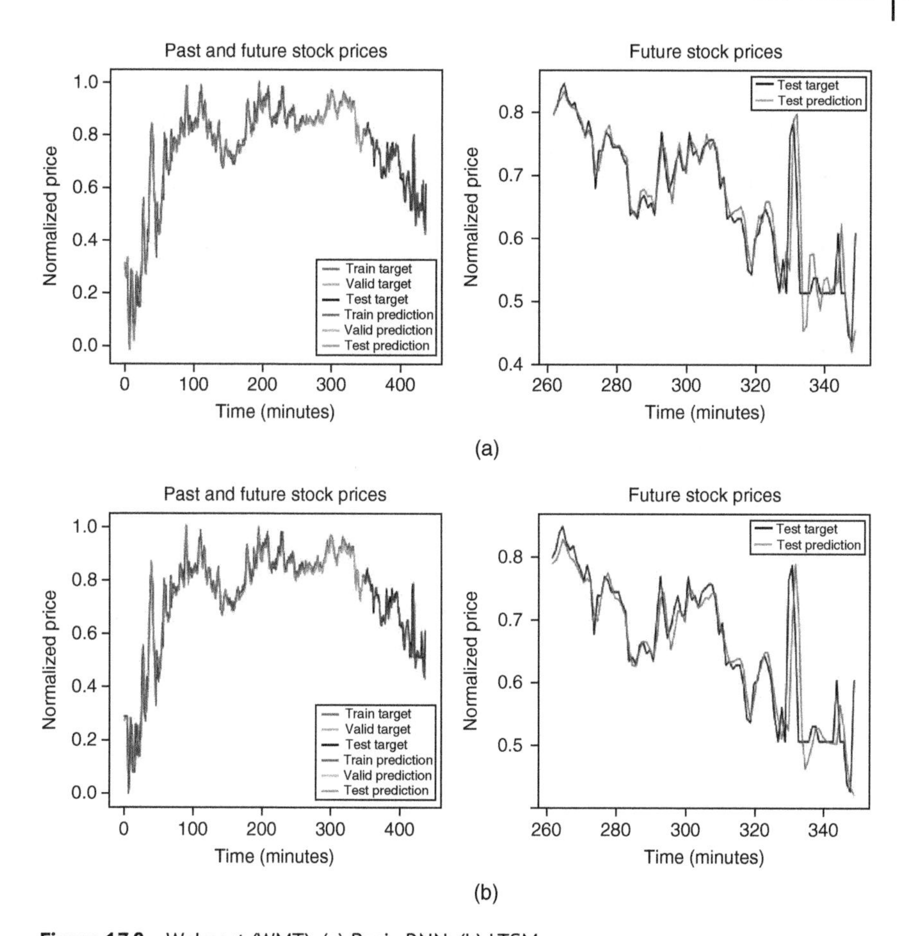

Figure 17.8 Walmart (WMT). (a) Basic RNN. (b) LTSM.

Calculate what will be the output value y of the unit for each of the following input patterns:

Pattern	P_1	P_2	P_3	P_4
x_1	1	0	1	0
x_2	0	0	1	1
x_3	1	1	1	1

4 Explain what effect will the following operations generally have on the bias and variance of your model.

(a) Increasing the size of the layers (more hidden units per layer).
(b) Getting more training data.

5 Discuss the problem(s) that will result from using a learning rate that's too high? How would you detect these problems?

6 Discuss the problem(s) that will result from using a learning rate that's too low? How would you detect these problems?

7 Distinguish between a feed forward neural network and a recurrent neural network.

8 Distinguish between a recurrent neural network and a long short-term neural network.

9 Suppose that the activation function $a() : R \Downarrow R$ is non constant, bounded, and continuous. Let I be a compact subset in $\mathbb{R}^p$ and $C(I)$ be the space of all real-valued continuous functions on I. Then, for any given $\epsilon > 0$ and any $h \in C(I)$, there $\exists M \in \mathbb{N}$, $w_m \in \mathbb{R}^p$, and constants ω_m, $b_m \in \mathbb{R}$ for $m = 1, \ldots, M$, prove that

$$|\overline{h(x) = h(x)}| < \epsilon, \forall x \in I,$$

$$\text{where, } \bar{h}(x) = \sum_{m=1}^{M} \omega_m.a(x^t w_m + b_m).$$

10 Explain and sketch a network diagram for a Feed-Forward Artificial Neural Network Architecture with two hidden layers.

18

Fourier Analysis

18.1 Introduction

In this chapter, we will discuss Fourier transforms which is one of the most popular transformation methodologies. Transformations are applied to signals to obtain further information from that signal that is not readily available in the raw signal. The main idea behind this technique is to transform data from the spatial domain to a frequency domain. The Fourier transform takes a time-based pattern, measures every possible cycle, and returns the amplitude, offset, and rotation speed for every cycle that was found. We will describe some properties of the Fourier transforms and discuss some applications to real financial data.

18.2 Definition

The Fourier transform of the function f is traditionally denoted by adding a circumflex: $\hat{f}$. There are several common conventions for defining the Fourier transform of an integrable function $f : \mathbb{R} \to \mathbb{C}$. It can be represented mathematically as

$$\hat{f}(\xi) = \int_{-\infty}^{\infty} f(x)\, e^{-2\pi i x \xi}\, dx, \tag{18.1}$$

for any real number ξ.

When the independent variable x represents time, the transform variable ξ represents frequency (e.g. if time is measured in seconds, then frequency is in hertz). Under suitable conditions, the function f is determined by $\hat{f}$ via the inverse transform:

$$f(x) = \int_{-\infty}^{\infty} \hat{f}(\xi)\, e^{2\pi i x \xi}\, d\xi,$$

for any real number x.

Data Science in Theory and Practice: Techniques for Big Data Analytics and Complex Data Sets, First Edition. Maria Cristina Mariani, Osei Kofi Tweneboah, and Maria Pia Beccar-Varela.

18.3 Discrete Fourier Transform

The discrete Fourier transform (DFT) is the equivalent of the continuous Fourier transform for signals known only at N instants separated by sample times T (i.e. a finite sequence of data). It is usually used in practical applications since we generally cannot observe a signal continuously. DFT converts a finite sequence of equally spaced samples of a function into a sequence of equally spaced samples of the discrete-time Fourier transform (DTFT), which is a complex-valued function of the frequency. A complex-valued function are functions F defined on the complex plane (or a subregion of the complex plane) which takes on complex values. A complex-valued function has a real part and an imaginary part.

The DFT is the most important discrete transform, used to perform Fourier analysis in several practical applications. In image processing, the samples can be the values of pixels along a row or column of an image. The DFT can also be used to efficiently solve partial differential equations, and to perform other operations such as convolutions or multiplying large integers.

The DFT of the sequence $f[k]$ is defined as:

$$F[n] = \sum_{k=0}^{N-1} f[k]e^{-j\frac{2\pi}{N}nk}, \quad n = 0, \dots, N-1. \tag{18.2}$$

Equation (18.2) can be written in matrix form as:

$$\begin{bmatrix} F[0] \\ F[1] \\ F[2] \\ F[3] \\ \vdots \\ F[N-1] \end{bmatrix} = \begin{bmatrix} 1 & 1 & 1 & 1 & \dots & 1 \\ 1 & W & W^2 & W^3 & \dots & W^{N-1} \\ 1 & W^2 & W^4 & W^6 & \dots & W^{N-2} \\ 1 & W^3 & W^6 & W^9 & \dots & W^{N-3} \\ \vdots & & & & & \\ 1 & W^{N-1} & W^{N-2} & W^{N-3} & \dots & W \end{bmatrix} \begin{bmatrix} f[0] \\ f[1] \\ f[2] \\ f[3] \\ \vdots \\ f[N-1] \end{bmatrix},$$

where $W = \exp(-j2\pi/N)$. Since $W^N = 1$ the columns are powers of the initial column. In linear algebra, this is a particular type of a Vandermonde matrix and multiplying vectors with it is a very fast operation.

The inverse DFT corresponding to (18.2) is defined as:

$$f[k] = \frac{1}{N}\sum_{n=0}^{N-1} F[n]e^{j\frac{2\pi}{N}nk}, \quad k = 0, \dots, N-1. \tag{18.3}$$

We can also rewrite Eq. (18.3) in matrix form as we did for Eq. (18.2). In this case, the inverse matrix will be $1/N$ times the complex conjugate of the original matrix.

In the process of taking the inverse transform, the terms $F[n]$ and $F[N-n]$ combine to produce two frequency components and only one is considered to be valid (the one at the lower of the two frequencies). From (18.3), the contribution to $f[k]$

of $F[n]$ and $F[N-n]$ is:

$$f_n[k] = \frac{1}{N}\left\{F[n]e^{j\frac{2\pi}{N}nk} + F[N-n]e^{j\frac{2\pi}{N-n}k}\right\}. \quad (18.4)$$

For all $f[k]$ real, $F[N-n] = \sum_{k=0}^{N-1} f[k]e^{-j\frac{2\pi}{N}(N-n)k}$.

The time taken to evaluate a DFT on a digital computer depends on the number of multiplications involved, since these are the computationally expensive operations. With the DFT, this number is directly related to the order of N^2 (matrix multiplication of a vector), where N is the length of the transform. For most problems, N is chosen to be at least 256 in order to get a reasonable approximation for the spectrum of the sequence under consideration, hence computational speed becomes a major consideration. Efficient computer algorithms for estimating DFTs are known as fast Fourier transform (FFT) algorithms (Section 18.4). These algorithms rely on the fact that the standard DFT involves a lot of redundant calculations.

For example, we can rewrite

$$F[n] = \sum_{k=0}^{N-1} f[k]e^{-j\frac{2\pi}{N}nk}, \quad n = 0, \dots, N-1$$

as

$$F[n] = \sum_{k=0}^{N-1} f[k]W_N^{nk}, \quad n = 0, \dots, N-1. \quad (18.5)$$

Where $W = \exp(-j2\pi/N)$. We see that in (18.5), the same values of W_N^{nk} are calculated many times as the computation proceeds.

18.4 The Fast Fourier Transform (FFT) Method

The work of Heston (1993) in stochastic volatility modeling, marks the first time Fourier inversion appears in option pricing literature. Since then, Fourier inversion has become more and more prevalent. This is due to the fact that analytic formulas are hard to calculate and can only be done in the simplest of models.

The Fast Fourier Transform (FFT) allows us to compute efficiently sums of the form:

$$W_m = \sum_{j=1}^{N} \exp\left(\frac{2\pi i(j-1)(m-1)}{N}\right) x_k, \quad (18.6)$$

with a complexity of $O(N \log N)$. The big O notation is a mathematical notation that describes the limiting behavior of a function when the argument tends toward a particular value. It is described in details in Chapter 7 of this book.

The Fourier transform methods arise naturally with the Lévy Khintchine formula (see Khintchine and Lévy 1936). In fact, the characteristic function of a distribution μ corresponds to the Fourier transform of the density function. The characteristic function of a distribution μ on $\mathbb{R}$ is defined as

$$\phi_\mu(u) = \hat{\mu}(u) = \int_{\mathbb{R}} e^{iux} \mu(dx). \tag{18.7}$$

In the case μ is a probability density function and its Fourier transform may be written in term of its moments. If we define the nth moment of μ by $\mu_n = \int x^n \mu(dx)$, then from (18.7) we obtain

$$\phi_\mu(u) = \int_{\mathbb{R}} \sum_{n=0}^{\infty} \frac{(iux)^n}{n!} \mu(dx) = \sum_{n=0}^{\infty} \frac{(iu)^n}{n!} \mu_n.$$

Fourier transform methods have become popular in finance in the context of exponential Lévy processes primarily due to lack of knowledge of the density functions in closed form and the impossibility to obtain closed form expressions for option prices.

As we stated before, the FFT methods have become a very important tool in finance. The fact that many characteristic functions can be expressed analytically usually makes the Fourier transform a natural approach to compute option prices.

Theorem 18.1 Let f be a function $f : \mathbb{R} \to \mathbb{R}$, continuous, bounded in $L^1(\mathbb{R})$, and so that $F(f) \in L^1(\mathbb{R})$. Then

$$E[f(X_t)] = F^{-1} [\tilde{f}](.)\phi_{X_t}(.)\Big]_0,$$

with F denoting the Fourier transform operator, and $\tilde{f}(x) = f(-x)$.

Proof: We observe that

$$\int_{\mathbb{R}} f(x)\mu(dx) = \int_{\mathbb{R}} f(y - x)\mu(dx)|_{y=0},$$

and the convolution product is integrable since f is bounded and continuous. Hence the theorem follows.

This theorem allows us to compute option values which are typically expressed as expectations of the final payoff f from the Fourier transform of the density of X_t (its characteristic function). When applied to the option pricing problem, the assumptions on f are usually not satisfied since most payoff functions are not in $L^1(\mathbb{R})$. The payoff function is usually replaced with a truncated version to satisfy the assumption.

Two Fourier transformations are possible, one with respect to the log strike price of the option, the other with respect to the log spot price at T. We will consider the first approach.

Theorem 18.2 Define the truncated time value option by

$$z_T(k) = E[(e^{X_T} - e^k)_+] - (1 - e^k)_+,$$

then for an exponential martingale e^{X_T} such that $E[e^{(1+\alpha)X_T}] < \infty$, the Fourier transform of the time value of the option is defined by

$$\zeta_T(v) = \int_{-\infty}^{\infty} e^{ivk} z_T(k)dk = e^{ivrT}\frac{\Phi_T(v-i) - 1}{iv(1+iv)}.$$

The price of the option is then given by

$$C(t,x) = x\left[(1 - e^{k_t}) + \frac{1}{2\pi}\int_{\mathbb{R}} \zeta_T(v)e^{-ivk_t}dv\right].$$

Proof: We sketch an outline of the proof: The discounted stock process $\tilde{S}_t = e^{X_t}$ is a martingale. Therefore,

$$\int_{-\infty}^{\infty} e^x q_{X_t}(x)dx = 1.$$

The stock process admits a moment of order $1 + \alpha$ so that there exists $\alpha > 0$, such that

$$\int_{-\infty}^{\infty} e^{(1+\alpha)x} q_{X_t}(x)dx < 0,$$

which can be used to show both integrals $\int_{-\infty}^{\infty} xe^x q_{X_t}(x)$ and $\int_{-\infty}^{\infty} e^x q_{X_t}(x)dx$ are finite. Using Fubini's theorem, we explicitly compute the Fourier transform z_T as follows: We write

$$\begin{aligned} z_T(k) &= E[(e^{X_T} - e^k)_+] - (1 - e^k)_+ \\ &= \int_{-\infty}^{\infty} [(e^x - e^k)1_{x\geq k} - (1 - e^k)1_{0\geq k}]q_{X_t}(x)dx \\ &= \int_{-\infty}^{\infty} [(e^x - e^k)(1_{x\geq k} - 1_{0\geq k})]q_{X_t}(x)dx, \end{aligned}$$

where in the second term we use the fact that $\tilde{S}_t$ is a martingale. Assuming we can interchange the integrals, the Fourier transform of z_T is given by

$$\begin{aligned} F(z_T)[v] &= \int_{-\infty}^{\infty} e^{ivk} z_T(k)dk \\ &= \int_{-\infty}^{\infty} e^{ivk} \int_{-\infty}^{\infty} (e^x - e^k)(1_{x\geq k} - 1_{0\geq k})q_{X_t}(x)dx\, dk \\ &= \int_{-\infty}^{\infty} q_{X_t}(x) \int_{-\infty}^{\infty} e^{ivk}(e^x - e^k)(1_{x\geq k} - 1_{0\geq k})dk\, dx \\ &= \int_{-\infty}^{\infty} q_{X_t}(x) \int_0^x e^{ivk}(e^x - e^k)dk\, dx. \end{aligned}$$

When applying the Fubini's theorem, we observe that when $x \geq 0$, $\int_0^x |e^{ivk}(e^x - e^k)| \leq e^x(x-1)dk$, and $\int_{-\infty}^{\infty} e^x(x-1)q_{X_t}(x)dx < \infty$. Also, when $x < 0$, $\int_0^x |e^{ivk}(e^x - e^k)| \leq (e^x - 1)dk$, and $\int_{-\infty}^{\infty}(e^x - 1)q_{X_t}(x)dx < \infty$. Thus,

$$\begin{aligned} F(z_t)[v] &= \int_{-\infty}^{\infty} q_{X_t}(x) \int_0^x e^{ivk}(e^x - e^k)dk\, dx \\ &= \int_{-\infty}^{\infty} q_{X_t}(x) \left[\frac{e^x}{iv}(e^{ivx} - 1) - \frac{1}{iv+1}(e^{(iv+1)x} - 1) \right] dx \\ &= \int_{-\infty}^{\infty} q_{X_t}(x) \frac{e^{(iv+1)x} - 1}{iv(iv+1)} dx. \end{aligned}$$

This approach has the advantage of not requiring an exact value of α in order to compute the option prices. The numerical implementation of this truncation function, which is not smooth, has the inconvenience of generating rather large truncation errors, making the convergence slow. Therefore, the Fourier transform of the time value of the option is

$$\zeta_T(v) = e^{ivrT} \frac{\Phi_T(v - i) - 1}{iv(1 + iv)} \sim |v|^{-2}.$$

18.5 Dynamic Fourier Analysis

A physical process can be described either in the time domain, by the values of some quantity f as a function of time t, for example, $f[t]$, or in the frequency domain, where the process is specified by giving its amplitude H (generally a complex number) as a function of frequency n, that is $F[n]$, with $-\infty < n < \infty$. For many applications, it is useful to think of $f[t]$ and $F[n]$ as being two different representations of the same function. Generally, these two representations are connected using Fourier transform formulas:

$$\begin{aligned} F[n] &= \int_{-\infty}^{\infty} f[t]e^{2\pi int}\, dt = \int_{-\infty}^{\infty} f[t](\cos(2\pi nt) + i\sin(2\pi nt))dt, \\ f[t] &= \int_{-\infty}^{\infty} F[n]e^{-2\pi int}\, dn, \end{aligned} \tag{18.8}$$

where, $F[n]$ is the Fourier transform and $f[t]$ represents the inverse Fourier transform. The Fourier transform $F[n]$ converts the data from the time domain into the frequency domain. The inverse Fourier transform $f[t]$ on the other hand converts the frequency domain components back into the original time-domain signal. A frequency-domain plot shows how much of the signal lies within each given frequency band over a range of frequencies.

In order to analyze a statistical time series using Fourier transforms, we need to assume that the structure of the statistical or stochastic process which generates the observations is time invariant. This assumption is summarized in the condition of stationarity which states that its finite dimensional distribution remains the same throughout time.

This condition is hard to verify and we generally enforce a weak stationarity condition. For example, a time series x_t is weak stationary, if its second-order behavior remains the same, regardless of the time t. Looking at the Fourier transform representation (18.8), we see that a stationary series is represented as a superposition of sines and cosines that oscillate at various frequencies. Therefore, a stationary time series can be matched with its sine and cosine series representation (these are the representative frequencies).

Next we introduce some definitions that will be used when applying the dynamic Fourier analysis.

18.5.1 Tapering

Generally, calculating the continuous Fourier transform and its inverse is a hard problem in practice. For this reason, a DFT is used when we observe discrete time series with a finite number of samples from a process that is continuous in time. We discussed the DFT in Section 18.3 however, applying it requires an extra step known as tapering.

When the original function (process) is discontinuous, the corresponding signal value abruptly jumps, yielding spectral leakage (that is, the input signal does not complete a whole number of cycles within the DFT time window). In this case to perform DFT, we need to multiply the finite sampled times series by a windowing function, or "a taper." The taper is a function that smoothly decays to zero near the ends of each window, and it is aimed at minimizing the effect of the discontinuity by decreasing the time series magnitude so it approaches zero at the ends of the window. Although spectral leakage cannot be prevented, it can be significantly reduced by changing the shape of the taper function in a way to minimize strong discontinuities close to the window edges. In particular in seismic data analysis, cosine tapers are often used since it is both effective and easy to calculate. The cosine taper can be written as:

$$c(t) = \begin{cases} \frac{1}{2}\left(1 - \cos \frac{\pi t}{a}\right), & 0 \le t \le a, \\ 1, & a \le t \le (1-a), \\ \frac{1}{2}\left(1 - \cos \frac{\pi}{a}(1-t)\right), & (1-a) \le t \le 1, \end{cases}$$

where t is time and a is the taper ratio. The cosine window represents an attempt to smoothly set the data to zero at the boundaries while not significantly reducing the level of the values within the window. This form of tapering reduces the leakage of the spectral power from a spectral peak to frequencies far away and it coarsens the spectral resolution by a factor $1/(1-a)$ for the above cosine tapers.

In general, the effects of applying a taper are:

1. Decrease the time series magnitude to zero or near zero at its start and end so that there is no sharp discontinuity between the first and last point in the periodic time series.
2. Change the weighting of samples in the time series so that those near the middle contribute more to the Fourier transform.
3. Reduce the resolution of the spectrum by averaging adjacent samples.

For nonstationary time series signals, tapering may bias the spectral amplitudes even if the taper is normalized. However, for these we should not use Fourier transform in the first place.

Generally, it is difficult to give precise recommendations on the kind of tapers to use in all specific situation. The work of Bingham et al. (1967) recommends to reduce the leakage by increasing the length N of the time window and at the same time decreasing the taper ratio a, such that the length (Na) of the cosine half-bells is kept constant.

In practice, a value of $a = 5\%$ for window lengths of 30 or 60 seconds is good enough for frequencies down to 0.2 Hz. See Pilz and Parolai (2012) for more details.

18.5.2 Daniell Kernel Estimation

In this section, we present a probabilistic approach used to estimate the frequency coefficients in a Fourier transform. A stationary process X_t may be defined by taking linear combinations of the form:

$$X_t = \sum_{j=1}^{m} (A_j \cos(2\pi\lambda_j t) + B_j \sin(2\pi\lambda_j t)), \tag{18.9}$$

where $0 \leq \lambda \leq \frac{1}{2}$ is a fixed constant and $A_1, B_1, A_2, B_2, \ldots, A_m, B_m$ are all uncorrelated random variables with mean zero and

$$\mathrm{Var}(A_j) = \mathrm{Var}(B_j) = \sigma_j^2.$$

We assume $\sum_{j=1}^{m} \sigma_j^2 = \sigma^2$ so that the variance of the process X_t is σ^2, and let the spectral density $f(\lambda)$ satisfy the equation

$$\int_{-\frac{1}{2}}^{\frac{1}{2}} f(\lambda) d\lambda = \sigma^2.$$

Then, the process given in (18.9) converges to a stationary process with spectral density f as $m \to \infty$.

In order to estimate the spectral density f, we define the estimators as a weighted average of periodogram values (I) for frequencies in the range $(j-m)/n$ to $(j+m)/n$. In particular we define:

$$\hat{f}(j/n) = \sum_{k=-m}^{m} W_m(k) I\left(\frac{j+k}{n}\right).$$

The set of weights $\{W_m(k)\}$ sum to one and the set is often referred to as a kernel or a spectral window. Essentially, this kernel with parameter m is a centered moving average which creates a smoothed value at time t by averaging all values between and including times $t-m$ and $t+m$.

We define:

$$W_m(k) = \frac{1}{2m+1} \quad \text{for} - m \leq k \leq m, \quad \sum_k W_m(k) = 1 \quad \text{and} \quad \sum_k k W_m(k) = 0.$$

The smoothing formula $\{u_t\}$ for a Daniell kernel with $m = 1$ corresponds to the three weights $(\frac{1}{3}, \frac{1}{3}, \frac{1}{3})$ is given by:

$$\hat{u}_t = \frac{1}{3}(u_{t-1} + u_t + u_{t+1}).$$

Applying the Daniell kernel again on smoothed values $\{\hat{u}_t\}$ produces a more extensive smoothing by averaging over a wider time interval. This is also defined as:

$$\hat{u}_t = \frac{\hat{u}_{t-1} + \hat{u}_t + \hat{u}_{t+1}}{3} = \frac{1}{9}u_{t-2} + \frac{1}{9}u_{t-2} + \frac{3}{9}u_t + \frac{2}{9}u_{t+1} + \frac{1}{9}u_{t+2}. \tag{18.10}$$

Consequently, applying the Daniell kernel transforms the spectral windows into a form of Gaussian probability density function.

18.6 Applications of the Fourier Transform

In this section, we briefly discuss some real-life applications of the Fourier transforms.

18.6.1 Modeling Power Spectrum of Financial Returns Using Fourier Transforms

We will use the dynamic Fourier transform technique to estimate the power spectrum of returns of daily and minute sampled financial stock market. Estimating the power spectrum is effective in characterizing the minute and daily-based data corresponding to different frequencies. This modeling techniques help to characterize some key variables of stationary time series that are very useful for making informed decisions in the stock market such as assessing financial risk in the market.

Table 18.1 Percentage of power for Discover data.

Frequency (Hz)	Daily data	Minute data
0–2	15.96	19.82
2–4	21.80	27.81
4–6	18.82	12.99
6–8	18.65	6.800
8–10	24.75	32.52

Table 18.2 Percentage of power for JPM data.

Frequency (Hz)	Daily data	Minute data
0–2	11.62	17.81
2–4	26.29	23.20
4–6	21.86	11.90
6–8	20.91	16.90
8–10	19.32	30.08

Table 18.3 Percentage of power for Microsoft data.

Frequency (Hz)	Daily data	Minute data
0–2	16.49	15.54
2–4	12.46	19.84
4–6	19.64	18.33
6–8	30.53	24.72
8–10	20.86	21.54

Table 18.4 Percentage of power for Walmart data.

Frequency (Hz)	Daily data	Minute data
0–2	15.96	11.90
2–4	21.80	24.00
4–6	18.82	18.799
6–8	18.65	18.95
8–10	24.75	26.24

The data studied comprises of the five trading days 10–14 March 2008 predating the merging announcement over the weekend as well as the two trading days 17 March and following the event. We analyzed four companies namely; Discover Financial Services company (Discover), Microsoft Corporation technology company (Microsoft), Walmart retail company (Walmart), and JPMorgan Chase financial services company (JPM). There is no particular reason for the specific choices company other than they are the largest companies judging the market capitalization numbers at that time. The data used consists of the financial return values. Next using dynamic Fourier analysis we model the power spectrum of financial returns.

The objective of the analysis is to estimate the power spectrum of returns of daily and minute sampled financial stock market. The technique is applied as follows. We begin by performing a dynamic Fourier analysis on a short segment of the

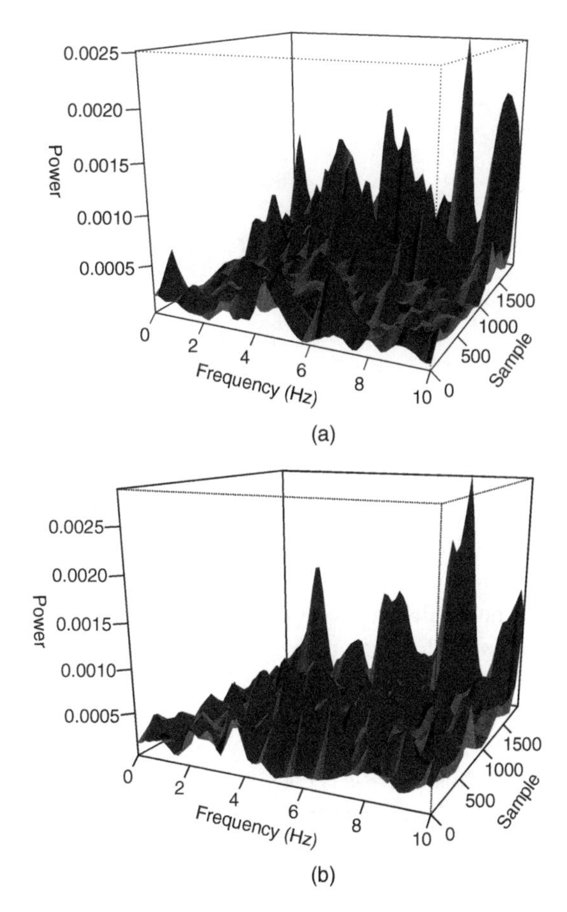

Figure 18.1 3D power spectra of the daily returns from the four analyzed stock companies. (a) Discover. (b) Microsoft. (c) Walmart. (d) JPM Chase.

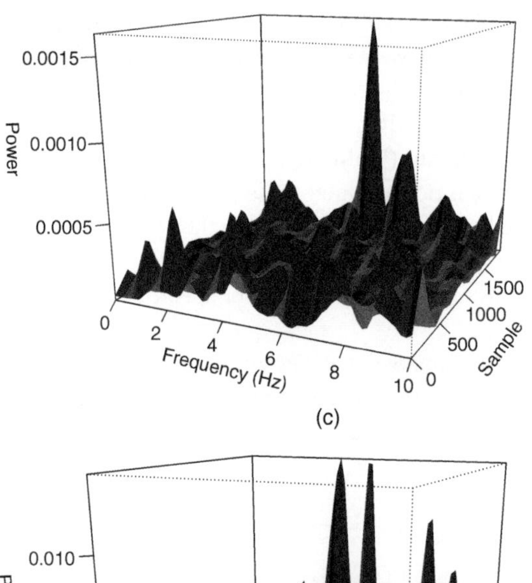

(c)

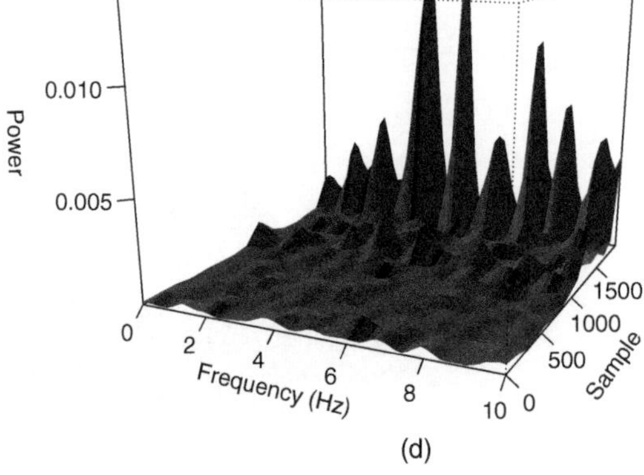

(d)

Figure 18.1 (*Continued*)

data. The section is then shifted, and the analysis is made on the new section. The process is repeated until the end of the time series.

In our analysis, the variable, X_t signifies the observations of financial time series, where $t = 1, \ldots, 2048$, and the analyzed data segments are $\{X_{t_k+1}, \ldots, X_{t_k+256}\}$, where $t_k = 128k$, and $k = 0, 1, \ldots, 14$. The analysis was carried out by developing an R statistical module.

The results are presented in Tables 18.1–18.4, for Discover, JPM, Microsoft, and Walmart stock companies. The tables report the percentage of power spectra for the minute and daily financial returns.

Based on the results in Tables 18.1–18.4, we observe that the power spectrum for both the minute and daily data is very high at lower levels (higher frequencies). This analysis is consistent with the results obtained in Beccar-Varela et al. (2017), where the authors analyzed high-frequency time series. Figures 18.1 and 18.2 reinforces the above trends. In these figures, the return series of the minute data has more spikes (peaks) in the power spectrum compared to the daily returns.

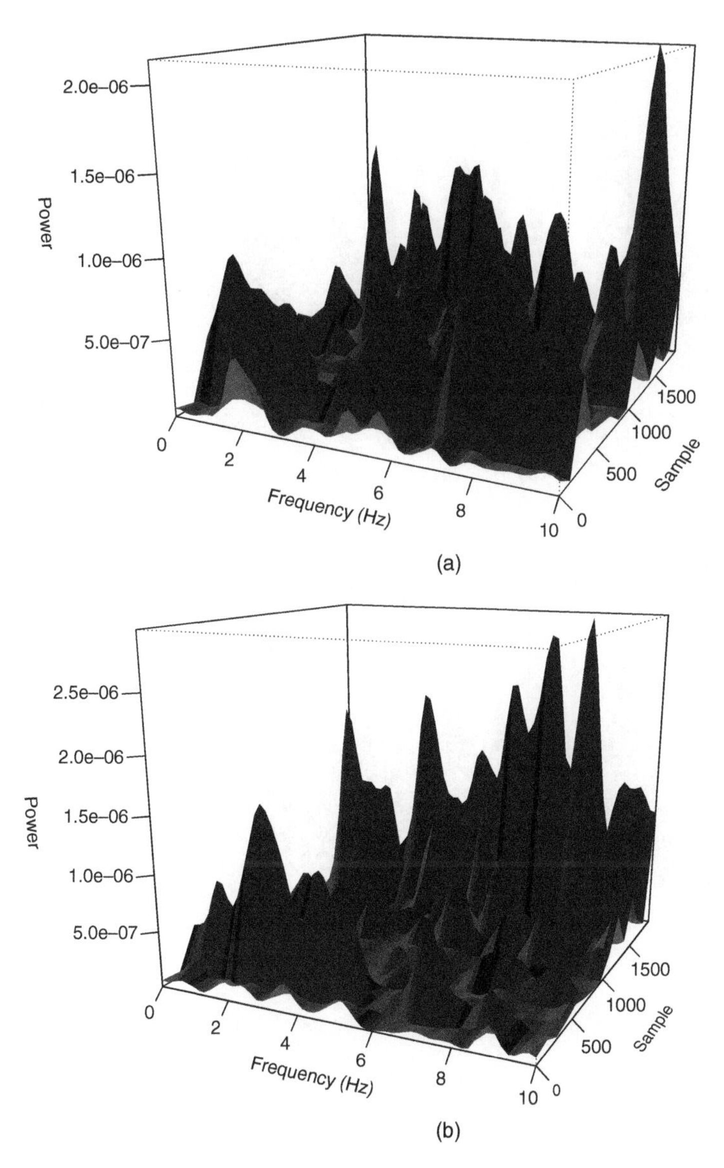

Figure 18.2 3D power spectra of the returns (generated per minute) from the four analyzed stock companies. (a) Discover. (b) Microsoft. (c) Walmart. (d) JPM Chase.

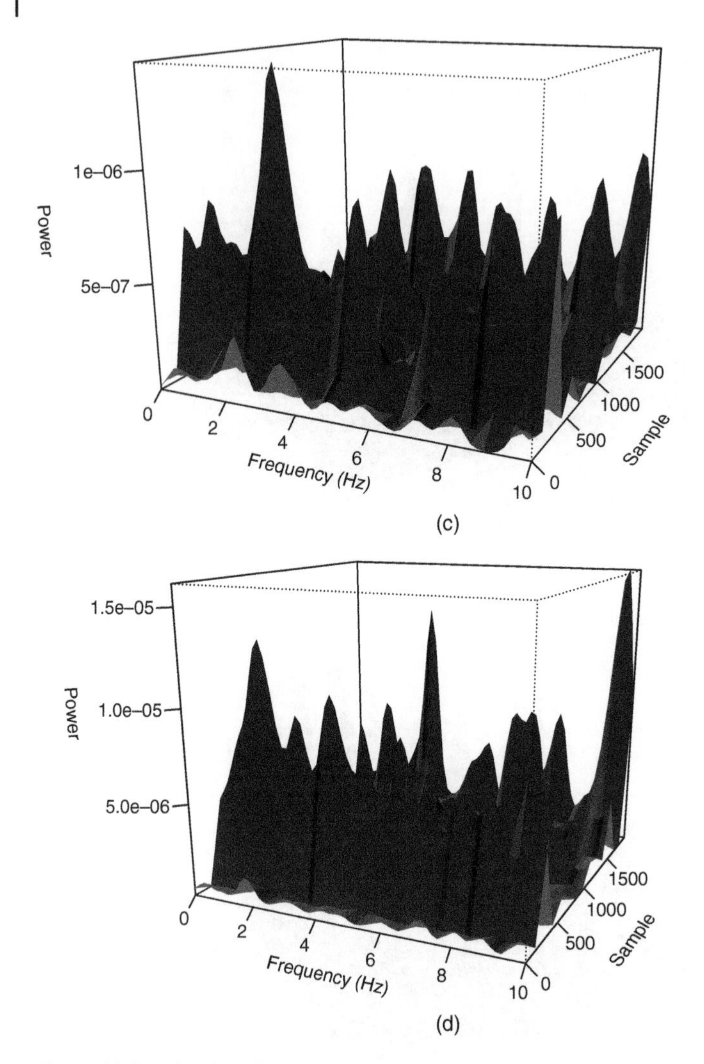

Figure 18.2 (*Continued*)

The significant number of peaks reflects how volatile the market is. The peak describes the amount of the variation explained by those frequencies. Thus the power spectrum for the minute and daily sampled data can be used to trace the market volatility, which reflects the state of the market, including financial gains or losses.

18.6.2 Image Compression

Another application of the Fourier transform is image compression and detection of edges in digital images. The Fourier transform decomposes an image into its sine and cosine components. The output of the transformation represents the image in the Fourier or frequency domain, while the input image is the spatial domain equivalent.

The Fourier transform is also used in a wide range of applications, such as image analysis, image filtering, and image reconstruction. The DFTs are often used to solve partial differential equations, where again the DFT is used as an approximation of the Fourier series (which is recovered in the limit of infinite N).

18.7 Problems

1 Express the inverse discrete Fourier transform in terms of the discrete Fourier transform.

2 Consider the signal $f(t) = \sin(8t)$. Use Matlab to obtain the discrete Fourier transform of this signal.

3 Let $f(\theta)$ be the 2π-periodic function determined by the formula

$$f(\theta) = \theta^2, \quad \text{for} - \pi \le \theta \le \pi.$$

Find the Fourier series for $f(\theta)$.

4 Fast Fourier transform (FFT) is an optimized faster equivalent to DFT. DFT has a computation complexity of N^2 and FFT reduces this to $N \log(N)$. How does FFT achieve this?

5 The discrete time transform shares all properties of the Fourier transform discussed except now some of these properties may take different forms. In the following, we always assume $\mathcal{F}[x[m]] = X(e^{j\omega})$ and $\mathcal{F}[y[m]] = Y(e^{j\omega})$.

(a) *Linearity*: Show that

$$\mathcal{F}[ax[m] + by[m]] = aX(e^{j\omega}) + bY(e^{j\omega}).$$

(b) *Time shifting*: Show that

$$\mathcal{F}[x[m - m_0]] = e^{-jm_0\omega}X(e^{j\omega}).$$

(c) *Differencing–Differencing is the discrete-time counterpart of differentiation*: Show that

$$\mathcal{F}[x[m] - x[m-1]] = (1 - e^{-j\omega})X(e^{j\omega}).$$

6 Download the monthly and daily equity data for the past five years at Yahoo finance. Please use any equity of your choice. For each time frequency, calculate a sequence of continuously compounded returns. Using the Fourier methodology, estimate the power spectrum of the monthly and weekly returns.

7 Investigate if the Fourier transform is distributive over multiplication.

8 Apply the *mean* filter operator to any image and compare its Fourier transform before and after the operation.

9 Write a program to estimate the power spectrum of any time series including extreme fluctuations.

19

Wavelets Analysis

19.1 Introduction

In this chapter, we will study another transformation technique, namely the wavelets transform. The wavelets technique is a more generalized version of the Fourier transform discussed in Chapter 18. We recall that the Fourier transform is only localized in the frequency domain. An advantage of the wavelets transform is the fact that is localized both in the frequency and time domain. Transformations are applied to signals to obtain further information from that signal that is not readily available in the raw signal. Most of the signals in practice are time-domain signals in their raw format and so transforming the signals helps to obtain a further information that is not readily available in the raw signal.

Recall that in order to perform Fourier decomposition the time series as well as the process needs to be stationary. From a regression point of view, we may imagine a system responding to various driving frequencies by producing linear combinations of sine and cosine functions. If a time series X_t is stationary, its mean, variance, and covariances remain the same regardless of the time t; therefore, we can match a stationary time series with sines and cosines because they behave the same indefinitely.

Nonstationary time series require a deeper analysis. The concept of wavelet analysis generalizes dynamic Fourier analysis, with functions that are better suited to capture the local behavior of nonstationary time series. These functions are called wavelet functions.

Wavelet transform is often compared with the Fourier transform, in the sense that we have a similar decomposition with wavelets replacing the sine functions. In fact, the Fourier transform may be viewed as a special case of a continuous wavelet transform with the choice of the wavelet function

$$\psi(t) = e^{-2\pi it}\psi(t) = e^{-2\pi it}.$$

Data Science in Theory and Practice: Techniques for Big Data Analytics and Complex Data Sets, First Edition. Maria Cristina Mariani, Osei Kofi Tweneboah, and Maria Pia Beccar-Varela.

The main difference between a wavelet transform and the Fourier transform is the fact that the wavelets analysis is a more general method which is localized in both time and frequency and has a faster computation speed, whereas the standard Fourier analysis is only localized in frequency (Shumway and Stoffer 2010).

As a mathematical tool, wavelets can be used to extract information from various kinds of data, including digital signals, images, and several others. Wavelets are well known for approximating data with sharp discontinuities (see Graps 1995).

19.1.1 Wavelets Transform

The wavelet transform of a function $f(t)$ with finite energy is defined as the integral transform for a family of functions

$$\psi_{\lambda,t}(u) \equiv \frac{1}{\sqrt{\lambda}}\psi\left(\frac{u-t}{\lambda}\right)$$

and is given as

$$Wf(\lambda, t) = \int_{-\infty}^{\infty} f(u)\psi_{\lambda,t}(u)du, \quad \lambda > 0. \tag{19.1}$$

The function ψ is called the wavelet function (mother wavelet). The λ is a scale parameter, t a location parameter, and the generated functions $\psi_{\lambda,t}(u)$ are called wavelets. In the case where $\psi_{\lambda,t}(u)$ is complex, we use complex conjugate $\overline{\psi}_{\lambda,t}(u)$ in the definition (19.1). The normalizing constant $\frac{1}{\sqrt{\lambda}}$ is chosen so that

$$|\psi_{\lambda,t}(u)|^2 \equiv \int |\psi_{\lambda,t}(t)|^2 du = \int |\psi(t)|^2 dt = 1$$

for all the scales λ.

The choice of the mother wavelet $\psi(t)$ is critical. Different choices lead to different decompositions, and the right choice very much depends on the particular data studied. The only property the mother wavelet function $\psi(t)$ has is unit energy (or variance in probability, i.e. $|\psi(t)|^2_{L^2} = \int |\psi(t)|^2 dt = 1$). However, to produce a good decomposition, it must possess the following properties:

- fast decay as $|t| \to \infty$ in order to obtain localization in space;
- zero mean, that is, $\int \psi(t)dt = 0$.

These two properties above typically make the function look like a wave reflected around 0 which is why the function $\psi(t)$ is called a wavelet. Please refer to Foufoula-Georgiou and Kumar (1994) for more details.

The wavelet function ψ is in effect a band-pass filter. When we scale it using λ, for each level the bandwidth is reduced. This creates the problem that in order to cover the entire spectrum, an infinite number of levels would be required. To deal with this in practice an extra decomposition is used. Wavelets are generated by a scaling function (father wavelet), ϕ in addition to a mother wavelet function, ψ. The scaling function is used to capture the smooth, low-frequency nature of the data, whereas the wavelet functions are used to capture the detailed, and high-frequency nature of the data.

The scaling function integrates to 1, and the wavelet function integrates to 0:

$$\int \phi(t)dt = 1 \quad \text{and} \quad \int \psi(t)dt = 0. \tag{19.2}$$

Generally, the analytic form of wavelets does not exist and they are typically generated using numerical schemes. Unlike in the Fourier transform case, here we talk about time and scale, rather than time and frequency. The departure from the periodic functions (sin and cos) means that frequency looses its precise meaning.

The orthogonal wavelet decomposition of a time series, $x_t, t = 1, \dots, n$ is defined as,

$$\begin{aligned} x_t = &\sum_k s_{J,k}\phi_{J,k}(t) + \sum_k d_{J,k}\psi_{J,k}(t) \\ &+ \sum_k d_{J-1,k}\psi_{J-1,k}(t) + \cdots + \sum_k d_{1,k}\psi_{1,k}(t), \end{aligned} \tag{19.3}$$

where the Jth level represents the number of scales (frequencies) for the orthogonal wavelet decomposition of the time series x_t. The index k ranges from 1 to the number of coefficients associated with the specified component. The functions $\phi_{J,k}(t), \psi_{J,k}(t), \psi_{J-1,k}(t), \dots, \psi_{1,k}(t)$ are generated from the scaling function, $\phi(t)$, and the wavelet function, $\psi(t)$, by shifting and scaling based on powers of 2:

$$\phi_{J,k}(t) = 2^{\frac{-J}{2}}\phi\left(\frac{t - 2^J k}{2^J}\right),$$

$$\psi_{j,k}(t) = 2^{\frac{-j}{2}}\psi\left(\frac{t - 2^j k}{2^j}\right), \quad j = 1, \dots, J,$$

where $2^j k$ is the shift parameter and 2^j is the scale parameter. The wavelet functions are spread out and shorter for larger values of j and narrow and taller for smaller values of the scale (Shumway and Stoffer 2010). The reciprocal of the scale parameter $(\frac{1}{2^j})$ in wavelet analysis is analogous to frequency $(\omega_j = \frac{j}{n})$ in Fourier analysis. This is as a result of the fact that larger values of the scale refer to slower, smoother movements of the signal, and smaller values of the scale refer to faster, finer movements of the signal.

The **discrete wavelet transform** of a time series data x_t, is given by the coefficients $s_{J,k}$ and $d_{J-1,k}$ for $j = J, J-1, \dots, 1$ in (19.3). To some degree of approximation, they are given by

$$s_{J,k} = n^{-1/2}\sum_{t=1}^{n} x_t\phi_{J,k}(t), \tag{19.4}$$

$$d_{j,k} = n^{-1/2}\sum_{t=1}^{n} x_t\psi_{j,k}(t), \quad j = J, J-1, \dots, 1. \tag{19.5}$$

The magnitudes of these coefficients measure the importance of the corresponding wavelet term in describing the behavior of the time series x_t. The $s_{J,k}$ are known as the smooth coefficients because they describe the smooth behavior of the data. The $d_{j,k}$ are known as the detail coefficients since they represent the finer and high frequency nature of the time series data.

A way to measure the importance of each level is to evaluate the proportion of the total power or energy explained by the phenomenon under study. The level corresponds to the number of scales (frequencies) for the orthogonal wavelet decomposition of the time series. For example, in seismic studies larger values of the scale correspond to slower and smoother movements of the seismogram, whereas smaller values of the scale correspond to faster and finer movements of the seismogram.

The total energy, P of a time series x_t, for $t = 1, \ldots, n$, is defined as:

$$P = \sum_{t=1}^{n} x_t^2. \tag{19.6}$$

The total energy associated to each level of scale is

$$P_J = \sum_{k=1}^{n/2^J} s_{J,k}^2 \tag{19.7}$$

and

$$P_j = \sum_{k=1}^{n/2^j} d_{j,k}^2, \quad j = J, J-1, \ldots, 1. \tag{19.8}$$

Thus, we can rewrite Eq. (19.6) as

$$P = P_J + \sum_{j=1}^{J} P_j. \tag{19.9}$$

The proportion of the total energy explained by each level is the ratios of the total energy associated with each coefficient of detail to the total energy of the time series given by:

$$P_j/P$$

for $j = J, J-1, \ldots, 1$. Please refer to Beccar-Varela et al. (2016b) for more details.

19.2 Discrete Wavelets Transforms

A discrete wavelet transform (DWT) is defined as any wavelet transform for which the wavelets are discretely sampled. Equations (19.4) and (19.5) exemplify such

DWT. A DWT decomposes a signal into a set of mutually orthogonal wavelet basis functions. One important feature of the DWT over other transforms, such as the Fourier transform, lies in its ability to offer temporal resolution, i.e. it captures both frequency and location (or time) information.

19.2.1 Haar Wavelets

The Haar DWT is one of the simplest possible wavelet transforms. One disadvantage of the Haar wavelet is the fact that it is not continuous, and therefore not differentiable. This property can, however, be an advantage for the analysis of signals with sudden transitions, such as monitoring of tool failure in machines (see e.g. Lee and Tarng 1999).

The Haar wavelet function $\psi(t)$ can be described as

$$\psi(t) = \begin{cases} 1, & 0 \leq t < \frac{1}{2} \\ -1, & \frac{1}{2} \leq t < 1 \\ 0, & \text{Otherwise} \end{cases}$$

and the scaling function $\phi(t)$ can be described as,

$$\phi(t) = \begin{cases} 1, & 0 \leq t < 1 \\ 0, & \text{Otherwise.} \end{cases}$$

The Haar functions in fact are very useful for demonstrating the basic characteristics of wavelets (see Shumway and Stoffer 2010).

19.2.1.1 Haar Functions

The family of N Haar functions $h_k(t)$, $(k = 0, \ldots, N-1)$ are defined on the interval $0 \leq t \leq 1$. The shape of the specific function $h_k(t)$ of a given index k depends on two parameters p and q:

$$k = 2^p + q - 1. \tag{19.10}$$

For any value of $k \geq 0$, p and q are uniquely determined so that 2^p is the largest power of 2 contained in k $(2^p < k)$ and $q - 1$ is the remainder $q - 1 = k - 2^p$.

Example 19.1 When $N = 20$, the index k with the corresponding p and q are shown in Table (19.1):

Thus, the Haar functions can be defined recursively as:

1. When $k = 0$, the Haar function is defined as a constant

$$h_0(t) = 1/\sqrt{N}.$$

Table 19.1 Determining p and q for $N = 20$.

k	0	1	2	3	4	5	6	7	8	9	10	11	12	13	14	15	16	17	18	19
p	0	0	1	1	2	2	2	3	3	3	3	3	3	3	3	3	4	4	4	4
q	0	1	1	2	1	2	3	4	1	2	3	4	5	6	7	8	1	2	3	4

2. When $k > 0$, the Haar function is defined by

$$\psi(t) = \frac{1}{\sqrt{N}} \begin{cases} 2^{p/2}, & (q-1)/2^p \leq t < (q-0.5)/2^p \\ -2^{p/2}, & (q-0.5)/2^p \leq t < (q-0.5)/2^p \\ 0, & \text{Otherwise.} \end{cases}$$

From the definition, it can be seen that p determines the amplitude and width of the nonzero part of the function, while q determines the position of the nonzero part of the function. We remark that to construct Table 19.1, if we consider for example the case $k = 19$, we can find unique values of p and q such that $k - 2^p = q - 1$. If $k = 19$ we need at the most $p = 4$ because in that case we have $k - 2^p = 19 - 16 = 3$ so then we can solve for $q = 1 + 3 = 4$.

19.2.1.2 Haar Transform Matrix

The N Haar functions can be sampled at $t = m/N$, where $m = 0, \ldots, N-1$ to form an N by N matrix for discrete Haar transform. For example, when $N = 2$, we have

$$\mathbf{H}_2 = \frac{1}{\sqrt{2}} \begin{bmatrix} 1 & 1 \\ 1 & -1 \end{bmatrix}$$

when $N = 4$, we have

$$\mathbf{H}_4 = \frac{1}{2} \begin{bmatrix} 1 & 1 & 1 & 1 \\ 1 & 1 & -1 & -1 \\ \sqrt{2} & -\sqrt{2} & 0 & 0 \\ 0 & 0 & \sqrt{2} & -\sqrt{2} \end{bmatrix}$$

and when $N = 8$, we have

$$\mathbf{H}_8 = \frac{1}{\sqrt{8}} \begin{bmatrix} 1 & 1 & 1 & 1 & 1 & 1 & 1 & 1 \\ 1 & 1 & 1 & 1 & -1 & -1 & -1 & -1 \\ \sqrt{2} & -\sqrt{2} & \sqrt{2} & -\sqrt{2} & 0 & 0 & 0 & 0 \\ 0 & 0 & 0 & 0 & \sqrt{2} & -\sqrt{2} & \sqrt{2} & -\sqrt{2} \\ 2 & -2 & 0 & 0 & 0 & 0 & 0 & 0 \\ 0 & 0 & 2 & -2 & 0 & 0 & 0 & 0 \\ 0 & 0 & 0 & 0 & 2 & -2 & 0 & 0 \\ 0 & 0 & 0 & 0 & 0 & 0 & 2 & -2 \end{bmatrix}.$$

One important property of the Haar transform matrix is the fact that it is real and orthogonal, i.e.

$$\mathbf{H} = \mathbf{H}^*, \quad \mathbf{H}^{-1} = \mathbf{H}^T, \quad \text{i.e.} \quad \mathbf{H}^T\mathbf{H} = \mathbf{I}, \tag{19.11}$$

where $\mathbf{I}$ is the identity matrix.

Example 19.2 When $N = 2$,

$$\mathbf{H}_2^{-1}\mathbf{H}_2 = \mathbf{H}_2^T\mathbf{H}_2 = \frac{1}{2}\begin{bmatrix}1 & 1\\1 & -1\end{bmatrix}\begin{bmatrix}1 & 1\\1 & -1\end{bmatrix}$$
$$= \begin{bmatrix}1 & 0\\0 & 1\end{bmatrix}.$$

19.2.2 Daubechies Wavelets

Daubechies wavelets defining the DWT are characterized by a maximal number of vanishing moments for the given support space. As in the Haar wavelets case, the Daubechies wavelets are defined recursively, with each resolution twice that of the previous scale. Daubechies wavelets are localized in the temporal domain, and approximately localized in the frequency domain.

There are several Daubechies transforms, but they are all very similar. In this section, we shall concentrate on the simplest one, the Daub4 wavelet transform. The Daub4 wavelet transform is defined in essentially the same way as the Haar wavelet transform. If a signal f has an even number N of values, then the 1-level Daub4 transform is the mapping $f \to (a^1|d^1)$ from the signal f to its first trend subsignal a^1 and first fluctuation subsignal d^1. Each value a_m of $a^1 = (a_1, \dots, a_{N/2})$ is equal to the scalar product:

$$a_m = f \cdot V_m^1 \tag{19.12}$$

of f with a 1-level scaling signal V_m^1. Similarly, the value d_m of $d^1 = (d_1, \dots, d_{N/2})$ is equal to the scalar product:

$$d_m = f \cdot W_m^1 \tag{19.13}$$

of f with a 1-level wavelet W_m^1. The Daub4 wavelet transform can be extended to multiple levels as many times as the signal length can be divided by 2. A signal is defined as a function that conveys information about the behavior or attributes of some phenomenon.

The main difference between the Daub4 transform and the Haar transform lies in the way that the wavelets and scaling signals are defined. We shall first discuss the scaling signals. The scaling numbers are defined as follows:

$$\alpha_1 = \frac{1+\sqrt{3}}{4\sqrt{2}},$$

$$\alpha_2 = \frac{3+\sqrt{3}}{4\sqrt{2}},$$
$$\alpha_3 = \frac{3-\sqrt{3}}{4\sqrt{2}},$$
$$\alpha_4 = \frac{1-\sqrt{3}}{4\sqrt{2}}.$$

The scaling signals have unit energy. This property is due to the following identity satisfied by the scaling numbers:

$$\alpha_1^2 + \alpha_2^2 + \alpha_3^2 + \alpha_4^2 = 1. \tag{19.14}$$

Another identity satisfied by the scaling function is

$$\alpha_1 + \alpha_2 + \alpha_3 + \alpha_4 = \sqrt{2}. \tag{19.15}$$

Equation (19.15) implies that each 1-level trend value $f \cdot V_m^1$ is an average of four values of f, multiplied by $\sqrt{2}$.

The Daub4 wavelet numbers are defined as follows:

$$\beta_1 = \frac{1-\sqrt{3}}{4\sqrt{2}},$$
$$\beta_2 = \frac{\sqrt{3}-3}{4\sqrt{2}},$$
$$\beta_3 = \frac{3+\sqrt{3}}{4\sqrt{2}},$$
$$\beta_4 = \frac{-1-\sqrt{3}}{4\sqrt{2}}.$$

The Daub4 wavelets have unit energy. This is evident in the 1-level wavelets since

$$\beta_1^2 + \beta_2^2 + \beta_3^2 + \beta_4^2 = 1. \tag{19.16}$$

The wavelet numbers and the scaling numbers are related by the equations:

$$\alpha_1 = -\beta_4, \quad \alpha_2 = \beta_3, \quad \alpha_3 = -\beta_2, \quad \alpha_4 = \beta_1.$$

Other forms of DWT includes the dual-tree complex wavelet transform, the undecimated wavelet transform, and the Newland transform. Please refer to Selesnick et al. (2005), Fowler (2005), and Newland (1993) for more details of the other forms of the DWT.

The DWT has practical applications in science, engineering, mathematics, and computer science.

19.3 Applications of the Wavelets Transform

In this section, we briefly discuss some-real life applications of the wavelet transforms.

19.3.1 Discriminating Between Mining Explosions and Cluster of Earthquakes

In this section, we discuss wavelets techniques used to analyze the seismograms of a set of mining explosions, reported in catalogs as earthquakes, and compared the results with natural earthquakes that occurred in the same region (within a radius of 10 km).

19.3.1.1 Background of Data

The earthquakes used in this study corresponds to a set of $M = 3.0 - 3.5$ aftershocks of a recent $M = 5.2$ intraplate earthquake that occurred on 26 June 2014 near the border between the states of Arizona and New Mexico, USA. These earthquakes were located near the town of Clifton, Arizona, where a large surface copper mine previously performed a set of explosions as part of a quarry blasts activity. We selected a group of explosions cataloged with similar magnitudes as the earthquakes ($M = 3.0 - 3.5$) and located in the same region (within a radius of 10 km).

19.3.1.2 Results

The objective of the analysis is to summarize the spectral behavior of the signal as it evolves over time. We used the seismograms containing the seismic waves generated by the earthquakes and explosions. The time series used were made up of 4096 data points. First, a wavelet analysis is performed on a short section of the time series data. The section is shifted, and the analysis is performed on a new section. The process is repeated until the end of the time series.

All the analyses were performed by developing a module using the R statistical software. The time-frequency images (Figures 19.1, 19.2, 19.5, and 19.6) were obtained by using (19.5), which is based on the local transforms of the time series data , x_t. The transforms are based on tapered cosines and sines that have been zeroed out over various regions in time (Shumway and Stoffer 2010). The three-dimensional graphics (Figures 19.3, 19.4, 19.7, and 19.8) were generated by adapting the wavelet module in R. The values in Tables 19.2 and 19.3 are the proportions of the ratios of the total energy associated with each coefficient of detail to the total energy of the time series. The values in the tables are rounded to two decimal places. Based on the values in the two tables, we observe that the significant values of the percentage of energy for the explosions are few relative to the significant values of the percentage of energy for the earthquake.

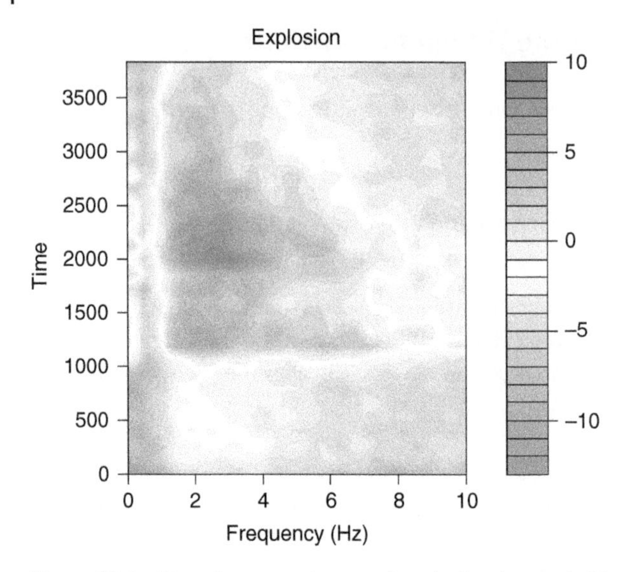

Figure 19.1 Time-frequency image of explosion 1 recorded by ANMO (Table 19.2).

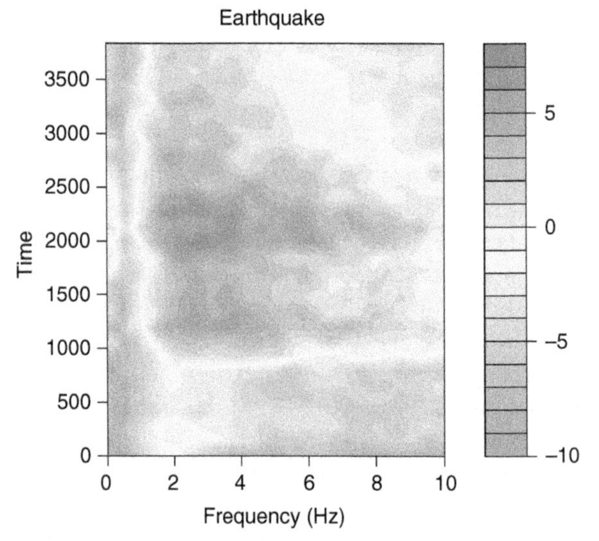

Figure 19.2 Time-frequency image of earthquake 1 recorded by ANMO (Table 19.2).

This observation is evident in Figures 19.3 and 19.7 which have fewer spikes compared to Figures 19.4 and 19.8 which have many spikes.

Also in the lower levels (i.e. level 1 and level 2) of Tables 19.2 and 19.3, the proportion of the total energy for the explosion is lower than the proportion of total energy for the earthquake. This observation is evident in Figures 19.1, 19.2, 19.5, and 19.6. In the figures, images in the darker areas correspond to higher energy

Table 19.2 Percentage of total power (energy) for Albuquerque, New Mexico (ANMO) seismic station.

Levels	Explosion (%)		Earthquake (%)	
	1	2	1	2
1	0	0.12	0	0.57
2	0	5.36	0.92	18.66
3	97.63	62.02	69.65	57.92
4	2.37	32.48	29.43	22.84
5	0	0.02	0	0.02
6	0	0.01	0	0

Table 19.3 Percentage of total power (energy) for Tucson, Arizona (TUC) seismic station.

Levels	Explosion (%)		Earthquake (%)	
	1	2	1	2
1	0.94	0.870	23.73	9.22
2	11.19	18.00	37.65	31.94
3	79.34	62.38	35.57	46.61
4	8.51	18.72	3.04	12.22
5	0.03	0.03	0.01	0.01

(Figures 19.7 and 19.8). Earthquakes show high energy at lower frequencies (levels), and the power remains strong for a long time. In contrast, explosions show energy at higher frequencies (levels) than the earthquake, and the energy of the signals does not last long as in the case of the earthquake.

The wavelets representation of seismograms in the amplitude–frequency–time is more efficient in enforcing the characteristic parameters of earthquake and explosions than the commonly used spectograms obtained using short-time Fourier transforms.

19.3.2 Finance

The wavelet transform has proved to be a very powerful tool for characterizing behavior, especially self-similar behavior, over a wide range of time scales. A self-similar behavior is a process that exhibits the same statistical properties at any temporal scale. We will discuss more about self-similarity processes in

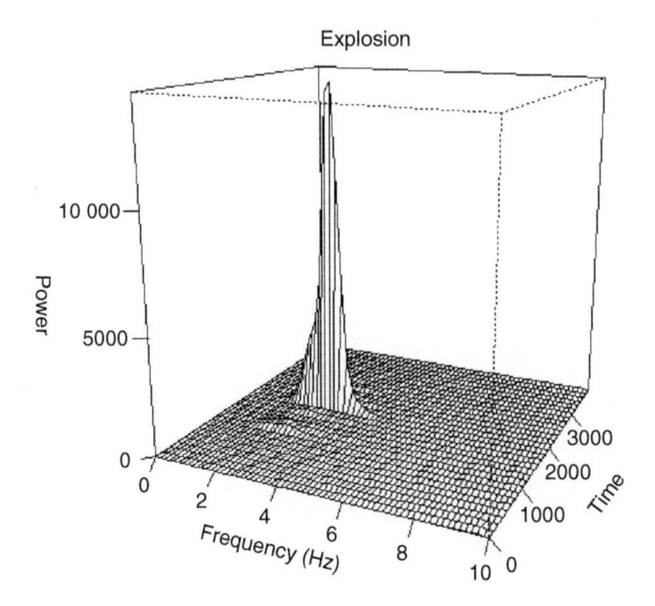

Figure 19.3 Three-dimensional graphic information of explosion 1 recorded by ANMO (Table 19.2).

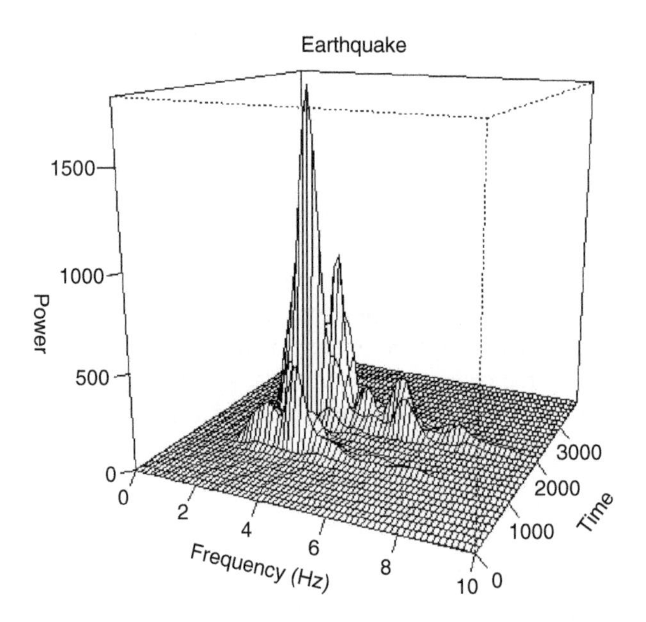

Figure 19.4 Three-dimensional graphic information of earthquake 1 recorded by ANMO (Table 19.2).

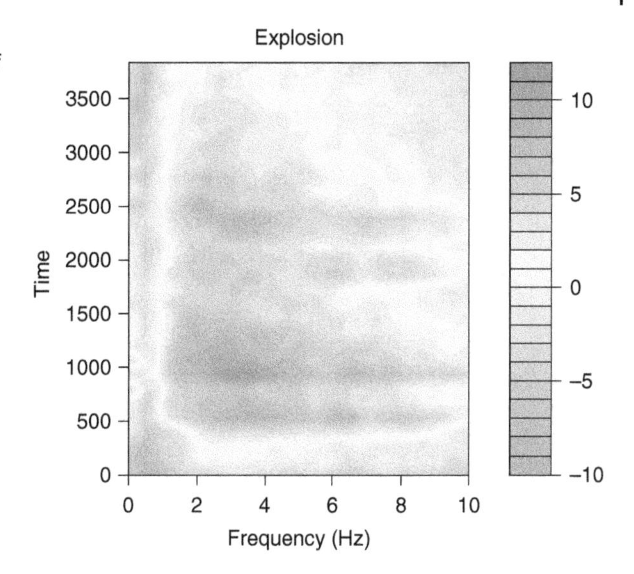

Figure 19.5 Time-frequency image of explosion 2 recorded by TUC (Table 19.3).

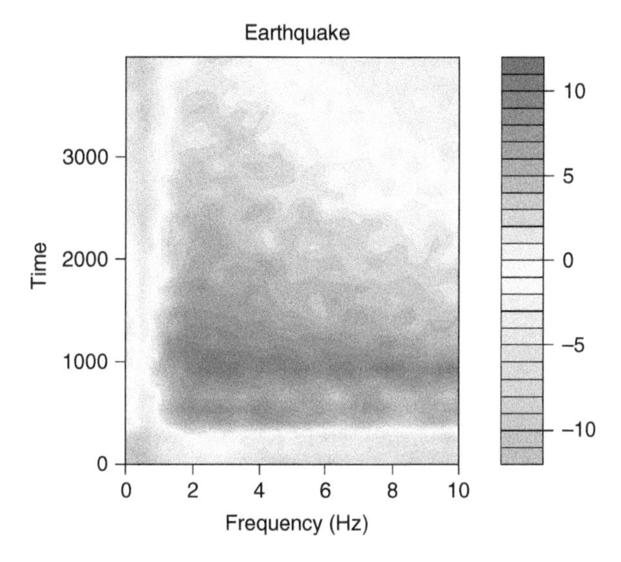

Figure 19.6 Time-frequency image of earthquake 2 recorded by TUC (Table 19.3).

Chapter 19 of this book. In the work of Beccar-Varela et al. (2017), the authors investigated the applicability of wavelet decomposition methods to determine if a market crash may be predictable. The premise of the work was that a predictable event produces a time series signal similar in nature with that recorded before a major earthquake which contains before-shock signals. A nonpredictable event would have a time series signal resembling data produced by an explosive event which is very sudden and not preceded by any prior signals.

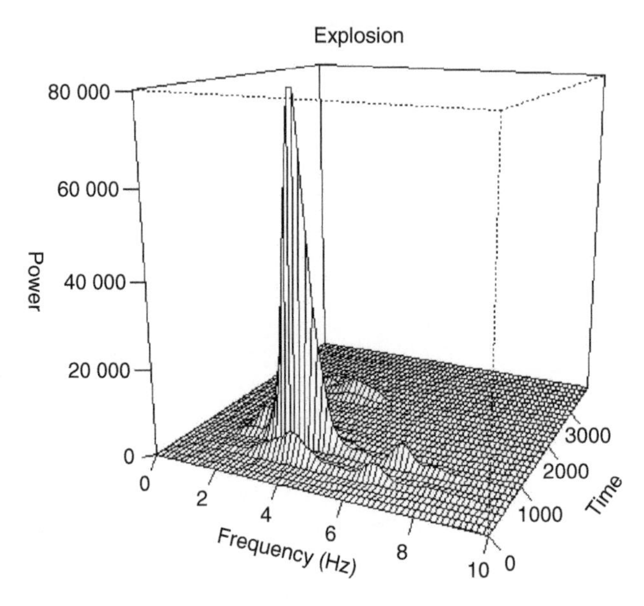

Figure 19.7 Three-dimensional graphic information of explosion 2 recorded by TUC (Table 19.3).

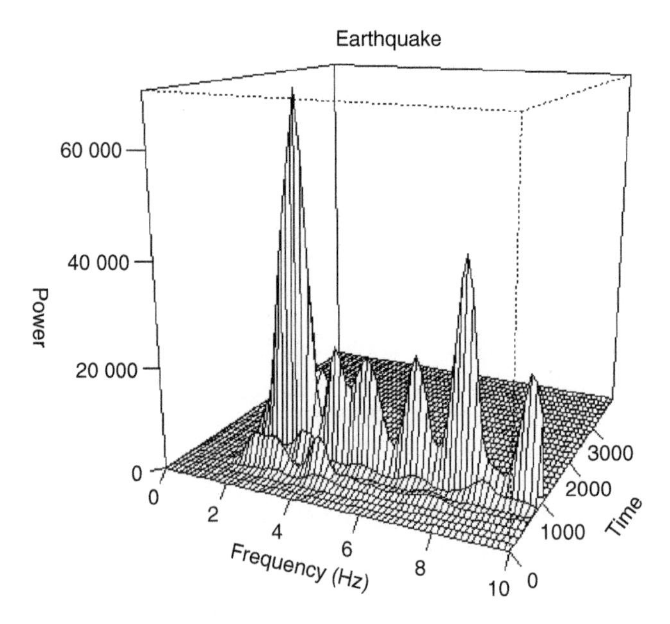

Figure 19.8 Three-dimensional graphic information of earthquake 2 recorded by TUC (Table 19.3).

The wavelets methodology was very useful in yielding a correct identification of the signal type. Correct identification of signal type using wavelets techniques help mitigate some of the potential effects of the events (Beccar-Varela et al. 2016b). Wavelets have also been used to investigate the dependence and interdependence between major stock markets (see Vuorenmaa 2005; Dajcman 2013).

19.3.3 Damage Detection in Frame Structures

The wavelet transform can be used to detect cracks in frame structures, such as beams and plane frames. The ability of wavelets to detect crack-like damage in structures can be demonstrated by means of several numerical examples. The method requires the knowledge of only the response of the damaged structure, i.e. no information about the original undamaged structure is required. This procedure can help detect the localization of the crack by using a response signal from static or dynamic loads. Moreover, the response needs to be obtained only at the regions where it is suspected that the damage may be present. The results of the simulation show that if a suitable wavelet is selected, the method is capable to extract damage information from the response signal in a simple, robust, and reliable way. Please refer to Ovanesova and Suárez (2004) and references there in for more details.

19.3.4 Image Compression

One major application of the wavelet transform is data compression and detection of edges in digital images.

Suppose one wants to send a digital image file to a friend via email. In order to expedite the process, the contents of the file needs to be compressed. There are two forms of image compression, lossless and lossy compression. Data compressed by lossless schemes can be recovered exactly without any loss of information. However for the lossy compression schemes, the data is altered. Good results can be obtained using lossy compression schemes, but the savings come at the expense of information lost from the original data (Fleet 2008).

The Haar wavelet compression is an efficient way to perform both lossless and lossy image compression. It relies on averaging and differencing values in an image matrix to produce a matrix which is sparse or nearly sparse. A sparse matrix is a matrix in which most of the elements are zero. Please refer to Fleet (2008) for more details of image compression.

19.3.5 Seismic Signals

Wavelet transforms have been applied to the decomposition of seismic signal energy into the time-scaling plane. For better representation, the scale is converted into frequency. The seismic signals are obtained from hitting a metal sphere against the lawn. After a number of experiments, both the expected value of the energy center frequencies and standard deviation are found and given in terms of the distance between the detector and place of excitation. The expected value and standard deviation decrease if the distance increases, so it is possible to estimate the distance using only one detector. These results are very useful for analyzing seismic signal whenever several excitations occur. This is demonstrated in experiments in which seismic signals are obtained from two sources. We refer to Stojanovic et al. (1999) for more details of the application of wavelets transform to seismic signals.

Wavelets are known to be strong methodologies because they can capture the nonstationary behavior of the time series and are localized in both frequency and time.

19.4 Problems

1 Define the Haar's wavelets function $\psi(t)$ and verify that for every t,

$$\psi(t) = \begin{cases} 1, & 0 \leq t < \frac{1}{2} \\ -1, & \frac{1}{2} \leq t < 1 \\ 0, & \text{Otherwise.} \end{cases}$$

2 Calculate the Haar wavelets transform for each pair s_{2k}, s_{2k+1} in the array $\vec{s} = (9, 4, 3, 2)$.

3 Calculate the Haar wavelets transform for the data $\vec{s} = (2, 4, 5, 2)$.

4 Discuss the differences between the wavelet transform and the Fourier transform.

5 For the scaling numbers,

$$\alpha_1 = \frac{2 + \sqrt{3}}{\sqrt{2}},$$

$$\alpha_2 = \frac{1 + \sqrt{3}}{\sqrt{2}},$$

$$\alpha_3 = \frac{1 - \sqrt{3}}{\sqrt{2}},$$

$$\alpha_4 = \frac{2-\sqrt{3}}{4\sqrt{2}}.$$

Verify the following identity:
(a) $\alpha_1^2 + \alpha_2^2 + \alpha_3^2 + \alpha_4^2 = 1$
(b) $\alpha_1 + \alpha_2 + \alpha_3 + \alpha_4 = \sqrt{2}$

6 Calculate the Daubechies wavelet transform of the data $\bar{a} = (a_0, a_1, a_2, a_3) = (3, 3, 3, 3)$.

7 Write a program for the signal $f(t) = \cos(4t)$ to obtain the discrete wavelets transform of this signal.

8 Explain the DWT to estimate power spectrum in high frequency time series data.

9 Find the Haar transform matrix for $N = 16$ and show its orthogonal property.

10 Discuss the difference between Haar wavelets and Daubechies wavelets.

20

Stochastic Analysis

20.1 Introduction

In this chapter, we introduce the reader to the fundamental ideas and some results of stochastic analysis up to the point that reader can acquire a working knowledge of this topic. Most of the concepts introduced will also serve as a prerequisite for the remaining chapters of the book.

Stochastic analysis is a basic tool in much of modern probability theory and is used in many applied fields such as biology, financial economics, physics, and especially statistical mechanics. It has become particularly well known via the Black–Scholes formula as a way of modeling financial markets and strategies, among others.

20.2 Necessary Definitions from Probability Theory

We begin the chapter with necessary definitions from probability theory. The following definitions can be found in many literature, for example Casella and Berger (2002), Mikosch (1998), Oksendal (2010a), Florescu and Tudor (2014), Florescu (2014), and references therein.

We will use the term "experiment" in a very general way to refer to some process that produces a random outcome. For completeness, we will recall some definitions from Section 2.3.

Definition 20.1 (Sample space) The set Ω, of all possible outcomes of a particular experiment is called the sample space for the experiment.

Definition 20.2 (σ-algebra) A σ-algebra is a collection of sets $\mathcal{F}$ of Ω satisfying the following condition:

Data Science in Theory and Practice: Techniques for Big Data Analytics and Complex Data Sets, First Edition. Maria Cristina Mariani, Osei Kofi Tweneboah, and Maria Pia Beccar-Varela.

1. $\emptyset \in \mathcal{F}$.
2. If $F \in \mathcal{F}$, then its complement $F^c \in \mathcal{F}$.
3. If $F_1, F_2, \ldots$ is a countable collection of sets in $\mathcal{F}$, then their union $\cup_{n=1}^{\infty} F_n \in \mathcal{F}$.

Definition 20.3 (Probability measure) Let $\mathcal{F}$ be a σ-algebra on Ω. A probability measure on $\mathcal{F}$ is a real-valued function P on $\mathcal{F}$ with the following properties:

1. $P(A) \geq 0$ for $A \in \mathcal{F}$.
2. $P(\Omega) = 1, P(\emptyset) = 0$.
3. If $A_n \in \mathcal{F}$ is a disjoint sequence of events, i.e. $A_i \cap A_j = \emptyset$, for $i \neq j$, then

$$P(\cup_{n=1}^{\infty} A_n) = \sum_{n=1}^{\infty} P(A_n).$$

Definition 20.4 (Probability space) A probability space is a triplet $(\Omega, \mathcal{F}, P)$, where Ω is a sample space, $\mathcal{F}$ is a σ-algebra on Ω, and P is a probability measure $P : \mathcal{F} \to [0, 1]$.

We recall in Chapter 2 that a random variable $\mathbf{X}$ is any measurable function defined on the probability space $(\Omega, \mathcal{F}, P)$ with values in $\mathbb{R}^n$.

For every random variable $\mathbf{X}$, we can associate a function called the cumulative distribution function (cdf) of $\mathbf{X}$ which is defined as follows:

Definition 20.5 (Cumulative distribution function) Given a random vector $\mathbf{X}$ with components $\mathbf{X} = (X_1, \ldots, X_n)$, its cdf is defined as follows:

$$F_{\mathbf{X}}(x) = P(\mathbf{X} \leq x) = P(X_1 \leq x_1, \ldots, X_n \leq x_n) \text{ for all } x.$$

Definition 20.6 (Continuous and discrete function) A random variable $\mathbf{X}$ is continuous if $F_{\mathbf{X}}(x)$ is a continuous function of x. A random variable $\mathbf{X}$ is discrete if $F_{\mathbf{X}}(x)$ is a step function of x.

Associated with a random variable $\mathbf{X}$ and its cdf $F_{\mathbf{X}}$ is another function called the probability density function (pdf) or probability mass function (pmf). The terms pdf and pmf were defined in Chapter 2.

20.3 Stochastic Processes

Definition 20.7 (Stochastic process) A stochastic process is a parametrized collection of random variables $\{X(t) : t \in \mathcal{I}\}$ defined on a probability space $(\Omega, \mathcal{F}, P)$ and assuming values in $\mathbb{R}^n$, where $\mathcal{I}$ is an index set.

The notations X_t and $X(t)$ are used interchangeably to denote the value of the stochastic process at index value t.

20.3.1 The Index Set $\mathcal{I}$

The set $\mathcal{I}$, that indexes the stochastic process determines the type of stochastic process. Below we give some examples:

- If the index set is defined as $\mathcal{I} = \{0, 1, 2 \dots\}$, we obtain the discrete-time stochastic processes. We shall denote the process as $\{X_n\}_{n \in \mathbb{N}}$ in this case.
- If the index set is defined as $\mathcal{I} = [0, \infty)$, we obtain the continuous-time stochastic processes. We shall also denote the process as $\{X_t\}_{t \geq 0}$. In most instances, t represents time.
- The index set can be multidimensional. For example, if $\mathcal{I} = [0, 1] \times [0, 1]$, we may be describing the structure of some surface, where, for instance $X(x, y)$ could be the value of some electrical field intensity at position (x, y).

20.3.2 The State Space $\mathcal{S}$

The state space $\mathcal{S}$ is the domain space of all the random variables X_t. Since we are discussing about random variables and random vectors, then necessarily $\mathcal{S} \subseteq \mathbb{R}$ or $\mathbb{R}^n$. This domain space can be defined using integers, real lines, n-dimensional Euclidean spaces, complex planes, or more abstract mathematical spaces. We present some examples as follows:

- If $\mathcal{S} \subseteq \mathbb{Z}$, then the process is integer valued or a process with discrete state space.
- If $\mathcal{S} = \mathbb{R}$, then X_t is a real-valued process or a process with a continuous state space.
- If $\mathcal{S} = \mathbb{R}^k$, then X_t is a k-dimensional vector process.

The state space $\mathcal{S}$ can be more general (for example, an abstract Lie algebra), in which case the definitions work very similarly except that for each t, we have X_t measurable functions.

20.3.3 Stationary and Independent Components

Definition 20.8 (Independent components) For any collection $\{t_1, t_2, \dots, t_n\}$ of elements in $\mathcal{I}$ if the corresponding random variables $X_{t_1}, X_{t_2}, \dots, X_{t_n}$ are independent, then the joint distribution $F_{X_{t_1}, X_{t_2}, \dots, X_{t_n}}$ is the product of the marginal distributions $F_{X_{t_i}}$, where $i = 1, \dots, n$.

Definition 20.9 (Strictly stationary) A stochastic process, X_t, is said to be strictly stationary if the joint distribution function of the vectors:

$$(X_{t_1}, X_{t_2}, \dots, X_{t_n}) \quad \text{and} \quad (X_{t_1+h}, X_{t_2+h}, \dots, X_{t_n+h})$$

are the same for all $h > 0$ and all arbitrary selection of index points $\{t_1, t_2, \dots, t_n\}$ in $\mathcal{I}$. Specifically: The process $\{x_t; t \in \mathbb{Z}\}$ is strongly stationary if

$$F_{t_1+k, t_2+k, \dots, t_s+k}(b_1, b_2 \dots, b_s) = F_{t_1, t_2, \dots, t_s}(b_1, b_2 \dots, b_s)$$

for any finite set of indices $\{t_1, t_2, \dots, t_s \subset \mathbb{Z}\}$ with $s \in \mathbb{Z}^+$ and any $k \in \mathbb{Z}$. Therefore, we can conclude that the process $\{x_t; t \in \mathbb{Z}\}$ is strongly stationary if the joint distribution function of the vector $(X_{t_1+k}, X_{t_2+k}, \dots, X_{t_s+k})$ is the same as the one of the vector $(X_{t_1}, X_{t_2}, \dots, X_{t_s})$ for any finite set of indices $\{t_1, t_2, \dots, t_s \subset \mathbb{Z}\}$ with $s \in \mathbb{Z}^+$, and any $k \in \mathbb{Z}$.

Definition 20.10 (Weak stationary) A stochastic process X_t, is said to be weak stationary if X_t has finite second moments for any t and if the covariance function $\text{Cov}(X_t, X_{t+h})$ depends only on h for all $t \in \mathcal{I}$.

Remarks 20.1 A strictly stationary process with finite second moments (so that covariance exists) is going to be automatically weak stationary. The reverse is not true.

The concept of weak stationarity was developed because of the practical way in which we observe stochastic processes. While strict stationarity is a very desirable concept, it is not possible to test it with real data. To show strict stationarity means, we need to test all joint distributions. However, in real life, the samples we gather are finite, so this is not possible. Instead, we can test the stationarity of the covariance matrix which only involves bivariate distributions.

Many phenomena can be described by stationary processes. In addition, many classes of processes eventually become stationary if observed for a long time. The white noise process is an example of a strictly stationary process. However, some of the most common processes encountered in practice – the Poisson process and the Brownian motion – are not stationary. However, they have stationary and independent increments. We define this concept next.

20.3.4 Stationary and Independent Increments

In order to discuss the increments for stochastic processes, we assume that the index set $\mathcal{I}$ has a total order, that is for any two elements a and b in $\mathcal{I}$ either $a \leq b$ or $b \leq a$. We note that a two-dimensional index set, for example $\mathcal{I} = [0,1] \times [0,1]$ does not have this property.

Definition 20.11 (Independent increments) A stochastic process X_t is said to have independent increments if the random variables:

$$X_{t_2} - X_{t_1}, X_{t_3} - X_{t_2}, \dots, X_{t_n} - X_{t_{n-1}}$$

are independent for any n and any choice of the sequence $\{t_1, t_2, \dots, t_n\}$ in $\mathcal{I}$ with $t_1 < t_2 < \dots < t_n$.

Definition 20.12 (Stationary increments) A stochastic process X_t is said to have stationary increments if for $s, t \in T$ with $s \leq t$, the increment $X_t - X_s$ has the same distribution as X_{t-s}.

This is not the same as stationarity of the process itself. In fact, with the exception of the constant process, there exists no process with stationary and independent increments which is also stationary.

Definition 20.13 (Quadratic variation for stochastic processes) Let X_t be a stochastic process on the probability space $(\Omega, \mathcal{F}, P)$ with filtration $\{\mathcal{F}_t\}_{t\in\mathcal{I}}$. Let $\pi_n = (0 = t_0 < t_1 < \cdots < t_n = t)$ be a partition of the interval $[0, t]$. We define the quadratic variation process

$$[X, X]_t = \lim_{\|\pi_n\|\to 0} \sum_{i=0}^{n-1} |X_{t_{i+1}} - X_{t_i}|^2,$$

where the limit of the sum is defined in probability.

The quadratic variation process is a stochastic process. The quadratic variation may be calculated explicitly only for some classes of stochastic processes. In fact, the stochastic processes used in finance have finite second-order variation. The third and higher order variations are all zero, while the first order is infinite. This is the fundamental reason why the quadratic variation has such a big role for stochastic processes used in finance.

20.3.5 Filtration and Standard Filtration

In the case, where index set $\mathcal{I}$ possesses a total order relationship, we can discuss about the information contained in the process $X(t)$ at some moment $t \in \mathcal{I}$. To quantify this information, we generalize the notion of sigma algebras by introducing a sequence of sigma algebras: the filtration.

Definition 20.14 (Filtration) A probability space $(\Omega, \mathcal{F}, P)$ is a filtered probability space if and only if there exists a sequence of sigma algebras $\{\mathcal{F}_t\}_{t\in\mathcal{I}}$ included in $\mathcal{F}$ such that $\mathcal{F}$ is an increasing collection, i.e.

$$\mathcal{F}_s \subseteq \mathcal{F}_t, \quad \forall s \leq t, \quad s, t \in \mathcal{I}.$$

A filtration is called *complete* if its first element contains all the null sets of $\mathcal{F}$.

Definition 20.15 (Right and left continuous filtrations) A filtration $\{\mathcal{F}_t\}_{t\in\mathcal{I}}$ is right continuous if and only if $\mathcal{F}_t = \mathcal{F}_{t+}$ for all t, and the filtration is left continuous if and only if $\mathcal{F}_t = \mathscr{F}_{t-}$ for all t.

Throughout this chapter, we shall assume that any filtration is right continuous.

Definition 20.16 (Adapted stochastic process) A stochastic process $\{X_t\}_{t\in I}$ defined on a filtered probability space $(\Omega, \mathscr{F}, P, \{\mathcal{F}_t\}_{t\in\mathcal{I}})$ is called adapted if and only if X_t is $\mathcal{F}_t$-measurable for any $t \in \mathcal{I}$.

This is an important concept since in general, $\mathcal{F}_t$ quantifies the flow of information available at any moment t. By requiring that the process be adapted, we ensure that we can calculate probabilities related to X_t based solely on the information available at time t. In addition, since the filtration by definition is increasing, this also means that we can calculate the probabilities at any later moment in time.

In some cases, we are only given a standard probability space (i.e. without a separate filtration defined on the space). This corresponds to the case where we assume that all the information available at time t comes from the stochastic process X_t itself. In this instance, we will be using the standard filtration generated by the process $\{X_t\}_{t\in\mathcal{I}}$ itself. Let

$$\mathcal{F}_t = \sigma(\{X_s : s \le t, s \in \mathcal{I}\}),$$

denote the sigma algebra generated by the random variables up to time t. The collection of sigma algebras $\{\mathcal{F}_t\}_t$ is increasing and the process $\{X_t\}_t$ is adapted with respect to it.

A stochastic process $\{Y_t\}$ is called a modification of a stochastic process $\{X_t\}$, if

$$\mathbb{P}[X_t = Y_t] = 1 \quad \text{for } t \in [0, \infty). \tag{20.1}$$

Two stochastic processes $\{X_t\}$ and $\{Y_t\}$ are identical in law, written as

$$\{X_t\} \overset{\mathrm{d}}{=} \{Y_t\}, \tag{20.2}$$

if the systems of their finite-dimensional distributions are identical. We discuss the concept of finite-dimensional distributions as follows:

Let $\{X_t\}_{t\in\mathcal{I}}$ be a stochastic process. For any $n \ge 1$ and for any subset $\{t_1, t_2, \dots, t_n\}$ of $\mathcal{I}$ we denote with $F_{X_{t_1}, X_{t_2}, \dots, X_{t_n}}$ the joint distribution function of the variables $X_{t_1}, X_{t_2}, \dots, X_{t_n}$. The statistical properties of the process X_t are completely described by the family of distribution functions $F_{X_{t_1}, X_{t_2}, \dots, X_{t_n}}$ indexed by the n and the t_i's. If we can describe these finite-dimensional joint distributions for all n and t's, we completely characterize the stochastic process.

20.4 Examples of Stochastic Processes

A Markov process is a simple type of stochastic process in which the time order in a sequence of events plays a significant role, i.e. the present state can influence the probability of what happens next. We present a formal definition as follows.

20.4.1 Markov Chains

In this section, we present some basic definitions that we will use later related to Markov chains, the Markov property, the Chapman–Kolmogorov equation, classification of states and limiting probabilities, and branching processes.

Suppose that

$$X = \{X_t : t \in T\}$$

is a random process on a probability space $(\Omega, \mathcal{F}, P)$ where X_t is a random variable taking values in S for each $t \in T$. We can think that X_t is the state of a system at time t. The state space S is usually either a countable set or a "nice" region of IR^k for some k. The time space T is either IN or $[0, \infty)$. For $t \in T$, let $\mathcal{F}_t$ denote the σ-algebra of events generated by $\{X_s : s \in T, s \leq t\}$.

Intuitively, $\mathcal{F}_t$ contains the events that can be defined in terms of X_s for $s \leq t$. In other words, if we are allowed to observe the random variables X_s for $s \leq t$, then we can tell whether or not a given event in $\mathcal{F}_t$ has occurred.

Definition 20.17 (Markov process) The random process X is a Markov process if the following property (known as the Markov property) holds: For every $s \in T$ and $t \in T$ with $s < t$, and for every $H \in \mathcal{F}_s$ and $x \in S$, the conditional distribution of X_t, given H and $X_s = x$ is the same as the conditional distribution of X_t just given $X_s = x$:

$$P(X_t \in A|H, X_s = x) = P(X_t \in A|X_s = x)$$

for all $A \subset S$.

In the statement of the Markov property, think of s as the present time and hence, t is a time in the future. Thus, x is the present state and H is an event that has occurred in the past. If we know the present state, then any additional knowledge of events in the past is irrelevant in terms of predicting the future (Shao, 2003).

The complexity of Markov processes depends greatly on whether the time space or the state space are discrete or continuous. We will assume that both are discrete, that is we assume that $T = \mathrm{IN}$ and that S is countable (and hence, the state variables have discrete distributions). In this chapter, Markov processes are known as Markov chains.

The Markov property for a Markov chain $X = (X_0, X_1, \dots)$ can be stated as follows: for any sequence of states $(x_0, x_1, \dots, x_{n-1}, x, y)$, we have that

$$P(X_{n+1} = y|X_0 = x_0, X_1 = x_1, \dots, X_{n-1} = x_{n-1}, X_n = x) = P(X_{n+1} = y|X_n = x).$$

In a more general way if $X_n = i$, then the process is said to be in state i at time n. Denote p_{ij} the fixed probability that the process will next be in state j, given that the process is in state i. That is, we suppose that

$$P(X_{n+1} = j | X_0 = i_0, X_1 = i_1, \dots, X_{n-1} = i_{n-1}, X_n = i) = P(X_{n+1} = j | X_n = i) = p_{ij},$$

$p_{ij} \geq 0$ and

$$\sum_{j=0}^{\infty} p_{ij} = 1.$$

If we write these equations in matrix form, we have that the matrix P is given by

$$P = \begin{pmatrix} p_{00} & p_{01} & \cdot & p_{0n} & \cdot \\ p_{10} & p_{11} & \cdot & p_{1n} & \cdot \\ \cdot & \cdot & \cdot & \cdot & \cdot \\ p_{i0} & p_{i1} & \cdot & p_{in} & \cdot \\ \cdot & \cdot & \cdot & \cdot & \cdot \end{pmatrix}.$$

To proceed, we briefly discuss transition probability matrices.

Definition 20.18 (Transition probability matrix or stochastic matrix) If P is a matrix on S, then P is a transition probability matrix or stochastic matrix if P is nonnegative and

$$\sum_{y \in S} P(x, y) = 1$$

for $x \in S$. Then, if P is a transition matrix, we have that $y \to P(x, y)$ is a pdf on S for each $x \in S$.

In matrix terminology, the row sums are 1.

Remarks 20.2 Suppose that P and Q are transition probability matrices on S and that f is a pdf on S. Then,

1. PQ is a transition probability matrix.
2. P^n is a transition probability matrix for each $n \in \mathrm{IN}$.

Next, we present some examples of Markov processes.

20.4.1.1 Examples of Markov Processes

Forecasting the Weather

Suppose that the probability of raining tomorrow depends on previous weather conditions only through whether or not it is raining today and not on past weather conditions. Assume also that if it rains today, then it will rain tomorrow with probability α and if it does not rain today, then it will rain tomorrow with probability β. If we say that the process is in state 0 when it rains and in state 1 when it does not rain, then we have a two state Markov chain whose transition probability P are given by the following matrix:

$$P = \begin{pmatrix} \alpha & 1-\alpha \\ \beta & 1-\beta \end{pmatrix}.$$

A Random Walk Model

A Markov chain whose state space is given by the integers $\{0, \pm 1, \pm 2, \dots\}$ is said to be a random walk if for some number $0 < p < 1$

$$p_{i,i+1} = 1 - p_{i,i-1}.$$

We may think of it as being a model for an individual walking on a straight line who at each point of time either takes one step to the right with probability p or one step to the left with probability $1-p$.

The Gambling Model

Consider a gambler who at each play of the game either wins 1 with probability p or loses 1 with probability $1-p$. If we suppose that our gambler quits playing either when he goes broke or he attains a fortune of \$ N, then the gambler fortune is a Markov chain having transition probabilities

$$p_{i,i+1} = p = 1 - p_{i,i-1}, \quad i = 1, 2, \dots, N-1,$$
$$p_{00} = p_{NN} = 1.$$

States 0 and N are called absorbing states since once entered they are never left. This is a finite state random walk with absorbing barriers (states 0 and N).

20.4.1.2 The Chapman–Kolmogorov Equation

Suppose that $X = (X_0, X_1, X_2, \dots)$ is a Markov chain with state space S. For $m, n \in$ IN with $m \leq n$, let

$$P_{m,n}(x, y) = P(X_n = y | X_m = x)$$

for $x \in S$, $y \in S$. The matrix $P_{m,n}$ is the transition probability matrix from time m to time n.

The following result gives the basic relationship between the transition matrices.

Theorem 20.1 ***(Chapman–Kolmogorov Equation)*** Suppose that k, m, n are nonnegative integers with $k \leq m \leq n$. Then

$$P_{k,m} P_{m,n} = P_{k,n}.$$

Corollary 20.1 It follows immediately that all of the transition probability matrices for X can be obtained from the one-step transition probability matrices: if m and n are nonnegative integers with $m \leq n$ then

$$P_{m,n} = P_{m,m+1} P_{m+1,m+2}, \dots, P_{n-1,n}.$$

Remarks 20.3 Combining Theorem 20.1 and Corollary 20.1, it follows that the distribution of X_0 (the initial distribution) and the one-step transition matrices determine the distribution of X_n for each n. Actually, these basic quantities determine the joint distributions of the process, a much stronger result.

In a more general notation, we now define the n-step transition probability $P^n_{i,j}$ to be the probability that a process in state i will be in state j after n additional transitions that is:

$$P^n_{i,j} = P(X_{n+k} = j|X_k = i) \quad n \geq 0, \quad i,j \geq 0.$$

Observe that $P^1_{i,j} = P_{i,j}$.

The Chapman–Kolmogorov equations provide a method for computing these n-step transition probabilities as follows:

$$P^{n+m}_{i,j} = \sum_{k=0}^{\infty} P^n_{i,k} P^m_{k,j} \quad n \geq 0, \quad i,j \geq 0.$$

Note that $P^n_{i,k} P^m_{k,j}$ represents the probability that starting in i, the process will go to state j in $n + m$ transitions through a path which takes it into state k at the n transition. Hence, summing over all intermediate states k yields the probability that the process will be in state j after $n + m$ transitions. Formally, we have

$$P^{n+m}_{i,j} = P(X_{n+m} = j|X_0 = i) = \sum_{k=0}^{\infty} P(X_{n+m} = j, X_n = k|X_0 = i) =$$

$$\sum_{k=0}^{\infty} P(X_{n+m} = j|X_n = k, X_0 = i) P(X_n = k|X_0 = i) = \sum_{k=0}^{\infty} P^n_{i,k} P^m_{k,j}.$$

If we let $P^{(n)}$ denote the matrix of the n-step transition probabilities $P^n_{i,j}$, then Chapman–Kolmogorov equation states that $P^{n+m} = P^n \cdot P^m$, where the dots represents matrix multiplication.

Next we present an example that revisits the weather forecasting. Suppose that in the previous example $\alpha = 0.8$ and $\beta = 0.4$. Then in order to calculate the probability that it will rain four days from today, given that it is raining today. The one-step transition probability matrix is given by

$$P = \begin{pmatrix} 0.8 & 0.2 \\ 0.4 & 0.6 \end{pmatrix}.$$

Then we have that

$$P^{(4)} = P^4 = \begin{pmatrix} 0.6752 & 0.3248 \\ 0.6496 & 0.3594 \end{pmatrix}$$

and the desired result is $P^{(4)}_{00} = 0.6752$.

20.4.1.3 Classification of States

We start this section with the following definitions.

Definition 20.19

1. State j is said to be accessible from state i if $P^n_{i,j} > 0$ for some $n \geq 0$.
2. Two states that are accessible to each other are said to communicate.
3. Two states that communicate are said to be in the same class.
4. The Markov chain is said to be irreducible if there is only one class.

Remarks 20.4 Note that each state communicates with itself because $P^0_{i,i} = P(X_0 = i|X_0 = i) = 1$. Also, if state j communicates with state i, then state i communicates with state j, and finally, if state i communicates with state j, and state j communicates with state k, then state i communicates with state k.

We now present an example.

Example 20.1

1. Consider the Markov chain consisting of the four states 0–3 and having transition probability matrix:

$$P = \begin{pmatrix} 1/3 & 2/3 & 0 & 0 \\ 2/3 & 1/3 & 0 & 0 \\ 1/4 & 1/4 & 1/4 & 1/4 \\ 0 & 0 & 0 & 1 \end{pmatrix}.$$

   The classes of this Markov chain are $\{0, 1\}$, $\{2\}$, $\{3\}$. Note that while state 0 or 1 is accessible from state 2, the reverse is not true. Since state 3 is an absorbing state, that is, $P_{33} = 1$ no other state is accessible from it.

Definition 20.20 For any state i, let f_i denote the probability that starting in state i the process will ever reenter state i. State i is said to be recurrent if $f_i = 1$, and transient if $f_i < 1$.

We state following theorem and corollaries without proof.

Theorem 20.2 State i is recurrent if:

$$\sum_{n=1}^{\infty} P^n_{ii} = \infty$$

and it is transient if

$$\sum_{n=1}^{\infty} P_{ii}^{n} < \infty.$$

Corollary 20.2 If state i is recurrent, and it communicates with state j, then state j is recurrent. If state i is transient and it communicates with state j, then state j is transient. All states of a finite irreducible Markov chain are recurrent.

20.4.1.4 Limiting Probabilities

In a previous example about forecasting the weather, we calculated

$$P^{(4)} = P^4 = \begin{pmatrix} 0.6752 & 0.3248 \\ 0.6496 & 0.3594 \end{pmatrix}.$$

If we now calculate

$$P^{(8)} = P^{(4)} \cdot P^{(4)} = \begin{pmatrix} 0.667 & 0.333 \\ 0.666 & 0.334 \end{pmatrix}.$$

So both matrices are almost identical, and we can guess that the entries have a limit. We will need the following:

Definition 20.21

1. State i is said to have a period d if $P_{ii}^{n} = 0$ whenever n is not divisible by d and d is the largest integer with this property. A state with period 1 is said to be aperiodic.
2. If state i is recurrent, it is said to be positive recurrent if starting in i the expected time until the process returns to state i is finite.
3. Positive recurrent aperiodic states are called ergodic.

Theorem 20.3 For an irreducible ergodic Markov chain, $\lim_{n\to\infty} P_{ij}^{n}$ exists, and it is independent of i. Furthermore, letting

$$\pi_j = \lim_{n\to\infty} P_{ij}^{n}, \quad j \geq 0$$

then π_j is the unique non negative solution of

$$\pi_j = \sum_{i=0}^{\infty} \pi_i P_{ij}, \quad j \geq 0$$

and

$$\sum_{j=0}^{\infty} \pi_j = 1.$$

We remark that

$$\sum_{i=0}^{\infty} P(X_{n+1} = j|X_n = i)P(X_n = i) = \sum_{i=0}^{\infty} P_{ij}P(X_n = i).$$

Letting $n \to \infty$ and assuming that we can bring the limit inside the summation we obtain

$$\pi_j = \sum_{i=0}^{\infty} P_{ij}\pi_i.$$

If the Markov chain is irreducible, it is possible to find a solution of

$$\pi_j = \sum_{i=0}^{\infty} P_{ij}\pi_i$$

$$\sum_{j=0}^{\infty} \pi_j = 1$$

if and only if the Markov chain is positive recurrent. If a solution exists, then it will be unique and π_j will be the long-run proportion of time that the Markov chain is in state j. If the chain is aperiodic, then π_j is also the limiting probability that the chain is in state j.

For example consider the previous application in which we assume that if it rains today, it will rain tomorrow with probability α and if it does not rains today, it will rain tomorrow with probability β. If we say that the state is 0 when it rains and it is 1 when it does not rain, then the limiting probability π_0 and π_1 are given by $\pi_0 = \alpha\pi_0 + \beta\pi_1$ $\pi_1 = (1-\alpha)\pi_0 + (1-\beta)\pi_1$ $\pi_0 + \pi_1 = 1$ which yields

$$\pi_0 = \frac{\beta}{1 + \beta - \alpha}$$

and

$$\pi_1 = \frac{1 - \alpha}{1 + \beta - \alpha}.$$

For example if $\alpha = 0.75$ and $\beta = 0.35$, then the limiting probability for rain is $\pi_0 = 0.35/(1 + 0.35 - 0.75) = 0.5833$.

20.4.1.5 Branching Processes

Consider a population consisting of individuals able to produce offspring of the same kind. Suppose each individual will produce j new offspring with probability $P_j, j \geq 0$ independently of the numbers produced by other individuals. A that $P_j < 1$ for all $j \geq 0$, then number of individuals initially present denoted by X_0 is called the size of the 0 generation. All offspring of the 0 generation constitute the first generation and their number is denoted by X_1, it follows that X_n for $n = 0, 1, \ldots$ is a Markov chain having as its state space the set of nonnegative integers (Ross, 2007).

Let

$$\mu = \sum_{j=0}^{\infty} jP_j$$

represent the mean number of offspring of a single individual and let

$$\sigma^2 = \sum_{j=0}^{\infty} (j-\mu)^2 P_j$$

represent the variance of number of offspring of a single individual.

Assuming that $X_0 = 1$, this means that initially there is a single individual present. We calculate $E(X_n)$ and $\mathrm{Var}(X_n)$ as follows. Let $X_n = \sum_{i=1}^{X_{n-1}} Z_i$, where X_n denotes the size of the nth generation, and Z_i represents the number of offspring of the ith individual of the $n-1$th generation, by conditioning on X_{n-1} we obtain

$$E(X_n) = E(E(X_n|X_{n-1})) = E\left(E\left(\sum_{i=1}^{X_{n-1}} Z_i|X_{n-1}\right)\right) =$$

$$E(X_{n-1}\mu) = \mu E(X_{n-1})$$

because $E(Z_i) = \mu$. Since $E(X_0) = 1$, we have that $E(X_1) = \mu$, $E(X_2) = E(X_1)\mu = \mu^2$, and $E(X_n) = \mu^n$. In the same way,

$$\mathrm{Var}(X_n) = E(\mathrm{Var}(X_n|X_{n-1})) + \mathrm{Var}(E(X_n|X_{n-1}).$$

We observe that given X_{n-1}, X_n is just the sum of X_{n-1} random variables each having distribution P_j. Then,

$$E(X_n|X_{n-1}) = X_{n-1}\mu$$

and

$$\mathrm{Var}(X_n|X_{n-1}) = X_{n-1}\sigma^2.$$

Then

$$\begin{aligned}
\mathrm{Var}(X_n) &= E(X_{n-1}\sigma^2) + \mathrm{Var}(X_{n-1}\mu) \\
&= \sigma^2\mu^{n-1} + \mu^2(\sigma^2\mu^{n-2} + \mu^2\mathrm{Var}(X_{n-2})) \\
&= \sigma^2(\mu^{n-1} + \mu^n) + \mu^4\mathrm{Var}(X_{n-2}) \\
&= \sigma^2(\mu^{n-1} + \mu^n) + \mu^4(\sigma^2\mu^{n-3} + \mu^2\mathrm{Var}(X_{n-3})) \\
&= \sigma^2(\mu^{n-1} + \mu^n + \mu^{n+1}) + \mu^6\mathrm{Var}(X_{n-3}) \\
&= \cdots \\
&= \sigma^2(\mu^{n-1} + \mu^n + \cdots + \mu^{2n-2}) + \mu^{2n}\mathrm{Var}(X_0) \\
&= \sigma^2(\mu^{n-1} + \mu^n + \cdots + \mu^{2n-2})
\end{aligned}$$

therefore,

$$\mathrm{Var}(X_n) = \sigma^2 \mu^{n-1} \left(\frac{1-\mu^n}{1-\mu} \right)$$

if $\mu \neq 1$, and

$$\mathrm{Var}(X_n) = n\sigma^2$$

if $\mu = 1$. Let π_0 denotes the probability that the population will die out under the assumption that $X_0 = 1$, then,

$$\pi_0 = \lim_{n\to\infty} P(X_n = 0|X_0 = 1).$$

We first note that $\pi_0 = 1$ if $\mu < 1$, because

$$\mu^n = E(X_n) = \sum_{j=1}^{\infty} jP(X_n = j) \geq \sum_{j=1}^{\infty} 1P(X_n = j) = P(X_n \geq 1)$$

as $\mu^n \to 0$ if $\mu < 1$, we have that $P(X_n \geq 1) \to 0$, and then, $P(X_n = 0) \to 1$. It is possible to show that $\pi_0 = 1$ even when $\mu = 1$. When $\mu > 1$, we have that $\pi_0 < 1$, and the following equation for determining π_0 can be obtained, see Ross (2007) for details:

$$\pi_0 = \sum_{j=0}^{\infty} \pi_0^j P_j,$$

where π_0^j is the probability that the population dies out given that $X_1 = j$.

For example, if $P_0 = 1/2$, $P_1 = P_2 = 1/4$, then as $\mu = 3/4 < 1$, it follows that $\pi_0 = 1$.

Similarly, if $P_0 = P_1 = 1/4$, $P_2 = 1/2$, then $\pi_0 = 1/4 + 1/4\pi_0 + 1/2\pi_0^2$. The smallest solution for this equation is $\pi_0 = 1/2$.

20.4.1.6 Time Homogeneous Chains

A Markov chain $X = (X_0, X_1, X_2, \ldots)$ is said to be time homogeneous if the transition matrix from time m to time n depends only on the difference $n - m$ for any nonnegative integers m and n with $m \leq n$. That is,

$$P(X_n = y|X_m = x) = P(X_{n-m} = y|X_0 = x)$$

for $x, y \in S$. It follows that there is a single one-step transition probability matrix P, given by

$$P(x, y) = P(X_{n+1} = y|X_n = x)$$

for $x, y \in S$, and all other transition matrices can be expressed as powers of P. Indeed, if $m \leq n$ then $P_{m,n} = P_{n-m,0}$, and the Chapman–Kolmogorov equation is simply the law of exponents for matrix powers.

Another example of a stochastic process is a Martingale.

20.4.2 Martingales

Definition 20.22 (Martingales) Let $(\Omega, \mathcal{F}, P)$ be a probability space. A martingale sequence of length n is a set of variables $X_1, X_2, \dots, X_n$ and corresponding σ-algebras $\mathcal{F}_1, \mathcal{F}_2, \dots, \mathcal{F}_n$ that satisfy the following relations:

1. Each X_i is an integrable random variable adapted to the corresponding σ-algebra $\mathcal{F}_i$.
2. The $\mathcal{F}_i$'s form a filtration.
3. For every $i \in [1, 2, \dots, n-1]$, we have

$$X_i = \mathbf{E}[X_{i+1} | \mathcal{F}_i].$$

This process has the property that the expected value of the future given the information we have today is going to be equal to the known value of the process today. In French, a martingale means a winning strategy. This is because for gamblers, a martingale is a betting strategy where the stake doubled each time the player loses. Players follow this strategy because, since they will eventually win, they argue they are guaranteed to make money. Examples of martingales are given below

1. Let $X_{t+1} = X_t \pm b_t$, where $+b_t$ and $-b_t$ occur with equal probability b_t is measurable $\mathcal{F}_t$, and the outcome $\pm b_t$ is measurable $\mathcal{F}_{t+1}$ (i.e. my "bet" b_t can only depend on what has happened so far and not on the future, but my knowledge $\mathcal{F}_t$ includes the outcome of all past bets). Then $\{X_t | \mathcal{F}_t\}$ is a martingale.
2. A random ± 1 walk is a martingale.

20.4.3 Simple Random Walk

A random walk is a stochastic sequence $\{X_n\}$, defined by

$$X_n = \sum_{t=1}^{n} X_t,$$

where X_t are independent and identically distributed (i.i.d) random variables.

The random walk is simple if $X_t = \pm 1$, with $P(X_t = 1) = p$ and $P(X_t = -1) = 1 - p$. A simple random walk is symmetric if the particle has the same probability for each of the neighbors. We recall that the simple random walk is both a martingale that is $E(X_{t+s} | X_t) = X_t$ and a stationary Markov process that is the distribution of $X_{t+s} | X_t = k_t, \dots, X_1 = k_1$ depends only on the value k_t.

20.4.4 The Brownian Motion (Wiener Process)

The Brownian motion also called the Wiener process is a continuous-time stochastic process.

Let $(\Omega, \mathcal{F}, P)$ be a probability space. A Brownian motion is a stochastic process B_t with the following properties:

1. $B_0 = 0$.
2. With probability 1, the function $t \to B_t$ is continuous in t.
3. The process B_t has stationary and independent increments.
4. The increment $B_{t+s} - B_s$ has a $N(0, t)$ distribution, where $N(0, t)$ denotes the normal distribution with mean 0 and variance t.

Other examples include the Lévy processes which will be discussed extensively in Chapter 21.

20.5 Measurable Functions and Expectations

Since Ω can be quite arbitrary, it is often convenient to consider a function (mapping) from Ω to a simpler space Λ (often $\Lambda = \text{IR}^k$). Let $B \subset \Lambda$, then the inverse image of B under f is $f^{-1}(B) = \{w \in \Omega : f(w) \in B\}$.

Let $(\Omega, \mathcal{F}, \nu)$ and $(\Lambda, \mathcal{G}, \mu)$ be measurable spaces and f a function from Ω to Λ. Then we recall that a function f is called a measurable function if and only if $f^{-1}(\mathcal{G}) \subset \mathcal{F}$. We mentioned in Chapter 2 that random variable is a measurable function.

Proposition 20.1 (Shao (2003)) Let $(\Omega, \mathcal{F}, \nu)$ be a measurable space, then

(i) $f : \Omega \to \text{IR}$ is Borel, or Borel measurable, if and only if $f^{-1}(a, \infty) \in \mathcal{F}$ for all $a \in \text{IR}$. This condition is equivalent to $f^{-1}(\mathcal{B}) \subset \mathcal{F}$, where $\mathcal{B}$ is the σ-algebra of Borel, that is, by definition, the least σ-algebra having the open sets of IR.
(ii) If f and g are Borel, then so are fg and $af + bg$ where a and b are real numbers. Also f/g provided that $g(\omega) \neq 0$.

Remarks 20.5 Let $(\Omega, \mathcal{F}, \nu)$ be a measurable space and f a measurable function from $(\Omega, \mathcal{F}, \nu)$ to $(\Lambda, \mathcal{G}, \mu)$. The induced measure by f denoted by $\nu \circ f^{-1}$, is a measure on $\mathcal{G}$ defined as $\nu \circ f^{-1}(B) = \nu(f^{-1}(B)), B \in \mathcal{G}$.

If $\nu = P$ is a probability measure and X is a random variable or a random vector, then $P \circ X^{-1}$ is called the law of distribution or the distribution of X and is denoted by F_X the c.d.f of P_X defined by

$$F(x) = P((-\infty, x])$$

or

$$F(x_1, \dots, x_k) = P((-\infty, x_1] \times \cdots (-\infty, x_k]), \quad x_i \in R.$$

Proposition 20.2 (Shao (2003)) Let $(\Omega, \mathcal{F}, \nu)$ be a measurable space, and f and g be Borel functions, then

i. f $\int f\, d\nu$ exists and $a \in \text{IR}$, then $\int (af) d\nu$ exists and is equal to $a \int f\, d\nu$.
ii. If both $\int f\, d\nu$ and $\int g\, d\nu$ exist, then $\int f\, d\nu + \int g\, d\nu$ is well defined and $\int (f + g) d\nu$ exist and is equal to $\int f\, d\nu + \int g\, d\nu$.

In order to define the expectation of a random variable (measurable function), we need the following result.

Theorem 20.4 **(Shao (2003) Change of variables)** Let f be measurable from $(\Omega, \mathcal{F}, \nu)$ to $(\Lambda, \mathcal{G}, \mu)$ and g be Borel on $(\Lambda, \mathcal{G}, \mu)$, that is, g is a measurable function from $(\Lambda, \mathcal{G}, \mu)$ to $(\text{IR}, \mathcal{B}, m)$, where m is the Lebesgue measure, then

$$\int g \circ f \, d\nu = \int g \, d(\nu \circ f^{-1}),$$

i.e. if either one of the integrals exist, then so does the other and the two are the same.

The proof of this theorem can be found in Billingley (1986, page 219) this result extend the change of variable for Riemann integrals, i.e.

$$\int g(y)dy = \int g(f(x))f'(x)dx, \quad y = f(x).$$

Let X be a random variable on a probability space $(\Omega, \mathcal{F}, P)$. If $E(X) = \int_\Omega X \, dP$ exists, then usually it is simpler to compute $E(X) = \int_{\text{IR}} x \, dP_X$ where $P_X = P \circ X^{-1}$ is the law of X. Let Y be a random vector from Ω to IR^k and g be Borel from IR^k to IR according to 1.16 $E(g(y))$ can be computed as $\int_{\text{IR}^k} g(y) d(P_Y)$ or $\int_{\text{IR}} x \, dP_{g(Y)}$ depending on which of P_Y and $P_{g(Y)}$ is easier to handle.

As an example consider $k = 2$, $Y = (X_1, X_2)$ and $g(Y) = X_1 + X_2$. Then using Proposition 20.2(ii), $E(X_1 + X_2) = E(X_1) + E(X_2)$ and hence $E(X_1 + X_2) = \int_{\text{IR}} x \, dP_{X_1} + \int_{\text{IR}} x \, dP_{X_2}$. So we need to handle two integrals. On the other hand, $E(X_1 + X_2) = \int_{\text{IR}} x \, dP_{X_1+X_2}$ which involves one integral with respect to $P_{X_1+X_2}$, which requires to know the joint c.d.f of (X_1, X_2).

20.5.1 Radon–Nikodym Theorem and Conditional Expectation

Before we state the Radon–Nikodym theorem, we present the following definitions:

Definition 20.23 Let ν, λ two measures in $(\Omega, \mathcal{F})$, we say that λ is absolutely continuous with respect to ν and write $\lambda \ll \nu$, if every time that $\nu(A) = 0$, we have $\lambda(A) = 0$, for every $A \in \mathcal{F}$.

As an example, we can consider

$$\lambda(A) = \int_A f \, dm,$$

where m is the Lebesgue measure and f is a nonnegative Lebesgue measurable and integrable function.

Definition 20.24 A measure ν in $(\Omega, \mathcal{F})$ is σ-finite if $\Omega = \cup_{i \in \text{IN}} \Omega_i$, with $\nu(\Omega_i) < \infty$.

In order to define the conditional expectation, we need the following result.

Theorem 20.5 (Radon– Nikodym) Let ν, λ two measures in $(\Omega, \mathcal{F})$, and ν be sigma finite. If $\lambda \ll \nu$, then there exists a nonnegative Borel function f on Ω such that

$$\lambda(A) = \int_A f \, d\nu$$

for every $A \in \mathcal{F}$.

Theorem 20.5 can be extended to the case in which ν is not positive, but a measure with values in IR (i.e. ν can take positive and negative values). Please refer to Billingsley (1995) for the proof of Theorem 20.5.

Definition 20.25 (Conditional expectations) Let X be an integrable random variable on $(\Omega, \mathcal{F}, P)$

1. Let $\mathcal{A}$ be a sub-σ-field (a sub-σ-field of $\mathcal{F}$ is a subset of $\mathcal{F}$ that is also a σ-field.) of $\mathcal{F}$. The conditional expectation of X given $\mathcal{A}$, denoted by $E(X|\mathcal{A})$ is the almost sure (i.e. happens with probability one) – unique random variable satisfying the following:
   (a) $E(X|\mathcal{A})$ is measurable from $(\Omega, \mathcal{A}, P)$ to $(\text{IR}, \mathcal{B}, m)$, where m is the Lebesgue measure.
   (b) $\int_A E(X|\mathcal{A})dP = \int_A X \, dP$ for any $A \in \mathcal{A}$.
2. Let $B \in \mathcal{F}$. The conditional probability of B given $\mathcal{A}$ is defined to be

$$P(B|\mathcal{A}) = E(I_B|\mathcal{A}).$$

3. Let Y be measurable from $(\Omega, \mathcal{F}, P)$ to $(\Lambda, \mathcal{G}, \mu)$. The conditional expectation of X given Y is defined to be

$$E(X|Y) = E(X|\sigma(Y)),$$

   where $\sigma(Y) = Y^{-1}(\mathcal{G})$.

Note that the existence of $E(X|\mathcal{A})$ follows from the Radon–Nikodym theorem, taking $\lambda(A) = \int_A X \, dP$, that is absolutely continuous with respect to P, and $f = E(X|\mathcal{A})$.

Remarks 20.6 If we take $A = \Omega$ in (b), we have that $\int_\Omega E(X|\mathcal{A})dP = \int_\Omega X \, dP$, that is $E[E(X|\mathcal{A})] = E(X)$, and $E[E(X|\mathcal{A})]$ is integrable.

Next, we recall the definition of covariance matrix in Chapter 3.

For a random k-vector $X = (X_1 \cdots X_k)^T$ its mean is $E(X) = (EX_1 \cdots EX_k)^T$. The extension of the variance is the variance-covariance matrix of X defined as follows:

$$\text{Var}(X) = E[(X - E(X))(X - E(X))^T].$$

The (i, j)th element of $\mathrm{Var}(X)$ for $i \neq j$ is $E[(X_i - E(X_i))(X_j - E(X_j))]$ which is called the covariance of X_i and X_j and is denoted by $\mathrm{Cov}(X_i, X_j)$. The (i, i)th element of $\mathrm{Var}(X)$ is the variance of X_i, denoted by $\mathrm{Var}(X_i)$.

Next, we present the convergence definition in relation to random variables.

Definition 20.26 A sequence of random variables $\{X_n, n \geq 1\}$ is said to converge almost surely to the random variable X if

$$P(\{\omega : \lim_{n\to\infty} X_n(\omega) \neq X(\omega)\}) = 0.$$

For example $\Omega = [0, 1]$, $X_n : \Omega \to \Omega$ with $X_n = \omega^n$. Then, $X_n \to 0$ on $[0, 1)$ and $X_n(1) = 1$. So $X_n \to 0$ a.s. The set in this case is only the number 1, that is why the measure is 0.

Example: Strong law of large numbers

Theorem 20.6 Let $\{X_n, n \geq 1\}$ be a sequence of i.i.d random variables with $E(X_i) = \mu$ and $\mathrm{Var}(X_i) = \sigma^2 < \infty$, and define $S_n = \frac{X_1+X_2+\cdots+X_n}{n}$. Then, for every $\varepsilon > 0$ we have that:

$$P(\lim_{n\to\infty} |S_n - \mu| < \varepsilon) = 1,$$

that is, S_n converges almost sure to μ. Please refer to Billingsley (1995) for the proof of Theorem 20.6.

Examples of the strong law of large numbers is the arrival times in a Poisson process and the return times of Markov chains.

Definition 20.27 A sequence of random variables $\{X_n, n \geq 1\}$ is said to converge in probability to the random variable X if given $\varepsilon > 0$,

$$\lim_{n\to\infty} P(|X_n - X| > \varepsilon) = 0$$

or equivalently,

$$\lim_{n\to\infty} P(|X_n - X| < \varepsilon) = 1.$$

Remarks 20.7 If a sequence of random variables $\{X_n, n \geq 1\}$ converges almost sure to the random variable X, then it converges in probability to the random variable X.

Let $X(s) = s$, then as $n \to \infty$, $P(|X_n - X| \geq \varepsilon)$ is the probability of an interval of s values whose length is going to 0. So X_n converges to X in probability. But X_n does not converges to X almost sure: there is no value of s for which $X_n(s) \to s = X(s)$. For every s, the value $X_n(s)$ alternates between the values s and $s + 1$ infinitely often. For example, If $s = 3/8$, $X_1(s) = 1 + 3/8$, $X_2(s) = 1 + 3/8$, $X_3(s) = 3/8$, $X_4(s) = 3/8$, $X_5(s) = 1 + 3/8$, $X_6(s) = 3/8$, etc.

Remarks 20.8 If a sequence converges in probability, it is always possible to find a subsequence that converges almost sure.

Remarks 20.9 In general, the convergence almost sure in probability theory, is the convergence almost everywhere in measure theory, and the convergence in probability in probability theory, is the convergence in measure in measure theory.

Note that convergence almost everywhere in measure theory does not imply convergence in measure, as an example, consider $X_n = X_{[-n,n]}$, then $X_n \to 1$ when $n \to \infty$, because $|X_n(x) - 1| = 0 < \varepsilon$, for all $\varepsilon > 0$ if $n \geq n_0 = n_0(\varepsilon, x) > |x|$, but for $0 < \lambda < 1$, we have that $m(\{x : |X_n(x) - 1| > \lambda\}) = m[(-\infty, -n) \cup (n, \infty)] = \infty$.
However, if the measure is finite, convergence almost everywhere in measure theory does imply convergence in measure; as the probability is a finite measure, then convergence almost sure implies convergence in probability.

Definition 20.28 A sequence of random variables $\{X_n, n \geq 1\}$ is said to converge in distribution to the random variable X if

$$\lim_{n\to\infty} F_n(x) = F(x)$$

for all $x \in C(F)$, where F_n, F are the distribution functions of X_n, X, respectively, and $C(F)$ is the continuity set of F. We remark that F is continuous in this set.

For example consider a sequence of random variables $\{X_n, n \geq 1\}$ with X_n having distribution $N(0, 1/n)$. Then

$$F_n(x) = P(X_n \leq x) = \sqrt{\frac{n}{2\pi}} \int_{-\infty}^{x} e^{-nz^2/2}\, dz = \frac{1}{\sqrt{2\pi}} \int_{-\infty}^{\sqrt{n}x} e^{-t^2/2}\, dt.$$

Therefore, as $n \to \infty$, $F_n(x) \to 0$ if $x < 0$, $F_n(x) \to 1/2$ if $x = 0$, $F_n(x) \to 1$ if $x > 0$, so $F(x) = I_{(0,\infty)}(x)$ if $x \neq 0$, and $F(0) = 1/2$; and we can conclude that $F_n(x) \to F(x) = I_{(0,\infty)}(x)$ in $C(F) = \mathrm{IR} - \{0\}$ the continuity set of F.

Remarks 20.10 If a sequence of random variables $\{X_n, n \geq 1\}$ converges in probability to the random variable X, then it converges in distribution to the random variable X.

20.6 Problems

1 Consider the Markov chain consisting of the three states 0–2 and having transition probability matrix:

$$P = \begin{pmatrix} 1/2 & 1/2 & 0 \\ 1/2 & 1/4 & 1/4 \\ 0 & 1/3 & 2/3 \end{pmatrix}.$$

Show that this Markov chain is irreducible.

2 Consider the Markov chain consisting of the four states 0–3 and having transition probability matrix:

$$P = \begin{pmatrix} 0 & 0 & 1/2 & 1/2 \\ 1 & 0 & 0 & 0 \\ 0 & 1 & 0 & 0 \\ 0 & 1 & 0 & 0 \end{pmatrix}.$$

Show that all the states communicate and are recurrent.

3 Let c be a constant and f and g two measurable real-valued functions defined on the same domain. Prove that the functions $cf, f + g$ and fg are also measurable.

4 Suppose that the price of an asset follows a Brownian motion:

$$dX = \mu X\ dt + \sigma X\ dz.$$

(a) What is the stochastic process for X^n?
(b) What is the expected value for X^n?

5 Discuss the difference between a stochastic process and an adapted stochastic process. Present examples.

6 Generate any single path of the process of Hull–White model where $\triangle t = 0.001, \theta_1 = 2, \theta = 0.7, \theta_3 = 0.8, a(t) = \theta_1 t, b(t) = \theta_t \sqrt{t}, \sigma(t) = \theta_3(t)$ and $dX_t = a(t)(b(t) - X_t)dt + \sigma(t)dW_t$.

7 If a process X_t has stationary and independent increments, prove that
(a) $E[X_t] = m_0 + m_1 t$ and
(b) $\text{Var}[X_t - X_0] = \text{Var}[X_1 - X_0]t$,
where, $m_0 = E[X_0]$ and $m_1 = E[X_1] - m_0$.

8 Explain the quadratic variation for stochastic process with examples.

9 Prove that for a deterministic function f which is differentiable with continuous first order derivative, all d-order variations with $d \geq 2$ are zero.

21

Fractal Analysis – Lévy, Hurst, DFA, DEA

21.1 Introduction and Definitions

In this chapter, we will discuss Fractal analysis: the main properties and examples of the Lévy stochastic processes and the truncated Lévy models which is a generalization of Lévy processes. We conclude this chapter with an introduction to the Range Scale Analysis (Hurst Analysis), Detrended Fluctuation Analysis (DFA) and Diffusion Entropy Analysis (DEA). The Hurst Analysis, DFA, and DEA methodologies are very important techniques for analyzing extreme events, like financial crashes and earthquakes. We begin our discussion with the Lévy processes and infinitely divisible distributions.

21.2 Lévy Processes

Lévy processes play a fundamental role in Mathematical Finance and in other fields such as Physics (turbulence), Engineering (telecommunications and dams), Actuarial Science (insurance risk), and several others.

Definition 21.1 (Lévy process) A stochastic process $\{X_t : t \geq 0\}$ on $\mathbb{R}^n$ is a *Lévy process* if the following conditions are satisfied,

1. For any choice of $n \geq 1$ and $0 \leq t_0 < t_1 < \cdots < t_n$, the random variables $X_{t_0}, X_{t_1} - X_{t_0}, X_{t_2} - X_{t_1}, \dots, X_{t_n} - X_{t_{n-1}}$ are independent. The process has independent increments.
2. $X_0 = 0$.
3. The distribution of $X_{s+t} - X_s$ does not depend on s, therefore, the process has stationary increments.

Data Science in Theory and Practice: Techniques for Big Data Analytics and Complex Data Sets, First Edition. Maria Cristina Mariani, Osei Kofi Tweneboah, and Maria Pia Beccar-Varela.

4. It is stochastically continuous.
5. There is $\Omega_0 \in \mathcal{F}$ (σ–algebra) with $\mathbb{P}[\Omega_0] = 1$ such that for every $\omega \in \Omega_0$, $X_t(\omega)$ is right-continuous on $t \geq 0$ and has left limits in $t > 0$.

Remarks 21.1 A Lévy process defined on $\mathbb{R}^n$ is known as an n-dimensional Lévy process. The law at time t of a Lévy process is completely determined by the law of X_1. The only degree of freedom we have when specifying a Lévy process is to define its distribution at a single time.

Definition 21.2 (Characteristic function) The *characteristic function* ϕ of a random variable X is the Fourier–Stieltjes transform of the distribution function $F(x) = \mathbb{P}(X \leq x)$:

$$\phi_X(u) = \mathbb{E}[e^{iuX}] = \int_{-\infty}^{\infty} e^{iux}\, dF(x), \tag{21.1}$$

where i is the imaginary number.

One important property of the characteristic function is the fact that for any random variable X, it always exists, it is continuous and it determines X unequivocally. If X and Y are independent random variables, then

$$\phi_{X+Y}(u) = \phi_X(u)\phi_Y(u). \tag{21.2}$$

Some functions related to the characteristic function that will be used throughout this chapter are as follows:

- The cumulant function:

$$k(u) = \log \mathbb{E}[e^{-uX}] = \log \phi(iu).$$

- The cumulant characteristic function or characteristic exponent:

$$\psi(u) = \log \mathbb{E}[e^{iuX}] = \log \phi(u). \tag{21.3}$$

Definition 21.3 (Infinitely divisible distribution) Suppose $\phi(u)$ is the characteristic function of a random variable X. If for every positive integer n, $\phi(u)$ is also the nth power of a characteristic function, we say that the distribution is infinitely divisible.

The normal distribution, Cauchy distribution, Poisson distribution, the negative binomial distribution, Gamma distribution, and all other members of the stable distribution family are examples of infinitely divisible distributions. We will discuss more about stable distributions in Section 21.2.4.

Theorem 21.1 describes the one-to-one relationship between Lévy processes and infinitely divisible distributions.

Theorem 21.1 ***(Infinite divisibility of Lévy processes)*** Let $X = \{X_t, t \geq 0\}$ be a Lévy process. Then X has infinitely divisible distributions F for every t. Conversely, if F is an infinitely divisible distribution, there exists a Lévy process X such that the distribution of X_1 is given by F.

Please refer to Sato (1999) for the proof of Theorem 21.1. We can write

$$\phi_{X_t}(z) = \mathbb{E}[e^{-izX_t}] = e^{t\psi(z)},$$

where $\psi(z) = \log(\phi(z))$ is the characteristic exponent as in Eq. (21.3). The characteristic exponent $\psi(z)$ of a Lévy process satisfies the following *Lévy–Khintchine formula* (see Khintchine and Lévy 1936; Bertoin 1996):

$$\psi(z) = -\frac{1}{2}\sigma^2 z^2 + i\gamma z + \int_{-\infty}^{\infty} (e^{izx} - 1 - izx\mathbb{I}_{\{|x|<1\}})\nu(dx), \tag{21.4}$$

where $\gamma \in \mathbb{R}, \sigma^2 \geq 0$ and ν is a measure on $\mathbb{R}\backslash\{0\}$ with

$$\int_{-\infty}^{\infty} \inf\{1, x^2\}\nu(dx) = \int_{-\infty}^{\infty} (1 \wedge x^2)\nu(dx) < \infty. \tag{21.5}$$

From Eq. (21.4), we observe that in general, a Lévy process consist of three independent parts, namely, linear deterministic part, Brownian part, and a pure jump part. The first two terms of Eq. (21.4) contains the deterministic and Brownian part and the third term contains the jump part.

We say that the corresponding infinitely divisible distribution has a Lévy triplet $[\gamma, \sigma^2, \nu(dx)]$. The measure ν is called the *Lévy measure* of X or sometimes referred to as the jump measure.

Definition 21.4 **(Lévy measure)** Let $X = \{X_t : t \geq 0\}$ be a Lévy process on $\mathbb{R}^n$. The measure ν on $\mathbb{R}^n$ defined by

$$\nu(A) = \frac{1}{t}\mathbb{E}\left(\sum_{0<s\leq t} \mathbb{I}_{\{\Delta X_s \in A\}}\right), \quad A \in \mathcal{B}(\mathbb{R}) \tag{21.6}$$

is called a *Lévy measure*, where $\nu(A)$ dictates how jumps occur and $\Delta X_s = X(s) - X(-s)$. Jumps of sizes in the set A occur according to a Poisson process with parameter $\nu(A) = \int_A \nu(dx)$. In other words, $\nu(A)$ is the expected number of jumps per unit time, whose size belongs to A.

A Lévy measure has no mass at the origin, but singularities may occur near the origin (small jumps). Special attention has to be considered on small jumps since the sum of all jumps smaller than some $\epsilon > 0$ may not converge. For example, consider the instance where the Lévy measure $\nu(dx) = \frac{dx}{x^2}$. As we move closer to the origin, there is an increasingly large number of small jumps and $\int_{-1}^{1} |x|\nu(dx) = +\infty$. But for the integral

$$\int_{-1}^{1} x^2 \nu(dx),$$

substituting $\nu(dx) = \frac{dx}{x^2}$ leads to the integral

$$\int_{-1}^{1} x^2 \frac{dx}{x^2} = \int_{-1}^{1} dx = 2$$

which is finite. However, as we move away from the origin, $\nu([-1,1]^c)$ is finite and so we do not experience any difficulties with the integral in Eq. (21.5) being finite. The Brownian motion (BM) has continuous sample paths with no jumps and as such $\Delta X_t = 0$.

A Poisson process (i.e. a stochastic process for modeling the times at which arrivals enter a system. We will discuss the Poisson process in the next subsection.) with rate parameter λ and jump size equal to 1 is a pure jump process with $\Delta X_t = 1$ and Lévy measure,

$$\nu(A) = \begin{cases} \lambda, & \text{if } \{1\} \in A \\ 0, & \text{if } \{1\} \notin A. \end{cases}$$

If the Lévy measure is of the form $\nu(dx) = u(x)dx$, then $u(x)$ is known as the Lévy density. The Lévy density has similar properties to a probability density. However, it does not need to be integrable and must have zero mass at the origin.

Consider the class of models with risk neutral dynamics of the underlying asset given by $S_t = \exp(rt + X_t)$, where X_t is a Lévy process and X_0 is set to zero. The characteristic function of X_t has the following Lévy–Khintchine representation

$$\mathbb{E}[e^{izX_t}] = \exp t\phi(z),$$

where

$$\phi(z) = -\frac{\sigma^2 z^2}{2} + i\gamma z + \int_{-\infty}^{\infty} (e^{izx} - 1 - izx1_{|x|\le 1})\nu(dx),$$

with $\sigma \ge 0$, γ are real constants and ν is a positive random measure on $\mathbb{R} - \{0\}$ verifying

$$\int_{-1}^{+1} x^2 \nu(dx) < \infty \quad \text{and} \quad \int_{|x|>1} \nu(dx) < \infty.$$

21.2.1 Examples of Lévy Processes

In this section, we present examples of some popular Lévy processes. We will begin with the so-called subordinators. A subordinator is a one-dimensional Lévy process that is nondecreasing almost surely (a.s.). They will be discussed in details in Section 21.2.3. Next, we will give some examples of Lévy processes defined on the real line. For each Lévy process defined, we will analyze their density function, characteristic function, and other important properties. We will also compute their respective moments, variance, skewness, and kurtosis if possible. We begin the examples with the Poisson process.

21.2.1.1 The Poisson Process (Jumps)

Definition 21.5 (Poisson process) A stochastic process $N = \{N_t, t \geq 0\}$ with intensity parameter $\lambda > 0$ is a *Poisson process* if it fulfills the following conditions:

1. $N_0 = 0$.
2. The process has independent and stationary increments.
3. For $s < t$, the random variable $N_t - N_s$ has a Poisson distribution with parameter $\lambda(t-s)$ such that,

$$\mathbb{P}[N_t - N_s = n] = \frac{\lambda^n (t-s)^n}{n!} e^{-\lambda(t-s)}.$$

The Poisson process is the simplest of all the Lévy processes. It is based on the Poisson distribution, which depends on the parameter λ and has the following characteristic function:

$$\phi_{\text{Poisson}}(z; \lambda) = \exp(\lambda(\exp(iz) - 1)).$$

The domain of the Poisson distribution is the set of nonnegative integers, $k = \{0, 1, 2, \ldots\}$ and the probability mass function at any given point k is given by

$$f(k; \lambda) = \frac{\lambda^k e^{-\lambda}}{k!}.$$

Since the Poisson distribution is infinitely divisible, we can define a Poisson process $N = \{N_t, t \geq 0\}$ with intensity parameter λ as the process which starts at zero, has independent and stationary increments and where the increments over a time interval of length $s > 0$ follow the Poisson distribution. The Poisson process is an increasing pure jump process with jump sizes equal to 1. That is, the time between two consecutive jumps follows an exponential distribution with mean λ^{-1}. This is referred to as the $\Gamma(1, \lambda)$ law since the exponential distribution is a special case ($n = 1$) of the gamma distribution. The moments of the Poisson distribution are given in Table 21.1.

21.2.1.2 The Compound Poisson Process

Definition 21.6 (Compound Poisson process) A *compound Poisson* process with intensity parameter λ and a jump size distribution L is a stochastic process $X = \{X_t, t \geq 0\}$ defined as follows:

$$X_t = \sum_{k=1}^{N_t} \chi_k, \tag{21.7}$$

where $N_t, t \geq 0$ is a Poisson process with intensity parameter λ and $(\chi_k, k = 1, 2, \ldots)$ is an independent identically distributed sequence.

The sample paths of X are piecewise constant (a function is said to be piecewise constant if it is locally constant in connected regions separated by a possibly

Table 21.1 Moments of the Poisson distribution with intensity λ.

	Poisson (λ)
Mean	λ
Variance	λ
Skewness	$\frac{1}{\sqrt{\lambda}}$
Kurtosis	$3 + \lambda^{-1}$

infinite number of lower-dimensional boundaries. An example of a 1D piecewise constant function is the Heaviside step function.), and the value of the process at time t, X_t is the sum of N_t random numbers with law L. The jump times have the same law as those of the Poisson process N. The ordinary Poisson process corresponds to the case where $\chi_k = 1$ for $k = 1, 2, \ldots$. The characteristic function of X_t is given by

$$\mathbb{E}[\exp(izX_t)] = \exp\left(t\int_{-\infty}^{\infty}(\exp(izx) - 1)\nu(dx)\right), \quad \forall u \in \mathbb{R}, \tag{21.8}$$

where ν is called the *Lévy measure* of the process X. The Lévy measure ν is a positive measure on $\mathbb{R}$ but not a probability measure since $\int \nu(dx) = \lambda \neq 1$.

21.2.1.3 Inverse Gaussian (IG) Process

The inverse Gaussian (IG) process $\{Y(t); t \geq 0\}$ is defined as the stochastic process satisfying the following properties:

- $Y(t)$ has independent increments, i.e. $Y(t_2) - Y(t_1)$ and $Y(s_2) - Y(s_1)$ are independent for all $t_2 > t_1 \geq s_2 > s_1$;
- $Y(t) - Y(s)$ follows an IG distribution $\mathcal{IG}(\Lambda(t) - \Lambda(s), \eta[\Lambda(t) - \Lambda(s)]^2)$, for all $t > s \geq 0$,

where $\Lambda(t)$ is a monotone increasing function and IG(a, b), $a, b > 0$, denotes the IG distribution with probability density function (PDF),

$$f_{\mathcal{IG}}(y; a, b) = \sqrt{\frac{b}{2\pi y^3}}\exp\left[-\frac{b(y-a)^2}{2a^2 y}\right], \quad y > 0.$$

Next, we define $T^{(a,b)}$ to be the first time a standard BM with drift $b > 0$ reaches a positive level $a > 0$. The first time, $T^{a,b}$ follows the IG law and has a characteristic function:

$$\phi(z; a, b) = \exp(-a(\sqrt{-2iz + b^2} - b)).$$

The IG distribution is infinitely divisible, and we can simply redefine the IG process $IG(a, b)$ to be a stochastic process X with parameters a, b as the process which starts at zero and has independent and stationary increments such that

$$\mathbf{E}[\exp(izX_t)] = \phi(z; at, b)$$
$$= \exp(-at(\sqrt{-2iz + b^2} - b)). \tag{21.9}$$

21.2.1.4 The Gamma Process

Definition 21.7 (Gamma process) A stochastic process $X = \{X_t, t \geq 0\}$ with parameters a and b is a *Gamma process* if it fulfills the following conditions:

1. $X_0 = 0$.
2. The process has independent and stationary increments.
3. For $s < t$ the random variable $X_t - X_s$ has a $\Gamma(a(t-s), b)$ distribution.

A random variable X has a gamma distribution $\Gamma(a, b)$ with rate and shape parameters, $a > 0$ and $b > 0$, respectively, if its density function is defined as:

$$f_X(x; a, b) = \frac{b^a}{\Gamma(a)} x^{a-1} e^{-bx}, \quad \forall x > 0. \tag{21.10}$$

The moments of the $\Gamma(a, b)$ distribution are presented in Table 21.2.

The Gamma process is a nondecreasing Lévy process and its characteristic function is given by

$$\phi(z; a, b) = \left(1 - \frac{iz}{b}\right)^{-a}.$$

21.2.2 Exponential Lévy Models

Suppose $(\Omega, F, F_t, \mathbb{P})$ is a filtered probability space and let $(S_t)_{t\in[0,T]}$ be the price of a financial asset modeled as a stochastic process on that space. Here F_t is taken to be the price history up to t. As mentioned in Cont and Voltchkova (2005), under the hypothesis of absence of arbitrage there exists a measure $\mathbb{Q}$ equivalent to $\mathbb{P}$ under which the discounted prices of all traded financial assets are $\mathbb{Q}$-martingales. In particular the discounted underlying $(e^{-rt}S_t)_{t\in[0,T]}$ is a martingale under $\mathbb{Q}$. A martingale is a sequence of a stochastic processes for which at any given time, the conditional expectation of the next value in the sequence is equal to the present value, regardless of all prior values.

Table 21.2 Moments of the $\Gamma(a, b)$ distribution.

	$\Gamma(a, b)$
Mean	$\frac{a}{b}$
Variance	$\frac{a}{b^2}$
Skewness	$2a^{\frac{1}{2}}$
Kurtosis	$3(1 + 2a^{-1})$

In exponential Lévy models the dynamics of S_t under $\mathbb{Q}$ is represented as the exponential of a Lévy process $S_t = S_0 e^{rt+X_t}$, where X_t is a Lévy process (under $\mathbb{Q}$) with characteristic triplet (σ, γ, ν), and the interest rate is given by r. As observed in Cont and Voltchkova (2005), the absence of arbitrage then imposes that $\hat{S}_t = S_t e^{-rt} = \exp X_t$ is a martingale.

21.2.3 Subordination of Lévy Processes

Subordination is a transformation of a stochastic process to a new stochastic process through random time change by increasing a stochastic process (subordinator) independent of the original process. The new process is a subordinate to the original one. The concept of subordination was introduced by Bochner in 1949 to model speculative price series, see Bochner (1949). In probability theory, a subordinator is an important concept that is related to stochastic processes. In order for a stochastic process to be a subordinator the process must be a Lévy process. A formal definition of a subordinator is as given below:

Definition 21.8 (Subordinator) A *subordinator* is a one-dimensional Lévy process that is nondecreasing a.s. As mentioned above, a subordinator can be thought of as a random model of time evolution. This is due to the fact that if $T = (T(t), t \geq 0)$ is a subordinator we have, $T(t) \geq 0$ a.s. for each $t > 0$ and $T(t_1) \leq T(t_2)$ a.s. whenever $t_1 \leq t_2$. Thus, a subordinator will determine the random number of time steps that occur within the subordinated process for a given unit of chronological time.

We proceed by stating the following theorem:

Theorem 21.2 If T is a subordinator, then its Lévy symbol takes the form

$$\eta(z) = ibz + \int_0^{\infty} (e^{izy} - 1)\lambda(dy), \tag{21.11}$$

where $b \geq 0$ and the Lévy measure λ satisfies the additional requirements

$$\lambda(-\infty, 0) = 0 \quad \text{and} \quad \int_0^{\infty} (y \wedge 1)\lambda(dy) < \infty.$$

Conversely, any mapping from $\mathbb{R}^d \rightarrow \mathbb{C}$ of the form (21.11) is the Lévy symbol of a subordinator. The function $y \wedge 1$ is defined as $y \wedge 1 = \inf\{1, y\}$.

The proof of Theorem 21.2 can be found in Bertoin (1996). The pair (b, λ) is the characteristic of the subordinator T.

Some classical examples of subordinators are the Poisson process, compound Poisson process (if and only if the $\chi_k, k = 1, 2, \ldots$ in Eq. (21.7) are all $\mathbb{R}^+$-valued) and the Gamma process. For more details of Lévy processes and subordinators, please refer to the following references: Bertoin (1996), Sato (1999), Schoutens (2003), Applebaum (2004), Schoutens and Cariboni (2009), Øksendal (2010a), and Tweneboah (2015).

In Section 21.2.4, we present some background of stable distributions.

21.2.4 Stable Distributions

Consider the sum of n independent identically distributed (i.i.d.) random variables x_i,

$$S_n = x_1 + x_2 + x_3 + \cdots + x_n = x(n\Delta t).$$

We observe that S_n can be regarded as the sum of n random variables or as the position of a single walker at time $t = n\Delta t$, where n is the number of steps performed and Δt the time required to perform one step. As the variables are independent, the sum can be obtained as the convolution, namely

$$P[x(2\Delta t)] = P(x_1) \otimes P(x_2)$$

or more generally,

$$P[x(n\Delta t)] = P(x_1) \otimes P(x_2) \cdots \otimes P(x_n).$$

Recall that for $n = 2$ if X_1 and X_2 are independent continuous random variables with pdfs $f_{X_1}(x_1)$ and $f_{X_2}(x_2)$ then if $Z = X_1 + X_2$, the pdf of Z is

$$f_Z(z) = \int_{-\infty}^{\infty} f_{X_1}(w) f_{X_2}(z - w)dw,$$

where we use $f_{Z,W}(z, w) = f_{X_1}(w) f_{X_2}(z - w)$ and integrating out w we obtain $f_Z(z)$, more details can be found in Casella and Berger 2002 (page 215).

We will say that the distribution is stable if the functional form of $P[x(n\Delta t)]$ is the same as the functional form of $P[x(\Delta t)]$. Specifically, given a random variable X, if we denote with Law(X) its PDF (for example, for a Gaussian random variable, we write Law(X) $= N(\mu, \sigma^2)$), then we will say that the random variable X is stable, or that it has a stable distribution if for any $n \geq 2$ there exists a positive number C_n and a number D_n so that

$$\text{Law}(X_1 + X_2 + \cdots + X_n) = \text{Law}(C_n X + D_n),$$

where $X_1, X_2, \ldots, X_n$ are independent random copies of X, this means that Law(X_i) = Law(X) for $i = 1, 2, \ldots, n$. If $D_n = 0$, then X is said to be a strictly stable variable. It can be shown (see Samorodnitsky and Taqqu 1994) that

$$C_n = n^{\frac{1}{\alpha}}$$

for some parameter α, $0 < \alpha \leq 2$. For example, if X is a Lorentzian random variable,

$$P(x) = \frac{\gamma}{\pi}\frac{1}{\gamma^2 + x^2}$$

and its characteristic function (that is, its Fourier transform) is given by

$$\varphi(q) = \int_{\mathrm{IR}} \exp(iqx)f(x)dx = E(\exp(iqx)),$$

where $f(x)$ is the PDF associated with the distribution $P(x)$. In this case, we obtain that

$$\varphi(q) = \exp(-\gamma|q|).$$

Now, if X_1, X_2 are two i.i.d. Lorentzian random variables, we have that

$$P[X(2\Delta t)] = P(X_1) \otimes P(X_2) = \frac{2\gamma}{\pi}\frac{1}{4\gamma^2 + x^2}.$$

Also, as the Fourier transform of a convolution is the product of the Fourier transforms, we obtain

$$\varphi_2(q) = \exp(-2\gamma|q|) = (\varphi(q))^2$$

and in general,

$$\varphi_n(q) = (\varphi(q))^n.$$

As another example, if X is a Gaussian random variable,

$$P(X) = \frac{1}{\sqrt{2\pi\sigma}} \exp\left(\frac{-x^2}{2\sigma^2}\right)$$

then its characteristic function is

$$\varphi(q) = \exp\left(-\frac{\sigma^2}{2}|q|^2) = \exp(-\gamma|q|^2\right),$$

where $\gamma = \frac{\sigma^2}{2}$, and again, we have that

$$\varphi_2(q) = (\varphi(q))^2.$$

By performing the inverse Fourier transform, we obtain that

$$P_2[X(2\Delta t)] = \frac{1}{\sqrt{8\pi\gamma}} \exp\left(\frac{-x^2}{8\gamma}\right) = \frac{1}{\sqrt{2\pi}(\sqrt{2}\sigma)} \exp\left(\frac{-x^2}{2(\sqrt{2}\sigma)^2}\right)$$

that is, the variance is now $\sigma_2^2 = 2\sigma^2$. Therefore, the two stable stochastic processes exist: Lorentzian and Gaussian, and in both cases, their Fourier transform has the form:

$$\varphi(q) = \exp(-\gamma|q|^{\alpha})$$

with $\alpha = 1$ for the Lorentzian, and $\alpha = 2$ for the Gaussian. Distributions with characteristic function

$$\varphi(q) = \exp(-\gamma |q|^{\alpha})$$

for $1 \leq \alpha \leq 2$ are stable.

In Section 21.3, we will see the form of all stable distributions also called Lévy flight models. This is a generalization of the Lévy processes.

21.3 Lévy Flight Models

Paul (1925) and Khintchine and Lévy (1936) discovered that the most general representation of a stable distribution is through the characteristic functions $\varphi(q)$. These characteristic functions are defined as follows:

$$\ln(\varphi(q)) = \mu q - \gamma \mid q \mid \left[1 + i\beta \frac{q}{\mid q \mid} \frac{2}{\pi} \log(q)\right], \quad \text{if } \alpha = 1 \tag{21.12}$$

and

$$\ln(\varphi(q)) = i\mu q - \gamma \mid q|^{\alpha} \left[1 - i\beta \frac{q}{\mid q \mid} \tan\left(\frac{\pi\alpha}{2}\right)\right], \quad \text{if } \alpha \neq 1, \tag{21.13}$$

where the scaling parameter of the Lévy: $0 < \alpha \leq 2$, γ is a positive scale factor ($\gamma^2 = \sigma^2$(variance)), μ is a real-valued constant known as the location parameter (or mean) and $-1 \leq \beta \leq 1$ is the skewness parameter.

Analytically, the stable Lévy distribution is only known for a few values of α and β. Consider the centered symmetric distribution ($\beta = 0$, $\mu = 0$), the characteristic function is

$$\varphi(q) = \exp(-\gamma \mid q|^{\alpha}). \tag{21.14}$$

Then, the stable Lévy distribution (Fourier transform of (21.14)) is defined as

$$P_L(x) = \frac{1}{\pi} \int_0^{\infty} \exp(-\gamma \mid q|^{\alpha}) \cos(qx) dq. \tag{21.15}$$

To avoid the complications that emerge in the infinite second moment (i.e. the fact that stable Lévy processes with $\alpha < 2$ have infinite variance), consider a stochastic process with finite variance that follows scale relations called truncated Lévy flight (TLF) (Mantegna and Stanley 1994).

The TLF distribution of a nonstable distribution with finite variance is defined by

$$T(x) = cP(x)\chi_{(-l,l)}(x), \tag{21.16}$$

where $P(x)$ is a symmetric Lévy distribution. However, depending on the magnitude of the cut-off length l, it relaxes convergence. For a smaller cut-off length l, convergence is faster which generates an instantaneous cut in its tails (Mantegna and Stanley 1994). To correct this problem, Koponen (1995) proposed a TLF with characteristic function shown in (21.17) below in which the cut function is a decreasing exponential function with parameter l.

$$\varphi(q) = \exp\left[c_0 - c_1 \frac{(q^2 + 1/l^2)^{\alpha/2}}{\cos(\pi\alpha/2)} \cos(\alpha \arctan(l \mid q \mid))\right] \tag{21.17}$$

and scale factors:

$$c_1 = \frac{2\pi \cos(\pi\alpha/2)}{\alpha\Gamma(\alpha)\sin(\pi\alpha)} At, \quad c_0 = \frac{2\pi}{\alpha\Gamma(\alpha)\sin(\pi\alpha)} Al^{-\alpha} t. \tag{21.18}$$

We obtain $T = N\Delta t$ if we discretize in time with steps Δt. Thus, we must calculate the sum of N stochastic variables that are independent identically distributed (i.i.d) at each interval.

Given a small sample of i.i.d random variables N, its probability distribution is very close to the stable Lévy distribution. The standardized model is given by

$$\ln \varphi_s(q) = \ln \varphi\left(\frac{q}{\sigma}\right) = c_0 - c_1 \frac{((q/\sigma)^2 + 1/l^2)^{\alpha/2}}{\cos(\pi\alpha/2)} \cos\left(\alpha \arctan\left(l \frac{\mid q \mid}{\sigma}\right)\right) \tag{21.19}$$

After substituting the values of c_1 and c_2 into (21.19), we obtain:

$$= \frac{2\pi A l^{-\alpha} t}{\alpha\Gamma(\alpha)\sin(\pi\alpha)} \left[1 - ((q/\sigma)^2 + 1)^{\alpha/2} \cos\left(\alpha \arctan\left(\frac{ql}{\sigma}\right)\right)\right] \tag{21.20}$$

if the variance equals 1. See Mariani and Liu (2007) for more details.

In Section 21.4, we present some techniques that help detect the presence of long-range correlations in nonstationary temporal series. We begin our discussion with the rescaled range analysis.

21.4 Rescaled Range Analysis (Hurst Analysis)

The idea of the Rescaled-Range analysis (R/S) was first presented by Hurst in the framework of his study on the long-run variations of the water level of the Nile river (Hurst 1951). It has become very popular since then and has been applied to a wide range of disciplines, including traffic analysis, bioengineering, physics, geology, biology, and geophysics (Graves et al., 2016).

The name H for the parameter derived from this technique was coined by Mandelbrot in tribute to the hydrologist Hurst and the mathematician Holder. The parameter H also known as index of dependence represents the relative trend of a time series and always lies between 0 and 1, it is equal to $\frac{1}{2}$ in the case of processes with independent increments. Of particular interest for our work is the case in which $0.5 < H < 1$ since it is an indicator of long-range correlations. Note that the case $H = 1/2$ represents the standard Brownian motion.

The numerical procedure to estimate the Hurst exponent by using the R/S analysis is presented next (for more details please see Shiryaev (2008) and references therein).

1. Let N be the length of the time series $(y_1, y_2, y_3 \dots, y_N)$. The logarithmic ratio of the time series is obtained. The length of the new time series $M(t)$ will be $N - 1$

   $$M(t) = \log\left(\frac{y_{t+1}}{y_t}\right)$$

   for $t = 1, 2, \dots, N - 1$.
2. The time series is then divided into m subseries of length n. n represents the number of elements in the series and m represents the number of subseries. Thus, $m \cdot n = N - 1$. Each subseries can be labeled as Q_a, where $a = 1, 2, \dots, m$ and each element in Q_a can be labeled as $L_{k,a}$ for $k = 1, 2, \dots, n$.
3. For each Q_a, the average value is calculated:

   $$Z_a = \frac{1}{n}\sum_{k=1}^{n} L_{k,a}.$$

4. The cumulative deviation in each Q_a is calculated:

   $$C_{k,a} = \sum_{j=1}^{k} (L_{j,a} - Z_a),$$

   $k = 1, 2, \dots, n$
5. Thus, the range of each subseries Q_a is given as follows:

   $$R(Q_a) = \max(C_{k,a}) - \min((C_{k,a})).$$

6. The standard deviation of each subseries Q_a is calculated as follows:

   $$S(Q_a) = \sqrt{\left(\frac{1}{n}\sum_{j=1}^{n}(L_{j,a} - Z_a)^2\right)}.$$

7. Each subseries is normalized by dividing the range, $R(Q_a)$ by the standard deviation, $S(Q_a)$. The average value of R/S for sub-series of length n is obtained by

   $$(R/S)_n = \frac{1}{m}\sum_{a=1}^{m}\frac{R(Q_a)}{S(Q_a)}.$$

8. Steps 2 through 7 are repeated for all possible values of n, thus obtaining the corresponding R/S values for each n. The relationship between length of the subseries, n and the rescaled range R/S is

   $$(R/S) = (cn)^H,$$

   where R/S is the rescaled range, n is the length of the subseries of the time series and H is the Hurst exponent. Taking logarithms yields

   $$\log(R/S) = H\log c + H\log n.$$

9. An ordinary least squares regression is performed using $\log(R/S)$ as a dependent variable and $\log(n)$ as an independent variable. The slope of the equation is the estimate of the Hurst exponent H.

If the Hurst exponent H for the investigated time series is 0.5, then it implies that the time series follows a random walk which is a process with independent increments, a value of $H = 0.5$ can indicate a completely uncorrelated series. For data series with long memory effects, H would lie between 0.5 and 1. It suggests all the elements of the observation are dependent. This means that what happens now would have an impact on the future. Time series that exhibit this property are called persistent time series, and this character enables prediction of any time series as it shows a trend. A high value in the series will probably be followed by another high value and that the future values for a long time will tend to be high. Lastly, if H lies between 0 and 0.5, it implies that the time-series possess anti-persistent behavior (negative autocorrelation), i.e. it is likely to reverse trend over the time frame considered. A high value may be followed by a low value and so on, and this tendency to switch between high and low values will continue for a long time.

We conclude this section with a short introduction to Fractional Brownian motion (FBM) which is a generalization of the BM and has some relationships with the Hurst index parameter H. The FBM is defined as follows:

Definition 21.9 (Fractional Brownian motion (FBM)) Let $B(t)$ be the ordinary Brownian motion, and Hurst index parameter H satisfy $0 < H < 1$. Then the FBM of the exponent H is a moving average of $dB(t)$ in which past increments of $B(t)$ are weighted by the kernel $(t - s)^{H-1/2}$.

Fractional Brownian motion provide useful models for several natural time series and below are some applications of FBMs. It has some applications in economics. Economic time series normally exhibit cycles of all orders of magnitude; the slowest cycles have periods of duration comparable to the total sample size. Another class of examples arises in the study of fluctuations in solids. Many such fluctuations are called $1/f$ noises because their sample spectral density takes the form λ^{1-2H}, where λ is the frequency and H is a number always satisfying $1/2 < H < 1$, and often close to 1 (see (Mandelbrot and Ness 1968). Other applications are in hydrology, where phenomena normally exhibit extremely long interdependencies.

There are numerous results that suggest that the price dynamics of financial products are consistent with FBM volatility models with Hurst index $H > 1/2$, which implies that the volatility has a long memory (see (Bollerslev and Mikkelsen 1996). Long memory refers to the slow decay of the autocorrelation function as a power-law with exponent less than 1. When $H = 1/2$, the FBM is the standard BM, which is a Markov process with short memory. A Markov is a random process whose future probabilities are determined by its most recent values. If $H < 1/2$, we have a power law decay (the increments of the process are negatively correlated).

21.5 Detrended Fluctuation Analysis (DFA)

The DFA method is an important technique in revealing long-range correlations in nonstationary time series. This method was developed by Peng et al. (1993), Peng et al. (1994), and has been successfully applied to the study of cloud breaking Ivanova and Ausloos (1999), DNA Peng et al. (1994), Buldyrev et al. (1995), Peng et al. (1995), cardiac dynamics Peng et al. (1993), Peng et al. (1995), climatic studies Koscienly-Bunde et al. (1998b), Koscienly-Bunde et al. (1998a), solid-state physics Kantelhardt et al. (1999), Vandewalle et al. (1999), and economic time series Liu et al. (1997), Cizeau et al. (1997), and Ausloos and Ivanova (2000). The advantages of DFA over conventional methods are that it permits the detection of intrinsic self-similarity embedded in a seemingly nonstationary time series (i.e. time series whose means, variances, and covariances change over time), and also avoids the spurious detection of apparent self-similarity, which may be an artifact of extrinsic trends. In general, self-similarity occurs when the shape of an object is similar, or approximately similar to the part of itself. That is each portion of the self-similar object can be considered as a reduced scale of the whole.

The numerical procedure to estimate the DFA exponent is presented below.

Let N be the length of time series $(y_1, y_2, y_3, \ldots, y_N)$. The logarithmic ratio of the time series is obtained and the length of the new time series $M(t)$ is $N-1$.

$$M(t) = \log\left(\frac{y_{t+1}}{y_t}\right), \quad t = 1, 2, \ldots, N-1. \tag{21.21}$$

The absolute value of $M(t)$ is integrated:

$$y(t) = \sum_{i=1}^{t} \mid M(i) \mid . \tag{21.22}$$

The integrated time series of length N is divided into m boxes of equal length n with no intersection between them. As the data is divided into equal length intervals, there may be some left over at the end. To take account of these leftover values, the same procedure is repeated but beginning from the end, obtaining $2N/n$ boxes.

A least square line is fitted to each box, representing the trend in each box, thus obtaining $y_n(t)$. Finally, the root mean square fluctuation (RMSF) is calculated using the formula

$$F(n) = \sqrt{\frac{1}{2N}\sum_{t=1}^{2N}[y(t) - y_n(t)]^2}. \tag{21.23}$$

This computation is repeated over all box sizes to characterize a relationship between the box size n and $F(n)$. A linear relationship between the $F(n)$ and n (i.e. box size) in a log–log plot reveals that the fluctuations can be characterized by a scaling exponent H, the slope of the line relating $\log F(n)$ to $\log n$. This generates the mathematical (power-law) relation:

$$F(n) \propto n^H. \tag{21.24}$$

For data series with no correlations or short-range correlation, α is expected to be 0.5. For data series with long-range power law correlations, α would lie between 0.5 and 1 and for power law anti-correlations α would lie between 0 and 0.5. This method was used to measure correlations in financial series of high frequencies and in the daily evolution of some of the most relevant indices.

Next, we discuss the DEA.

21.6 Diffusion Entropy Analysis (DEA)

The DEA can be used to analyze and detect the scaling properties of low- and high-frequency time series. Using the DEA, one can determine if the characterization of a time series follows a Gaussian or Lévy distribution, as well as establish the existence of long-range correlations in the time series.

DEA is a PDF scaling method which perceives the numbers in a time series as the trajectory of a diffusion process (Huang et al. 2012). The scaling property for the stationary time series takes the form

$$p(x,t) = \frac{1}{t^{\delta}} F\left(\frac{x}{t^{\delta}}\right), \tag{21.25}$$

where x denotes the diffusion variable, $p(x,t)$ is its PDF at time t, and $0 < \delta < 1$ is the scaling exponent.

The scaling property for the nonstationary time series takes the form

$$p(x,t) = \frac{1}{t^{\delta(t)}} F\left(\frac{x}{t^{\delta(t)}}\right), \tag{21.26}$$

where $\delta(t) = \delta_0 + \eta \log(t)$. As derived in Scafetta (2011); Scafetta and Grigolini (2002), a diffusion process generated by Lévy walk is characterized by the following relation:

$$\delta = \frac{1}{3 - 2(H,\alpha)}. \tag{21.27}$$

If $\delta = (H,\alpha)$, the time series can be characterized by FBM, since the variance methods are based subtly on the Gaussian assumption. However if $\delta \neq (H,\alpha)$, and the Eq. (21.27) holds true, the noise can be characterized by Lévy statistics. (H,α) in Eq. (21.27) refers to the scaling exponent derived from the variance scaling methods. We recall that Lévy walk is a mathematical construction for describing random patterns of movement with unusual fractal properties that seem to have no place in biology. Movement patterns resembling Lévy walks have been observed at scales ranging from the microscopic to the ecological in the molecular machinery operating within cells during intracellular trafficking, in the movement patterns of T cells within the brain, in DNA, and in some insects, fish, birds and mammals.

21.6.1 Estimation Procedure

In this section, we describe the estimation technique for the scaling exponent, δ. We first present a brief background on the Shannon Entropy that is used for estimating δ.

21.6.1.1 The Shannon Entropy

The concept of entropy was developed by Rudolph Clausius in 1865, a few years after he stated the laws of thermodynamics. The entropy is an indicator of the lack of information about the measure of an event that occurs with propability p. Other types of entropies are the Kolmogorov–Sinai entropy, the Renyi entropy, and the Tsallis entropy (Scafetta 2011, Scafetta and Grigolini 2002). The Shannon entropy (named by Shannon) measures information of a probability distribution as follows:

$$S(t) = -\sum_{1}^{N} p_i \log p_i. \tag{21.28}$$

The summation is replaced by the integral in the case of continuous probability distributions. The above equation is used to derive the log equation that will be used to determine the DEA δ scaling. See below the technique for estimating δ:

- The time series data is first transformed into a diffusion process.
- Shannon's entropy of the diffusion process is calculated. A log-linear equation or log-quadratic equation is derived from the Shannon entropy by substituting Eqs. (21.25) and (21.26), respectively. Simplifying the result from the substitutions, we have the following relation for stationary time series:

  $$S(t) = A + \delta \ln(t). \tag{21.29}$$

  For the nonstationary series, the relation is as follows:

  $$S(t) = A + \delta(t)\tau, \tag{21.30}$$

  where $\delta(t) = \delta_0 + \eta \log(t)$ and $\tau = \log(t)$ with $\eta \log(t) < 1 - \delta_0$. After some simplifications, Eq. (21.30) becomes

  $$S(t) = A + (\delta_0 - K)\log(t) + (1 - \delta_0)(\log(t))^2, \tag{21.31}$$

  where $K < 0$ and $\delta_0 \equiv \delta$ from the stationary PDF. Thus, by fitting a log-quadratic model in the nonstationary series and a log-linear model in the stationary series, we are able to determine the δ (δ_0) scaling. At $t = 1$, it is clear that the constant A in both Eqs. (21.29) and (21.30) is given by $S(1)$.

  Thus, δ (or δ_0) is derived by an estimation of the slope of the above linear-log equation or by the coefficients from the quadratic-log equation. For details of the algorithm used when transforming the series into a diffusion process, we refer the reader to Scafetta (2011).

The DEA technique is able to determine the correct scaling exponent even when the statistical properties of the time series, as well as the dynamic properties, are anomalous.

In Section 21.6.2, we discuss some relationship between the Hurst exponent H and the α obtained from the TLF. We begin this discussion by first defining self-similarity.

Definition 21.10 A stochastic process is said to be *self-similar* if there exists a constant $H > 0$ such that for any scaling factor $a > 0$, the process $\{X_{at}\}_{t\geq 0}$ and $\{a^H X_t\}_{t\geq 0}$ have the same law in the sense of finite dimensional distributions. The constant H is called the self-similarity exponent or the Hurst exponent of the process X.

As observed in Cont and Tankov (2004), from the definition we obtain that for any c, t positive, X_{ct} and $c^H X_t$ have the same distribution. Choosing $c = 1/t$ yields that $X_t = t^H X_1$ in distribution, for any positive t. Therefore,

$$F_t(x) = P(X_t \leq x) = P(t^h X_1 \leq x) = F_1\left(\frac{x}{t^H}\right).$$

If F_t has a density ρ_t, that is, if ρ_t denotes the density or distribution function of X_t, then we obtain, by differentiating the previous formula that the Definition 21.10 implies that

$$\rho_t(x) = \frac{1}{t^H}\rho_1\left(\frac{x}{t^H}\right). \tag{21.32}$$

If $\Phi_t(q)$ is the characteristic function of X_t and furthermore $\Phi_t(q)$ is an even function, then

$$\rho_t(x) = \frac{1}{\pi}\int_0^\infty \Phi_t(q)\cos(qx)dq \tag{21.33}$$

and

$$\rho_1\left(\frac{x}{t^H}\right) = \frac{1}{\pi}\int_0^\infty \Phi_1(q)\cos\left(q\frac{x}{t^H}\right)dq. \tag{21.34}$$

Thus, in that case (21.32) gives

$$\int_0^\infty \Phi_t(q)\cos(qx)dq = \frac{1}{t^H}\int_0^\infty \Phi_1(q)\cos\left(q\frac{x}{t^H}\right)dq. \tag{21.35}$$

The Lévy process (where the increments are independent and follow the Lévy distribution) is self-similar. For this process, $\Phi_t(q) = \exp(-t|q|^\alpha)$, and we obtain from (21.35) that

$$\int_0^\infty \exp(-t|q|^\alpha)\cos(qx)dq = \frac{1}{t^H}\int_0^\infty \exp(-|q|^\alpha)\cos\left(\frac{qx}{t^H}\right)dq$$

by using the change of variable: $q = t^H y$ the second integral becomes

$$\int_0^\infty \exp(-t^{\alpha H}|q|^\alpha)\cos(qx)dq$$

so the equality holds only if

$$\alpha = \frac{1}{H}.$$

This result is well known and has been used widely in literature.

21.6.2 The $H-\alpha$ Relationship for the Truncated Lévy Flight

We now consider the standardized truncated Lévy model discussed in Section 21.3. This model is not self-similar due to truncation and standardization. However, we would like to find out how close this model is from a self-similar model. We will do this by looking at the relationship between the α characterizing the *TLF* and the resulting H parameter characterizing the self-similar property.

We recall that the standardized *TLF* model is

$$\Phi_t(q) = \varphi_S(q) = \varphi\left(\frac{q}{\sigma}\right)$$
$$= \exp\left[G(\alpha)t\left[1-\left(\left(\frac{q}{\sigma}\right)^2+1\right)^{\frac{\alpha}{2}}\cos\left(\alpha\arctan\left(\frac{ql}{\sigma}\right)\right)\right]\right], \quad (21.36)$$

where

$$G(\alpha) = \frac{2\pi A l^{-\alpha}}{\alpha\Gamma(\alpha)\sin(\pi\alpha)}. \quad (21.37)$$

Therefore,

$$\rho_t(x) = \frac{1}{\pi}\int_0^\infty \exp\left[G(\alpha)t\left[1-\left(\left(\frac{q}{\sigma}\right)^2+1\right)^{\frac{\alpha}{2}}\right.\right.$$
$$\left.\left.\times\cos\left(\alpha\arctan\left(\frac{ql}{\sigma}\right)\right)\right]\right]\cos(qx)dq. \quad (21.38)$$

Observe that when $\frac{q}{\sigma}$ is large enough (in other words q is large enough) $\cos(\alpha\arctan(\frac{ql}{\sigma})) \approx \cos(\alpha\frac{\pi}{2})$ because $\arctan(\infty) = \frac{\pi}{2}$ and $((\frac{q}{\sigma})^2+1)^{\frac{\alpha}{2}} \approx (\frac{q}{\sigma})^\alpha$. Choose B large enough so that when $q > \frac{B}{t^{1/\alpha}}$,

$$1-\left(\left(\frac{q}{\sigma}\right)^2+1\right)^{\frac{\alpha}{2}}\cos\left(\alpha\arctan\left(\frac{ql}{\sigma}\right)\right) \approx -\left(\frac{q}{\sigma}\right)^\alpha\cos\left(\alpha\frac{\pi}{2}\right).$$

Then

$$\rho_t(x) = \frac{1}{\pi}\int_0^{\frac{B}{t^{1/\alpha}}} \exp\left[G(\alpha)t\left[1-\left(\left(\frac{q}{\sigma}\right)^2+1\right)^{\frac{\alpha}{2}}\right.\right.$$

$$\times \cos\left(\alpha \arctan\left(\frac{ql}{\sigma}\right)\right)\Bigg]\Bigg] \cos(qx)dq$$

$$+\frac{1}{\pi}\int_{\frac{B}{t^{1/\alpha}}}^{\infty} \exp\left[G(\alpha)t\left[1-\left(\left(\frac{q}{\sigma}\right)^2+1\right)^{\frac{\alpha}{2}}\right.\right.$$

$$\times \cos\left(\alpha \arctan\left(\frac{ql}{\sigma}\right)\right)\Bigg]\Bigg] \cos(qx)dq$$

$$\approx I_0^{\frac{B}{t^{1/\alpha}}}(t,x) + \frac{1}{\pi}\int_{\frac{B}{t^{1/\alpha}}}^{\infty} \exp\left[-G(\alpha)t\cos\left(\alpha\frac{\pi}{2}\right)\left(\frac{q}{\sigma}\right)^{\alpha}\right]\cos(qx)dq$$

$$= I_0^{\frac{B}{t^{1/\alpha}}}(t,x) + \frac{1}{t^{1/\alpha}}\frac{1}{\pi}\int_{B}^{\infty} \exp\left[-G(\alpha)\cos\left(\alpha\frac{\pi}{2}\right)\left(\frac{u}{\sigma}\right)^{\alpha}\right]$$

$$\times \cos\left(u\frac{x}{t^{1/\alpha}}\right)du, \tag{21.39}$$

where

$$I_0^{\frac{B}{t^{1/\alpha}}}(t,x) = \frac{1}{\pi}\int_{0}^{\frac{B}{t^{1/\alpha}}} \exp\left[G(\alpha)t\left[1-\left(\left(\frac{q}{\sigma}\right)^2+1\right)^{\frac{\alpha}{2}}\right.\right.$$

$$\times \cos\left(\alpha \arctan\left(\frac{ql}{\sigma}\right)\right)\Bigg]\Bigg] \cos(qx)dq \tag{21.40}$$

and the change of variable $q = \frac{u}{t^{\frac{1}{\alpha}}}$ was done in the second term to obtain the last equality.

So we obtain the formula:

$$\rho_t(x) = I_0^{\frac{B}{t^{1/\alpha}}}(t,x) + \frac{1}{t^{1/\alpha}}\frac{1}{\pi}\int_{B}^{\infty} \exp\left[-G(\alpha)\cos\left(\alpha\frac{\pi}{2}\right)\left(\frac{u}{\sigma}\right)^{\alpha}\right]\cos\left(u\frac{x}{t^{1/\alpha}}\right)du$$

Taking $t = 1$ in the last formula, we obtain

$$\rho_1(x) \approx I_0^B(1,x) + \frac{1}{\pi}\int_{B}^{\infty} \exp\left[-G(\alpha)\cos\left(\alpha\frac{\pi}{2}\right)\left(\frac{u}{\sigma}\right)^{\alpha}\right]\cos(ux)du.$$

Thus, in the last formula, evaluating in $\frac{x}{t^{\frac{1}{\alpha}}}$ and multiplying by $t^{\frac{1}{\alpha}}$ both sides we have

$$\frac{1}{t^{1/\alpha}}\rho_1\left(\frac{x}{t^{1/\alpha}}\right)$$

$$\approx \frac{1}{t^{1/\alpha}}I_0^B\left(1,\frac{x}{t^{1/\alpha}}\right) + \frac{1}{t^{1/\alpha}}\frac{1}{\pi}\int_{B}^{\infty} \exp\left[-G(\alpha)\cos\left(\alpha\frac{\pi}{2}\right)\left(\frac{u}{\sigma}\right)^{\alpha}\right]$$

$$\times \cos\left(u\frac{x}{t^{1/\alpha}}\right)du.$$

Therefore,

$$\rho_t(x) - \frac{1}{t^{1/\alpha}}\rho_1\left(\frac{x}{t^{1/\alpha}}\right) \approx I_0^{\frac{B}{t^{1/\alpha}}}(t,x) - \frac{1}{t^{1/\alpha}}I_0^B\left(1,\frac{x}{t^{1/\alpha}}\right). \tag{21.41}$$

So when the quantity $\left| I_0^{\frac{B}{t^{1/\alpha}}}(t,x) - \frac{1}{t^{1/\alpha}} I_0^B(1, \frac{x}{t^{1/\alpha}}) \right|$ becomes small, the standardized truncated Lévy model tends to show the self-similar structure with corresponding Hurst exponent H approximately equals to $\frac{1}{\alpha}$. The choice of B in (21.41) depends on the numerical computation.

Remarks 21.2 For the case of the present analysis, empirical evidence presented in Barany et al. (2011) shows that when $T = 1$, $H \approx \frac{1}{\alpha}$. In general because of the truncation and standardization of the Lévy model, we should not expect $H = \frac{1}{\alpha}$.

Please refer to Barany et al. (2011) for more details of the H–α relationship for the TLF model.

21.7 Application – Characterization of Volcanic Time Series

In this application, we consider several geophysical time series data and analyze their long-range correlations using R/S analysis, the DFA, and the DEA. The continuous time-varying Lévy process is effective for capturing the stochastic volatility (SV) and fat tails of data distribution. It is known that the volatilities of high-frequency data are correlated, and they vary stochastically over time. We seek to determine the characterization of the time series data (i.e. whether it follows the Gaussian or Lévy distribution) by comparing the relation between the scaling exponent derived with the R/S and DFA against that of the DEA.

21.7.1 Background of Volcanic Data

The volcanic data used was recorded by seismic stations belonging to the Bezymianny Volcano Campaign Seismic Network (PIRE). Data was requested for 10 days before and 5 days after the published time of the volcanic eruptions. The seismic stations used were BEZB and BELO. Volcanic eruptions 1 and 2 were from BEZB and Volcanic eruptions 3–8 were from BELO.

21.7.2 Results

Next, we explain the analysis of the volcanic time series when the models are applied to the data sets. Table 21.3 shows the scaling exponents derived from applying the three scaling methods. The δ, H, and α exponents are used to obtain $\delta_{\text{Levy}}(R/S)$ and $\delta_{\text{Levy}}(\text{DFA})$. The Hurst analysis of volcanic eruptions 1 and 2 are shown in Figure 21.1. The slope of the best straight line fitted on the

Table 21.3 Scaling exponents of Volcanic Data time series.

Eruption number	$R/S(H)$	DFA (α)	DEA (δ)	$\delta_{Levy}(R/S)$	δ_{Levy} (DFA)
1	0.45	0.74	0.6837	0.4756	0.6547
2	0.51	0.92	0.6837	0.5093	0.8682
3	0.38	0.85	0.6837	0.4472	0.7636
4	0.39	0.66	0.6837	0.4509	0.5957
5	0.39	0.76	0.6837	0.4513	0.6729
6	0.37	0.67	0.6837	0.4433	0.6002
7	0.42	0.81	0.6837	0.4634	0.7194
8	0.504	0.75	0.6837	0.5018	0.6684

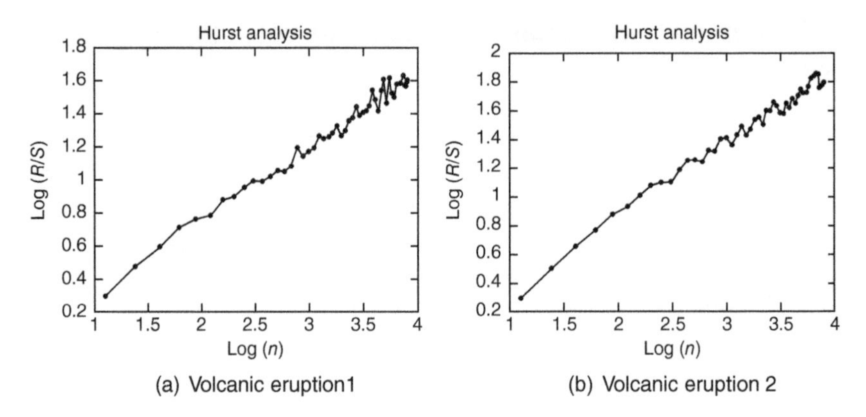

Figure 21.1 R/S for volcanic eruptions 1 and 2.

logarithmic plot of rescaled range (R/S) versus time is the Hurst exponent (see Table 21.3). Figure 21.2 summarizes the DFA analysis of volcanic eruptions 1 and 2, showing the linear trend when plotting (n) and $F(n)$ on a log–log scale. A linear relationship on a double log graph indicates that there is a scaling or self-similarity in the graph, and the fluctuations can be characterized by scaling exponent. Table 21.3 and Figure 21.2 shows that the scaling exponent (α) is less than 1, which confirms the presence of long-range correlations, that is, the large values are likely to be followed by large values and vice versa. So the DFA allows us to study the correlations in data, without disturbance of seasonality or trend. In Figure 21.3, we observe that there is a considerable difference between the DEA analysis of volcanic eruptions data. In addition, $S(t) - S(1)$ of the volcanic eruption data is increased almost exponentially with the logarithm of time scale.

Based on the results obtained, the R/S analysis is unable to correctly detect the existence of long-range correlations since the volcanic data is nonstationary. However, the DEA and DFA correctly detect long-range correlations. Equation

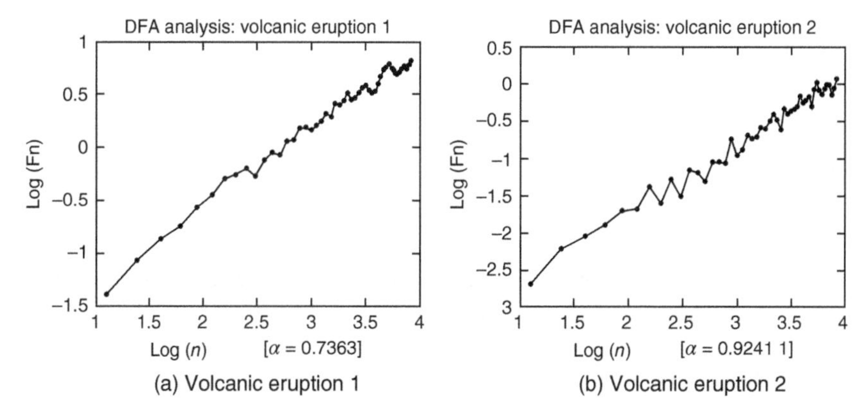

Figure 21.2 DFA for volcanic eruptions 1 and 2.

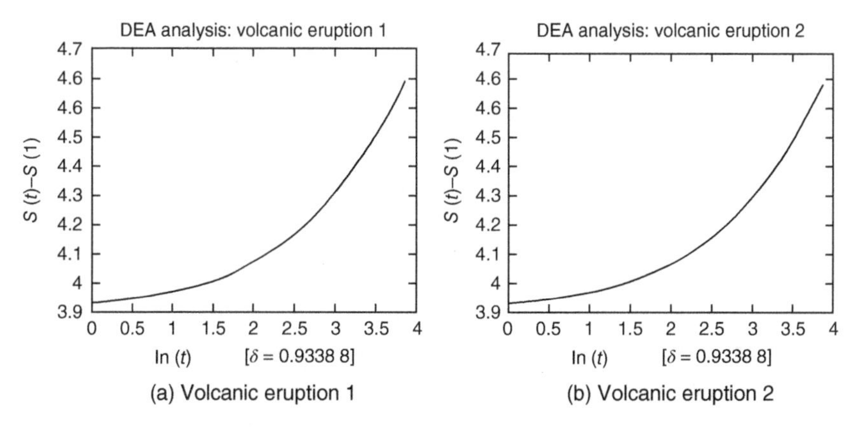

Figure 21.3 DEA for volcanic eruptions 1 and 2.

(21.27) is however not satisfied and clearly $\delta \neq (H, \alpha)$. Hence, the volcanic series can neither be characterized by FBM nor Lévy walk. The volcanic time series is thus characterized by a Lévy flight (i.e. it has an infinite variance). Lévy flights and Lévy walks serve as two paradigms of random walks. One of the main difference between the two is the discontinuity versus continuity of their trajectories and infinite versus finite propagation velocity.

21.8 Problems

1 Discuss examples of stochastic processes that are not Lévy processes.

2 Discuss examples of stochastic processes that are Lévy processes.

3 Discuss examples of Lévy processes that are subordinators.

4 Discuss examples of Lévy processes that are not subordinators.

5 Prove directly that, for each $t \geq 0$,

$$\mathbb{E}(e^{-zT(t)}) = \int_0^\infty e^{-zs} f_{T(t)}(s)ds = e^{-tz^{1/2}},$$

where $(T(t), t \geq 0)$ is the Lévy subordinator.

6 Present examples of exponential Lévy models that have infinite number of jumps in every compact interval.

7 Discuss and compare the DFA and Rescaled Range Analysis.

8 Discuss why the DEA technique is able to detect the scaling exponent of a time series when statistical properties of that particular time series are irregular.

9 What is a unit root test and what is its effects.

10 Discuss the relations between the DFA, Hurst analysis, and the DEA.

11 Implement the Hurst analysis and DFA algorithm discussed in Sections 21.4 and 21.5, respectively, to detect long-range correlations in the 1997 Asian financial crisis. Please download the historical data (02 July 1997 to 25 October 2001) from Yahoo finance. Please use any equity of your choice.

12 Repeat Problem 11, using data corresponding to the Lehman Brothers Collapse. Please download the historical data (01 September 2001 to 30 September 2008) from Yahoo finance. Please use any equity of your choice.

13 Download the monthly and daily equity data for the past five years at Yahoo finance. Please use any equity of your choice. For each time frequency, calculate a sequence of continuously compounded returns. Apply R/S range analysis and obtain an estimate for the H parameter. What do you observe about the estimate when the sampling frequency increases. Repeat the analysis for the DFA estimator.

22

Stochastic Differential Equations

22.1 Introduction

A stochastic differential equation (SDE) is a differential equation (DE) in which one or more of the terms are a stochastic process, resulting in a solution which is also a stochastic process. Stochastic differential equations (SDEs) are used to model various phenomena such as stock prices or physical systems subject to thermal fluctuations. SDEs can also be used to model dynamic phenomena, where the exact dynamics of the system are uncertain. Typically, SDEs contain a variable which represents random white noise calculated as the derivative of Brownian motion or the Wiener process. However, other types of random behavior are possible, such as jump processes. In this chapter, we also present an overview of numerical approximation methods for SDEs. For a complete introduction and study of SDEs, we refer to the reader to Øksendal (2010) and Tweneboah (2020).

22.2 Stochastic Differential Equations

The theory of DEs is the origin of classical calculus and motivated the creation of differential and integral calculus. A DE is an equation involving an unknown function and its derivative. Typically, a DE is a functional relationship

$$f(t, x(t), x'(t), x''(t), \dots) = 0, \quad 0 \leq t \leq T \tag{22.1}$$

involving the time t, an unknown function $x(t)$ and its derivative. The solution of the DE is to find a function $x(t)$ which satisfies Eq. (22.1).

Now consider the deterministic DE:

$$dx(t) = a(t, x(t))dt, \quad x(0) = x_0. \tag{22.2}$$

A simple way to introduce randomness in this equation is to randomize the initial condition. The solution $x(t)$ then becomes a stochastic process $(X_t, t \in [0, T])$

Data Science in Theory and Practice: Techniques for Big Data Analytics and Complex Data Sets, First Edition. Maria Cristina Mariani, Osei Kofi Tweneboah, and Maria Pia Beccar-Varela.

defined as:

$$dX_t = a(t, X_t)dt, \quad X_0(\omega) = Y(\omega). \tag{22.3}$$

Equation (22.3) is called a random DE. Random DEs can be considered as a deterministic equations with a perturbed initial condition. Note this is not a full SDE.

Definition 22.1 A SDE is defined as a deterministic DE which is perturbed by random noise.

In general, an SDE is given as

$$dX(t, \omega) = f(t, X(t, \omega))dt + g(t, X(t, \omega))dW(t, \omega), \tag{22.4}$$

where ω denotes that $X = X(t, \omega)$ is a random variable and possesses the initial condition $X(0, \omega) = X_0$ with probability 1.

As an example, we have

$$dY(t, \omega) = \mu(t)dt + \sigma(t)dW(t, \omega),$$

where $dW(t, \omega)$ is the white noise process (a process that does not exist). White noise can be thought of as the derivative of a Brownian motion. We recall that a Brownian motion is a continuous stationary stochastic process $W(t)$ having independent increments and for each t we have that $W(t)$ is a Gaussian random variable with mean 0 and variance t. We recall that stationary increments means that for any $0 < s, t < 1$ the distribution of the increment $W_{t+s} - W_s$ has the same distribution as $W_t - W_0 = W_t$; we also recall that a stochastic process with stationary, independent increments is called a Lévy process. The standard Wiener process is the intersection of the class of Gaussian processes with the Lévy processes.

22.2.1 Solution Methods of SDEs

In this section, we present some necessary steps in order to find an analytic solution for a class of SDEs that arise in population models. This method can also be applied to other models.

Consider the following nonlinear SDE for the stochastic process $X = \{X_t : t \in T\}$:

$$dX_t = rX_t(k - X_t^m)dt + \beta X_t \, dB_t, \tag{22.5}$$

where the constant $k > 0$ is called the carrying capacity of the environment, the constant $r \in R$ is a measure of the quality of the environment and the constant $\beta \in R$ is a measure of the size of the noise in the system. Equation (22.5) arises in modeling the growth of a population of size X_t in a stochastic, crowded, environment. We will present our analysis in five parts.

Part 1:

It is possible to find an integrating factor F_t for Eq. (22.5). Indeed, the integrating factor F_t is given by:

$$F_t = \exp\left(-\beta B_t + \frac{1}{2}\beta^2 t\right).$$

More generally, we have the following result:

Theorem 22.1 Consider the following SDE

$$dX_t = f(t, X_t)dt + c(t)X_t\, dB_t,$$

where

$$f : \mathrm{I\,R} \times \mathrm{I\,R} \rightarrow \mathrm{I\,R}$$

is a continuous deterministic function, $f(t,x)$ and $c : \mathrm{I\,R} \rightarrow \mathrm{I\,R}$ is also a continuous deterministic function. Then

$$F_t = \exp\left(-\int_0^t c(s)dB_s + \frac{1}{2}\int_0^t c(s)^2 ds\right)$$

is an integrating factor for the SDE, that transform it into a deterministic DE:

$$d(F_t X_t) = F_t f(t, X_t)dt.$$

Before we present the proof of Theorem 22.1, we state the following remark.

Remarks 22.1 In the case of Eq. (22.5), $c(s) = \beta$ constant therefore,

$$F_t = \exp\left(-\int_0^t \beta\, dB_s + \frac{1}{2}\int_0^t \beta^2\, ds\right) = \exp\left(-\beta B_t + \frac{1}{2}\beta^2 t\right).$$

Proof: We will split the proof in four steps

In the first step, we present the following generalization of the integration by parts for stochastic integrals of a deterministic function:

Theorem 22.2

$$\int_0^t c(s)dB_s = c(t)B_t - \int_0^t c'(s)B_s\, ds$$

Proof: In order to prove Theorem 22.2, we can use the one dimensional Itô's formula which is presented in the theorem below.

Theorem 22.3 One dimensional Itô's formula.

If Z_t is an Itô's process given by

$$dZ_t = \mu\, dt + v\, dB_t$$

and $Y_t = g(t, Z_t)$, where g is a C^2 function, then Y_t is also an Itô's process and

$$dY_t = \frac{\partial g}{\partial t}(t, Z_t)dt + \frac{\partial g}{\partial x}(t, Z_t)dZ_t + \frac{1}{2}\frac{\partial^2 g}{\partial x^2}(t, Z_t)(dZ_t)^2.$$

In this case, we can take

$$Z_t = c(t)B_t \tag{22.6}$$

and

$$g(t,x) = c(t)x \tag{22.7}$$

then differentiating Eq. (22.7) partially with respect to t, we obtain:

$$\frac{\partial g}{\partial t} = c'(t)x$$

and similarly, differentiating Eq. (22.7) partially with respect to x, we get:

$$\frac{\partial g}{\partial x} = c(t) \tag{22.8}$$

and finally the second derivative of Eq. (22.8) with respect to x is:

$$\frac{\partial^2 g}{\partial x^2} = 0 \tag{22.9}$$

therefore

$$dZ_t = d(c(t)B_t) = c'(t)B_t\, dt + c(t)dB_t. \tag{22.10}$$

Integrating Eq. (22.10), we obtain:

$$\int_0^t \frac{d(c(s)B_s)}{ds} ds = \int_0^t c'(s)B_s\, ds + \int_0^t c(s)dB_s$$

now, using that $\int_0^t \frac{d(c(s)B_s)}{ds} ds = c(t)B_t - c(0)B_0 = c(t)B_t$ because $B_0 = 0$ we obtain:

$$c(t)B_t = \int_0^t c'(s)B_s\, ds + \int_0^t c(s)dB_s$$

or equivalently,

$$\int_0^t c(s)dB_s = c(t)B_t - \int_0^t c'(s)B_s\, ds$$

Now, in the case of our integrating factor, we have that:

$$\begin{aligned} F_t &= \exp\left(-\int_0^t c(s)dB_s + \frac{1}{2}\int_0^t c(s)^2 ds\right) \\ &= \exp\left(-c(t)B_t + \int_0^t c'(s)B_s\, ds + \frac{1}{2}\int_0^t c(s)^2 ds\right). \end{aligned}$$

For the second step in order to compute dF_t, we will use again the one dimensional Itô's formula above. In this case,

$$g(t,x) = \exp\left(-c(t)x + \int_0^t c'(s)x(s)ds + \frac{1}{2}\int_0^t c(s)^2 ds\right). \tag{22.11}$$

Therefore, we obtain the partial derivatives of $g(t, x)$ in Eq. (22.11) as follows:

$$\frac{\partial g}{\partial t} = g\left(-c'(t)x + c'(t)x + \frac{1}{2}c(t)^2\right) = \frac{1}{2}c(t)^2 g,$$

$$\frac{\partial g}{\partial x} = g(-c(t)),$$

and

$$\frac{\partial^2 g}{\partial x^2} = g(-c(t))(-c(t)) = g(c(t))^2.$$

Then

$$dF_t = \frac{\partial g}{\partial t}dt + \frac{\partial g}{\partial x}dB_t + \frac{1}{2}\frac{\partial^2 g}{\partial x^2}(dB_t)^2$$
$$= g\frac{1}{2}c(t)^2 dt - gc(t)dB_t + g\frac{1}{2}c(t)^2 dt = gc(t)^2 dt - gc(t)dB_t = F_t c(t)^2 dt - F_t c(t)dB_t,$$

where we use the fact that $(dB_t)^2 = dt$

In the third step, to compute $d(F_t X_t)$ we need the following result in Theorem 22.4, which is the generalization of the integration by parts for two stochastic processes:

Theorem 22.4 Let X_t, Y_t be two Itô's processes, then

$$d(X_t Y_t) = X_t\, dY_t + Y_t\, dX_t + dX_t\, dY_t. \tag{22.12}$$

This formula is also called integration by parts, or derivative of a product for two stochastic processes.

The proof of Theorem 22.4 is as follows:

Proof: The proof of this identity is done by using the two dimensional Itô's formula below, and $(dX_t)^2$, $(dY_t)^2$, $dX_t\, dY_t$ are computed by using that $dt\, dt = dt\, dB_t = dB_t\, dt = 0$, and $dB_t\, dB_t = dt$.

We state the two dimensional Itô's formula.

Theorem 22.5 Two dimensional Itô's formula.

If X_t and Y_t are Itô's processes given by

$$dX_t = \mu_1\, dt + v_1\, dB_t,$$
$$dY_t = \mu_2\, dt + v_2\, dB_t,$$

and $Z_t = g(t, X_t, Y_t)$, where g is a C^2 function, then Z_t is also an Itô's process and

$$d(Z_t) = d(g(t, X_t, Y_t)) = \frac{\partial g}{\partial t}(t, X_t, Y_t)dt + \frac{\partial g}{\partial x}(t, X_t, Y_t)dX_t + \frac{\partial g}{\partial y}(t, X_t, Y_t)dY_t$$
$$+ \frac{1}{2}\frac{\partial^2 g}{\partial x^2}(t, X_t, Y_t)(dX_t)^2 + \frac{1}{2}\frac{\partial^2 g}{\partial y^2}(t, X_t, Y_t)(dY_t)^2$$
$$+ \frac{\partial^2 g}{\partial x \partial y}(t, X_t, Y_t)dX_t\, dY_t.$$

Now, we can prove the formula (22.12), i.e.

$$d(X_t Y_t) = X_t\, dY_t + Y_t\, dX_t + dX_t\, dY_t.$$

We apply Itô's formula with $g(t, x, y) = g(x, y) = xy$, then we get:

$$d(Z_t) = d(X_t Y_t) = d(g(X_t, Y_t)) = \frac{\partial g}{\partial t}(X_t, Y_t)dt + \frac{\partial g}{\partial x}(X_t, Y_t)dX_t + \frac{\partial g}{\partial y}(X_t, Y_t)dY_t$$
$$+ \frac{1}{2}\frac{\partial^2 g}{\partial x^2}(X_t, Y_t)(dX_t)^2 + \frac{1}{2}\frac{\partial^2 g}{\partial y^2}(X_t, Y_t)(dY_t)^2$$
$$+ \frac{\partial^2 g}{\partial x \partial y}(X_t, Y_t)dX_t\, dY_t.$$

We have that as $g(x, y) = xy$, then, the derivative of $g(x, y)$ partially with respect to x is

$$\frac{\partial g}{\partial x} = y$$

hence

$$\frac{\partial g}{\partial x}(X_t, Y_t) = Y_t$$

and similarly the derivative of $g(x, y)$ partially with respect to y is

$$\frac{\partial g}{\partial y} = x$$

hence

$$\frac{\partial g}{\partial y}(X_t, Y_t) = X_t$$

$$\frac{\partial g}{\partial t} = \frac{\partial^2 g}{\partial x^2} = \frac{\partial^2 g}{\partial y^2} = 0$$

and

$$\frac{\partial^2 g}{\partial x \partial y} = 1.$$

Therefore,

$$d(X_t Y_t) = X_t\, dY_t + Y_t\, dX_t + dX_t\, dY_t.$$

In the final step, we will compute

$$d(X_tF_t) = X_t\, dF_t + F_t\, dX_t + dX_t\, dF_t,$$

where

$$dX_t = f(t, X_t)dt + c(t)X_t\, dB_t \tag{22.13}$$

and

$$dF_t = F_tc(t)^2dt - F_tc(t)dB_t. \tag{22.14}$$

Multiplying through Eqs. (22.13) and (22.14) by F_t and X_t respectively, we have

$$F_tdX_t = F_tf(t, X_t)dt + c(t)F_tX_t\, dB_t$$

and

$$X_t\, dF_t = X_tF_tc^2(t)dt - X_tF_tc(t)dB_t$$

thus,

$$dX_t\, dF_t = (f(t, X_t)dt + c(t)X_t\, dB_t)(F_tc(t)^2dt - F_tc(t)dB_t) = -F_tc^2(t)X_t\, dt,$$

where we did use the fact that: $dB_t\, dt = 0$, $dt\, dB_t = 0$, $(dt)^2 = 0$ and $(dB_t)^2 = dt$. Then finally,

$$d(X_tF_t) = F_tf(t, X_t)dt + c(t)F_tX_t\, dB_t + X_tF_tc^2(t)dt - X_tF_tc(t)\, dB_t$$

$$-F_tc^2(t)X_t\, dt = F_tf(t, X_t)dt.$$

And this completes the proof of Theorem 22.1.

Remarks 22.2 We observe that in this last equation, we do not have anymore dB_t, so the integrating factor transformed the original equation in an equation that is a deterministic equation.

Part 2 :

Now, we go back to Eq. (22.5), and we will compute dF_t for our case. In order to compute dF_t, we will use the one dimensional Itô's formula below as we did before.

In this case, as F_t is given by:

$$F_t = \exp\left(-\beta B_t + \frac{1}{2}\beta^2 t\right)$$

then we can take $Z_t = B_t$ in the one dimensional Itô's formula, that is, $\mu = 0, \upsilon = 1$, and $Y_t = F_t = g(t, B_t)$, with

$$g(t, x) = \exp\left(-\beta x + \frac{1}{2}\beta^2 t\right)$$

therefore,

$$\frac{\partial g}{\partial t} = g\frac{1}{2}\beta^2,$$

$$\frac{\partial g}{\partial x} = g(-\beta),$$

and

$$\frac{1}{2}\frac{\partial^2 g}{\partial x^2} = \frac{1}{2}\frac{\partial g(-\beta)}{\partial x} = \frac{1}{2}(-\beta)\frac{\partial g}{\partial x} = \frac{1}{2}(-\beta)g(-\beta) = \frac{1}{2}\beta^2 g.$$

Since $g = F_t$, we have that:

$$dF_t = \frac{1}{2}F_t\beta^2\, dt + F_t(-\beta)dB_t + \frac{1}{2}F_t\beta^2(dB_t)^2 = -\beta F_t\, dB_t + F_t\beta^2\, dt.$$

Part 3:
Now, we will compute

$$d(X_tF_t) = X_t\, dF_t + F_t\, dX_t + dX_t\, dF_t,$$

where

$$dX_t = rX_t(k - X_t^m)dt + \beta X_t\, dB_t,$$

$$dF_t = -\beta F_t\, dB_t + \beta^2 F_t\, dt.$$

We have that:

$$F_t\, dX_t = F_t rX_t(k - X_t^m)dt + F_t\beta X_t\, dB_t,$$

$$X_t\, dF_t = -\beta X_tF_t\, dB_t + \beta^2 X_tF_t\, dt,$$

and

$$\begin{aligned} dX_t\, dF_t &= (rX_t(k - X_t^m)dt + \beta X_t\, dB_t)(-\beta F_t\, dB_t + \beta^2 F_t\, dt) \\ &= -\beta^2 F_tX_t(dB_t)^2 = -\beta^2 F_tX_t\, dt. \end{aligned}$$

Then finally,

$$\begin{aligned} d(X_tF_t) &= F_t rX_t(k - X_t^m)dt + F_t\beta X_t\, dB_t - \beta X_tF_t\, dB_t \\ &\quad + \beta^2 X_tF_t\, dt - \beta^2 F_tX_t\, dt = rF_tX_t(k - X_t^m)dt. \end{aligned} \tag{22.15}$$

Remarks 22.3 We observe that the last equationfv is the corresponding one to the general case that we proved first, with $f(t, X_t) = rX_t(k - X_t^m)$. Again we see that in this last equation, we do not have anymore dB_t, so the integrating factor transformed the original equation in an equation that is not only exact but also a deterministic equation.

Now, setting $Y_t = F_t X_t$, we have that:

$$dY_t = d(F_t X_t) = rF_t X_t (k - X_t^m)dt = rY_t \left(k - \frac{Y_t^m}{F_t^m} \right) dt$$

or

$$\frac{dY}{dt} = rY \left(k - \frac{Y^m}{F_t^m} \right) = rkY - r\frac{Y^{m+1}}{F_t^m},$$

where we call $Y = Y_t$.

Remarks 22.4 We observe that the last equation is no more a SDE, it is a deterministic one. That is why we will call $Y = Y_t$, in order to use the classical notation for ordinary DEs.

Part 4:

By using change of variables, Eq. (22.15) can be transformed in a linear equation by setting

$$z = -\frac{Y^{1-(m+1)}}{1-(m+1)} = \frac{Y^{-m}}{m}$$

then

$$mz = Y^{-m}$$

and

$$mdz = -mY^{-(m+1)}\, dY$$

therefore

$$dz = -Y^{-(m+1)}\, dY$$

or

$$dY = -Y^{m+1}\, dz.$$

Hence, the equation

$$dY = \left(rkY - r\frac{Y^{m+1}}{F_t^m} \right) dt$$

becomes

$$-Y^{m+1} dz = \left(rkY - r\frac{Y^{m+1}}{F_t^m} \right) dt$$

or, equivalently,

$$dz = \left(-rkY^{1-m-1} + r\frac{Y^{m+1-m-1}}{F_t^m} \right) dt.$$

Hence,

$$dz = \left(-rkY^{-m} + \frac{r}{F_t^m}\right) dt = \left(-rkmz + \frac{r}{F_t^m}\right) dt$$

that can be written as:

$$\frac{dz}{dt} = -rkmz + \frac{r}{F_t^m}$$

or

$$\frac{dz}{dt} + rkmz = \frac{r}{F_t^m}.$$

Part 5:

The solution to the DE

$$\frac{dz}{dt} + P(t)z = Q(t)$$

is given by:

$$z(t) = \frac{1}{\mu(t)} \left[\int_0^t \mu(s)Q(s)ds + C\right]$$

with $\mu(t) = \exp(\int_0^t P(s)ds)$, because if we want to find an integrating factor $\mu(t)$ for the equation

$$dz + P(t)z\, dt - Q(t)dt = 0$$

then

$$\mu(t)dz + \mu(t)(P(t)z - Q(t))dt$$

needs to be exact, that is, there exists F so that

$$dF = \mu(t)dz + \mu(t)(P(t)z - Q(t))dt$$

hence,

$$\frac{\partial \mu}{\partial t} = \frac{\partial[\mu(t)(P(t)z - Q(t))]}{\partial z}$$

and therefore,

$$\frac{d\mu}{dt} = \mu(t)P(t)$$

or equivalently, $\frac{1}{\mu(t)}\frac{d\mu}{dt} = P(t)$. So, integrating we have that:

$$\mu(t) = \exp\left(\int_0^t P(s)ds\right).$$

In this case, $P(t) = rkm$, so $\mu(t) = \exp(rkmt)$, and

$$z(t) = \exp(-rkmt)\left[\int_0^t \exp(rkms)\frac{r}{F_s^m}ds + C\right].$$

As $Z = \frac{Y^{-m}}{m}$, where $Y = Y_t$ and $Y_t = X_t F_t$ we obtain that $Y = (mz)^{-\frac{1}{m}}$ and finally,

$$X_t = \frac{Y_t}{F_t} = \frac{(mz)^{-\frac{1}{m}}}{F_t}.$$

Now, as

$$F_t = \exp\left(-\beta B_t + \frac{1}{2}\beta^2 t\right).$$

If $m = 1$ we obtain that:

$$X_t = \frac{z^{-1}}{F_t}$$

and hence

$$X_t = \frac{1}{\exp(-\beta B_t + \frac{1}{2}\beta^2 t)}\frac{1}{\exp(-rkt)}\frac{1}{\int_0^t \exp(rks)\frac{r}{\exp(-\beta B_s + \frac{1}{2}\beta^2 s)}ds + C},$$

where C is a constant that will be fixed with the initial condition, or equivalently,

$$X_t = \frac{\exp(\beta B_t - (\frac{1}{2}\beta^2 - rk)t)}{r\int_0^t \exp((-\frac{1}{2}\beta^2 + rk)s + \beta B_s)ds + C}.$$

Similarly, if $m = 2$ we obtain that:

$$X_t = \frac{(2z)^{-1/2}}{F_t}$$

and hence:

$$X_t = \frac{2^{-1/2}\exp(\beta B_t - (\frac{1}{2}\beta^2 - rk)t)}{[r\int_0^t \exp((-\beta^2 + 2rk)s + 2\beta B_s)ds + C]^{1/2}}.$$

22.3 Examples

In this section, we present two examples.

22.3.1 Modeling Asset Prices

In this case, the SDE that models the evolution of an asset price is:

$$dX_t = \mu X_t\, dt + \sigma X_t\, dB_t,$$

where X_t is the asset price at time t, B_t is a standard Wiener process, μ is the drift, and σ is the diffusion coefficient. The first term models the deterministic trends and the second term models the random, unpredictable events occurring during this motion. The SDE modeling asset price that is contained in the class of SDEs that we have studied before:

$$dX_t = f(t, X_t)dt + c(t)X_t\, dB_t$$

taking $f(t, X_t) = \mu X_t$ and $c(t) = \sigma$. Then we have

$$F_t = \exp\left(-\int_0^t \sigma\, dB_s + \frac{1}{2}\int_0^t \sigma^2\, ds\right) = \exp\left(-\sigma B_t + \frac{1}{2}\sigma^2 t\right)$$

is an integrating factor for the SDE, that transform it into a deterministic DE; setting $Y_t = F_t X_t$ we have that:

$$dY_t = d(F_t X_t) = F_t f(t, X_t)dt = \mu F_t X_t\, dt = \mu Y_t\, dt.$$

Therefore,

$$\frac{dY_t}{Y_t} = \mu\, dt$$

and integrating both sides of the above equation, we obtain:

$$\log Y_t = \mu t + C$$

or equivalently,

$$Y_t = e^{\mu t + C} = Ae^{\mu t}$$

and from this last equation, we can obtain X_t as:

$$X_t = \frac{Ae^{\mu t}}{F_t} = Ae^{\mu t}e^{\sigma B_t - \frac{1}{2}\sigma^2 t}.$$

22.3.2 Modeling Magnitude of Earthquake Series

The SDE (Tweneboah 2020) that models the evolution of the magnitude of earthquakes is given as

$$dX(t) = -\lambda X(t)dt + dZ(t), X_0 > 0, \quad \lambda \in \mathbb{R}^+, \tag{22.16}$$

where $Z = \{Z_t, t \geq 0\}$ is a Lévy process and the rate parameter λ is a positive number.

For $\lambda > 0$ and $t > 0$, we can find an explicit solution to Eq. (22.16). Define

$$g(t, X_t) = e^{\lambda t}X_t. \tag{22.17}$$

Applying Itô's lemma to Eq. (22.17), we obtain

$$\begin{aligned} d(e^{\lambda t}X_t) &= [\lambda X_t e^{\lambda t} - \lambda X_t e^{\lambda t}]dt + e^{\lambda t}\, dZ_t \\ &= e^{\lambda t} dZ_t. \end{aligned} \tag{22.18}$$

Integrating both sides of Eq. (22.18) and dividing through by $e^{\lambda t}$ yields the solution of our SDE,

$$X_t = e^{-\lambda t}X_0 + \int_0^t e^{-\lambda(t-s)}\, dZ_s. \tag{22.19}$$

Equation (22.19) models the evolution of earthquake sequences. Please see Mariani and Tweneboah (2016), Tweneboah (2015), and Tweneboah (2020) for more details of the model for the evolution of earthquake magnitudes. More applications of SDEs can be found in Øksendal (2010).

22.4 Multidimensional Stochastic Differential Equations

If a finite number of SDEs are given in a model, then the multidimensional case should be considered. This situation can arise for the modeling of several phenomena, e.g. the price evolution of multiple stocks, interest rates, volatilities, and several others see Oksendal (2010) and Lamberton and Lapeyre (2007).

For the case of random fluctuation in higher dimensions (i.e. $m \geq 2$), let $\mathbf{B}_t = (B_t^1, B_t^2, \ldots, B_t^m)^T$ denotes an m-dimensional Brownian motion at time t. In this case, the deterministic (drift) part $b : \mathbb{R}^d \times \mathbb{R}^+ \to \mathbb{R}^d$, namely $\mathbf{b}(X_t, t)$ is a measurable vector process and the diffusion part $\sigma : \mathbb{R}^d \times \mathbb{R}^+ \to \mathbb{R}^{d \times m}$, namely $\boldsymbol{\sigma}(X_t, t)$ is a measurable matrix-valued process.

For the underlying probability space $(\Omega, \mathbf{F}, P)$, adapted to the filtration $(\mathbf{F})_{t \geq 0}$, a d-dimensional stochastic process $\mathbf{X} = (\mathbf{X}_t : t \in [0, \infty))$ is represented by d DEs, d-dimensional initial vector and suitable conditions.

If we write a coupled system of SDEs with d processes of states and m-dimensional Brownian motions, we obtain:

$$d\mathbf{X}_t = b(\mathbf{X}_t, t)dt + \sigma(\mathbf{X}_t, t)d\mathbf{B}_t, \tag{22.20}$$

where

$$\mathbf{X}_t = \begin{bmatrix} X_1(t) \\ X_2(t) \\ \vdots \\ X_d(t) \end{bmatrix}, \quad b(\mathbf{X}_t, t) = \begin{bmatrix} b_1(\mathbf{X}_t, t) \\ b_2(\mathbf{X}_t, t) \\ \vdots \\ b_d(\mathbf{X}_t, t) \end{bmatrix}, \quad d\mathbf{B}_t = \begin{bmatrix} dB_1(t) \\ dB_2(t) \\ \vdots \\ dB_m(t) \end{bmatrix},$$

$$\sigma(\mathbf{X}_t, t) = \begin{bmatrix} \sigma_{11}(\mathbf{X}_t, t) & \sigma_{12}(\mathbf{X}_t, t) & \ldots & \sigma_{1m}(\mathbf{X}_t, t) \\ \sigma_{21}(\mathbf{X}_t, t) & \sigma_{22}(\mathbf{X}_t, t) & \ldots & \sigma_{2m}(\mathbf{X}_t, t) \\ \vdots & \vdots & \ddots & \vdots \\ \sigma_{d1}(\mathbf{X}_t, t) & \sigma_{d2}(\mathbf{X}_t, t) & \ldots & \sigma_{dm}(\mathbf{X}_t, t) \end{bmatrix}.$$

In Section 22.4.1,wediscuss the multidimensional Ornstein–Uhlenbeck process.

22.4.1 The multidimensional Ornstein–Uhlenbeck Processes

Consider a stochastic process $\mathbf{X}(t)$ driven in terms of two DE. From these two SDEs, the first one $\{X_1(t)\}$ describes the physical process which is affected by location of an event, while the second SDE $\{X_2(t)\}$ describes the location of an event which is affected by the physical process.

The above coupled system satisfies the SDE:

$$\begin{cases} dX_1(t) = -\lambda_1 X_1(t)dt + \sigma_{11}\ dZ_1(t) + \sigma_{12}\ dZ_2(t), & \lambda_1 \in \mathbb{R}^+ \\ dX_2(t) = -\lambda_2 X_2(t)dt + \sigma_{21}\ dZ_1(t) + \sigma_{22}\ dZ_2(t), & \lambda_2 \in \mathbb{R}^+ \end{cases} \tag{22.21}$$

with $\mathbf{X}(0) = (X_0^1, X_0^2)^T$, where both $X_0^1 > 0$ and $X_0^2 > 0$ denotes the initial condition for $X_1(t)$ and $X_2(t)$ respectively, $Z_1(t) = Z(\lambda_1 t)_{t\geq 0}, Z_2(t) = Z_1(\lambda_2 t)_{t\geq 0}$ are Lévy processes and λ_1, λ_2 are the intensity parameters. The parameters σ_{11} and σ_{22} determines the volatility of the system and σ_{12} and σ_{21} describes the correlation of the system. The processes Z_1 and Z_2 are called the background driving Lévy processes (BDLP). The intensity parameters describe the velocity at which the time series returns toward its mean value. The volatility parameters are used to tune the effect of $\mathbf{Z}_t$ on $\mathbf{X}_t$. A higher value implies more randomness in the system. Choosing $\sigma_{11} = \sigma_{22} = \sigma_{12} = \sigma_{21} = 0$ reduces (22.21) to a deterministic DE. The processes $X_1(t)$ and $X_2(t)$ are correlated if $\sigma_{12}=\sigma_{21} \neq 0$.

In matrix notation, we can rewrite (22.21) as:

$$d\mathbf{X}(t) = A\mathbf{X}(t)dt + \sum_{i=1}^{2} B_i(t) d\mathbf{Z}(\lambda t), \tag{22.22}$$

where

$$\mathbf{X} = \begin{pmatrix} X_1 \\ X_2 \end{pmatrix}, \quad A = \begin{pmatrix} -\lambda_1 & 0 \\ 0 & -\lambda_2 \end{pmatrix}, \quad B_1(t) = \begin{pmatrix} \sigma_{11} & 0 \\ 0 & \sigma_{21} \end{pmatrix},$$

$$B_2(t) = \begin{pmatrix} \sigma_{12} & 0 \\ 0 & \sigma_{22}, \end{pmatrix} \quad \text{and} \quad \mathbf{Z}(\lambda t) = \begin{pmatrix} Z_1(\lambda t) \\ Z_2(\lambda t) \end{pmatrix}.$$

Equation (22.21) is an example of a multidimensional SDE. Thus, we have a two-dimensional SDE. This particular system can be solved easily by finding the solution for X_1 and X_2. However we will present a more general theory.

22.4.2 Solution of the Ornstein–Uhlenbeck Process

From (22.22), we have the following system of SDEs:

$$d\mathbf{X}(t) = A\mathbf{X}(t)dt + B_1\ d\mathbf{Z}(\lambda t) + B_2\ d\mathbf{Z}(\lambda t) \tag{22.23}$$

for an n-dimensional process $\mathbf{X}(t)$, where A, B_1, and B_2 are matrices.

We rewrite (22.23) as

$$e^{-At}\ d\mathbf{X}(t) - e^{-At}A\mathbf{X}(t)dt = e^{-At}B_1\ d\mathbf{Z}(\lambda t) + e^{-At}B_2\ d\mathbf{Z}(\lambda t), \tag{22.24}$$

where for any general $n \times n$ matrix A, we define e^A to be an exponential matrix of the form:

$$e^A = \sum_{n=0}^{\infty} \frac{1}{n!} A^n, \tag{22.25}$$

where $A^0 = I$ is the identity matrix.

From (22.24), we observe that the left-hand side is related to

$$d(e^{-At}\, dX(t)).$$

To achieve this, we apply the two-dimensional version of the Itô formula (Theorem 22.5) to the two coordinate functions f_1, f_2 of

$$f : [0, \infty) \times \mathbb{R}^2 \to \mathbb{R}^2 \text{ given by } f(t, x_1, x_2) = e^{-At} \begin{pmatrix} x_1 \\ x_2 \end{pmatrix},$$

we obtain that

$$d(e^{-At}\, d\mathbf{X}(t)) = e^{-At}\, d\mathbf{X}(t) - e^{-At} A\mathbf{X}(t) dt. \tag{22.26}$$

Substituting (22.26) into (22.24) and taking into account the timing in the BDLP, we obtain the solution:

$$\mathbf{X}(t) = e^{At}\mathbf{X}(0) + \int_0^t e^{A(t-s)} B_1\, d\mathbf{Z}(\lambda s) + \int_0^t e^{A(t-s)} B_2\, d\mathbf{Z}(\lambda s) \tag{22.27}$$

by integration by parts. We applied the integration by part to (22.24) after substituting (22.26).

Remarks 22.5 *The issue when calculating the solution is to calculate the exponential e^{At}, when A is a $n \times n$ matrix. The idea is to write the matrix A as $A = PDP^{-1}$, where P is an invertible matrix and D is a diagonal matrix. We can find P and D by diagonalizing the matrix.*

Definition 22.2 (Diagonalizable) A square matrix A is said to be diagonalizable if A is similar to a diagonal matrix, i.e. if $A = PDP^{-1}$ where P is invertible and D is a diagonal matrix.

We state the following theorem without proof.

Theorem 22.6 Diagonalizable theorem An $n \times n$ matrix A is diagonalizable if and only if A has n linearly independent eigenvectors. In fact $A = PDP^{-1}$ with D a diagonal matrix, if and only if the columns of P are n linearly independent eigenvectors of A. In this case, the diagonal entries of D are eigenvalues of A that correspond, respectively, to the eigenvectors in P.

From Theorem 22.6, if we can find n linearly independent eigenvectors for an $n \times n$ matrix A, then A diagonalizable. In addition, we can use the eigenvectors and their corresponding eigenvalues to find an invertible matrix P and a diagonal matrix D required to show that A is diagonalizable.

For a detailed study of the multidimensional SDE and its solution methods, see Mikosch (1998), Oksendal (2010), Lamberton and Lapeyre (2007), and references therein.

22.5 Simulation of Stochastic Differential Equations

In this section, we will discuss the simulation of SDEs focusing mainly on the Euler–Maruyama method Maruyama (1955) and Euler–Milstein scheme Milstein (1974) for approximating the sample path of SDEs. Paul Glasserman in his book Glasserman (2010) presents other techniques for generating sample paths of SDEs. We begin our discussion with the Euler–Maruyama method, which is also known as the Euler method.

22.5.1 Euler–Maruyama Scheme for Approximating Stochastic Differential Equations

The Euler–Maruyama method is a technique used to approximate the numerical solution of a SDE. It is a generalization of the Euler method for approximating ordinary DEs to SDEs. Consider the following SDE:

$$dX_t = b(t, X_t)dt + \sigma(t, X_t)dB_t, \tag{22.28}$$

with initial condition $X_0 = x_0$ and B_t a one-dimensional standard Brownian motion. The solution of (22.28) is given as:

$$X_t = x_0 + \int_0^t b(s, X_s)ds + \int_0^t \sigma(s, X_s)dB_s. \tag{22.29}$$

Therefore, approximating the path of X_t is equivalent to approximating the integral. There are several ways to approximate the first integral, the second however has to be approximated using the Euler scheme. The Euler method uses a simple rectangular rule. First assume that the interval $[0, t]$ is divided into n equal subintervals. This implies that the increment is $\Delta t = t/n$ and that the points are $t_0 = 0, t_1 = \Delta t, \dots, t_i = i\Delta t, \dots, t_n = n\Delta t = t$. Thus, using X_i to denote X_{t_i} we have:

$$\begin{cases} X_0 = x_0 \\ X_i = X_{i-1} + b(t_{i-1}, X_{i-1})\Delta t + \sigma(t_{i-1}, X_{i-1})\Delta B_i, \quad \forall i \in \{1, 2, \dots, n\}, \end{cases}$$

where ΔB_i is the increment of a standard Brownian motion over the interval $[t_{i-1}, t_i]$. We recall that the Brownian motion has independent and stationary increments, so it follows that each of such increment is independent of all others and is distributed as a normal (Gaussian) random variable with mean 0 and variance the length of the time sub-interval (i.e. Δt). Therefore, the standard deviation of the increment is $\sqrt{\Delta t}$.

The Euler–Maruyama algorithm for generating the sample paths of example (22.29) using a fixed number of paths, n, and discretization interval, Δt are presented in Algorithm 22.1.

Algorithm 22.1 Generating a sample path using Euler–Maruyama's method to estimate $\theta = E[f(X_t)]$.

for $j = 1$ to n **do**
  $t = 0; \hat{X} = X_0$
  for $k = 1$ to $[T/\Delta t] =: m$ **do**
    generate $Z \sim N(0, 1)$
    set $\hat{X} = \hat{X} + b(t, \hat{X})\Delta t + \sigma(t, \hat{X})\sqrt{\Delta t}Z$
    set $t = t + \Delta t$
  end for
  set $f_j = f(\hat{X})$
end for
set $\hat{\theta}_n = (f_1 + \ldots + f_n)/n$
set $\hat{\sigma}_n^2 = \sum_{j=1}^{n}(f_j - \hat{\theta}_n)^2/(n-1)$
set approximately $100(1-\alpha)\%$ $\quad \text{CI} = \hat{\theta}_n \pm z_{1-\alpha/2}\frac{\hat{\sigma}_n}{\sqrt{n}}$

This algorithm can be extended to general b and σ functions by creating separate functions. The scheme can also be generalized to approximate multidimensional SDEs. In the multidimensional case, $X_t \in \mathbb{R}^d$, $B_t \in \mathbb{R}^p$, and $b(t, X_t) \in \mathbb{R}$ are vectors and $\sigma(t, X_t) \in \mathbb{R}^{d \times p}$ is a matrix. The Euler–Maruyama method gives a first-order approximation for the stochastic integral. Please refer to Haugh (2017) and Chapter 6 of the book by Glasserman (2010) for more details of simulating a stochastic differential equations using the Euler–Maruyama Scheme.

The Euler–Milstein method discussed in Section 22.5.2 provides an improvement by including second-order terms.

22.5.2 Euler–Milstein Scheme for Approximating Stochastic Differential Equations

The Euler–Milstein scheme is a technique for approximating the numerical solution of a SDE. The idea in this scheme is to consider expansions on the coefficients b and σ. This method is applied when the coefficients of the process are functions of only the main process, i.e. do not depend on time. The scheme is designed to work with SDEs of the type

$$dX_t = b(X_t)dt + \sigma(X_t)dB_t,$$

with initial conditions $X_0 = x_0$. We consider expansions on the coefficients $b(X_t)$ and $\sigma(X_t)$ using Itô's lemma. We then obtain:

$$db(X_t) = b'(X_t)dX_t + \frac{1}{2}b''(X_t)(dX_t)^2$$

and

$$d\sigma(X_t) = \left(b'(X_t)b(X_t) + \frac{1}{2}b''(X_t)\sigma^2(X_t)\right)dt + b'(X_t)\sigma(X_t)dB_t.$$

Writing out the integral form from t to u for any $u \in (t, t + \Delta t]$, we obtain

$$b_u = b_t + \int_t^u \left(b'_s b_s + \frac{1}{2}b''_s \sigma_s^2\right) ds + \int_t^u b'_s \sigma_s \, dB_s,$$

$$\sigma_u = \sigma_t + \int_t^u \left(\sigma'_s b_s + \frac{1}{2}\sigma''_s \sigma_s^2\right) ds + \int_t^u \sigma'_s \sigma_s \, dB_s,$$

where we used the notation $b_u = b(X_u)$. Substituting these expressions in the original SDE, we obtain

$$X_{t+\Delta t} = X_t + \int_t^{t+\Delta t} \left(b_t + \int_t^u \left(b'_s b_s + \frac{1}{2}b''_s \sigma_s^2\right) ds + \int_t^u b'_s \sigma_s \, dB_s\right) du$$
$$+ \int_t^{t+\Delta t} \left(\sigma_t + \int_t^u \left(\sigma'_s b_s + \frac{1}{2}\sigma''_s \sigma_s^2\right) ds + \int_t^u \sigma'_s \sigma_s \, dB_s\right) dB_u.$$

In this expression, we eliminate all terms which will produce higher orders than Δt after integration. That means eliminating terms of the type $ds\, du = O(\Delta_t^2)$ and $du\, dB_s = O(\Delta t^{\frac{3}{2}})$. The only terms remaining other than simply du and ds are the ones involving $dB_u\, dB_s$ since they are of the right order. Thus, after eliminating the terms, we obtain:

$$X_{t+\Delta t} = X_t + b_t \int_t^{t+\Delta t} du + \sigma_t \int_t^{t+\Delta t} dB_u + \int_t^{t+\Delta t} \int_t^u \sigma'_s \sigma_s \, dB_s \, dB_u. \tag{22.30}$$

For the last term, we apply Euler discretization in the inner integral (this follows from Theorem 22.2) as follows:

$$\begin{aligned}\int_t^{t+\Delta t} \left(\int_t^u \sigma'_s \sigma_s \, dB_s\right) dB_u &\approx \int_t^{t+\Delta t} \sigma'_t \sigma_t (B_u - B_t) dB_u \\ &= \sigma'_t \sigma_t \left(\int_t^{t+\Delta t} B_u \, dB_u - B_t \int_t^{t+\Delta t} dB_u\right) \\ &= \sigma'_t \sigma_t \left(\int_t^{t+\Delta t} B_u \, dB_u - B_t B_{t+\Delta t} + B_t^2\right).\end{aligned} \tag{22.31}$$

For the integral term inside, recall that

$$\int_0^t B_u \, dB_u = \frac{1}{2}(B_t^2 - t).$$

Therefore, applying for t and $t + \Delta t$ and taking the difference, we obtain

$$\int_t^{t+\Delta t} B_u \, dB_u = \frac{1}{2}(B_{t+\Delta t}^2 - t - \Delta t) - \frac{1}{2}(B_t^2 - t).$$

Therefore, substituting back into (22.31), we have

$$\begin{aligned}\int_t^{t+\Delta t} \left(\int_t^u \sigma_s' \sigma_s \, dB_s \right) dB_u &\approx \sigma_t' \sigma_t \left(\frac{1}{2}(B_{t+\Delta t}^2 - B_t^2 - \Delta t) - B_t B_{t+\Delta t} + B_t^2 \right) \\ &= \sigma_t' \sigma_t \left(\frac{1}{2} B_{t+\Delta t}^2 + \frac{1}{2} B_t^2 - B_t B_{t+\Delta t} - \Delta t \right) \\ &= \sigma_t' \sigma_t \left(\frac{1}{2}(B_{t+\Delta t} - B_t)^2 - \Delta t \right).\end{aligned}$$

We recall that $B_{t+\Delta t} - B_t$ is the increment of the Brownian motion which we know is $N(0, \Delta t)$ or $\sqrt{\Delta t} Z$, where $Z \sim N(0, 1)$. In summary, for the SDE

$$dX_t = b(X_t)dt + \sigma(X_t)dB_t, \quad X_0 = x_0,$$

the Euler–Milstein scheme starts with $X_0 = x_0$ and for each successive point, we first generate $Z \sim N(0, 1)$ and then calculate the next point as

$$X_{t+\Delta t} = X_t + b(X_t)\Delta t + \sigma(X_t)\sqrt{\Delta t} Z + \frac{1}{2}\sigma'(X_t)\sigma(X_t)\Delta t(Z^2 - 1), \tag{22.32}$$

where σ' denotes the derivative of $\sigma(x)$ with respect to x.

Remarks 22.6 When $\sigma'(X_t) = 0$, i.e. the diffusion term does not depend on X_t, the Euler–Milstein scheme is equivalent to the Euler–Maruyama method.

The Euler–Milstein method can also be generalized to approximate multidimensional SDEs. Please refer to Alnafisah (2018).

22.6 Problems

1 Prove the two-dimensional Itô formula.

2 Calculate the solution to the following mean reverting Ørnstein–Uhlenbeck SDE:

$$dX_t = \alpha X_t \, dt + \beta \, dB_t$$

with $X_0 = x$.

3 Solve the following SDE:

$$dX_t = \frac{1}{2} x_t \, dt + X_t \, dB_t$$

with $X_0 = 1$.

4 Calculate the solution to the following SDE:

$$dX_t = \mu(m - X_t)dt + \sigma\, dB_t$$

with $X_0 = x$.

5 Write a python script to simulate B_t, a standard Brownian motion on the interval $t \in [0, 3]$.

6 Give the Euler and Euler–Milstein approximation scheme for the following SDE:

$$dX_t = \mu_X X_t\, dt + \sigma X_t^{\beta}\, dB_t,$$

where $\beta \in (0, 2]$. Generate five paths and plot them for the following parameter values:

$$\mu_X = 0.2, \quad \sigma = 0.8, \quad \beta = \frac{1}{4}, \quad X_0 = 200.$$

7 Given the SDE,

$$dX_t = a(t)(b(t) - X_t)dt + \sigma(t)dW_t,$$

let $a(t) = \theta_1 t$, $b(t) = \theta_2\sqrt{t}$, and $\sigma(t) = \theta_3 t$, where $\theta_1 = 2, \theta_2 = 0.8$, $\theta_3 = 0.9$, and $\Delta t = 0.001$.

(a) Use the Euler scheme to generate a single path of the process by choosing a $\Delta t = 0.001$ from $t = 0$ to $t = 1$.

(b) Repeat part (a) with the Euler–Milstein approximation scheme. Write a conclusion based on the results obtained.

8 Consider the following nonlinear stochastic DE arises in population growth:

$$dX_t = rX_t(k - X_t^m)dt + \beta X_t\, dB_t, \tag{22.33}$$

where the constant $k > 0$ is called the carrying capacity of the environment, the constant $r \in R$ is a measure of the quality of the environment, and the constant $\beta \in R$ is a measure of the size of the noise in the system.
Write a program in R to simulate data corresponding to the above population model on the interval $t \in [0, 2]$ and discuss your findings.

9 Using the Ornstein–Uhlenbeck SDE:

$$dX(t) = -\lambda X(t)dt + dZ(t), X_0 > 0, \quad \lambda \in \mathbb{R}^+, \tag{22.34}$$

where $Z = \{Z_t, t \geq 0\}$ is a Lévy process and the rate parameter λ is a positive number, simulate data corresponding to the above SDE on the interval $t \in [0, 1]$.

10 Explain a SDE that violates the Lipschitz condition.

23

Ethics: With Great Power Comes Great Responsibility

23.1 Introduction

In recent years, science and technology keep evolving very fast and to a large extent have improved our daily life. Technology has revealed some questions regarding the way people relate and interact with others. For example, physical human contact and interactions are decreasing every day due to the virtual world, and there is a considerable increase in virtual interaction. In addition, online social networking is replacing real face-to-face and physical contact and as a result, alienation can increase. Alienation describes the separation between oneself and other that properly belong together. Other problems such as cyber bullying, online stalking, and cybercrime, which are related to the anonymity of the Internet are also increasing rapidly. With the enormous availability of data, practitioners and researchers make decisions based on data and as a result are confronted with a set of ethical questions. These questions does not only concern the misuse of data but also questions on how to preserve data privacy, avoid bias in data selection, ways to prevent disruption and hacking of data, issues of transparency in data collection, research, and dissemination. Some individuals and organizations uses data it collects from its own merchants to compete against them. The questions often asked by people are: Who owns the data? Who has the right of access to it and under what conditions? Quoting Peter Parker's Uncle Ben in Spiderman, "With great power comes great responsibility"; if we have the ability to make decisions using data, as practitioners and researchers, we must ensure we do that for the good of others. In fact, learning and complying with ethical data norms is a symbol of accountability.

In order to systematically and effectively apply data science techniques in our real life, it is important to observe and follow some ethical guidelines. These guidelines offer a set of principles that ensure that the steps of data preparation, modeling, deployment, and model monitoring are performed in an ethical manner. Data science involves interdisciplinary teams where discussions around ethics are challenging in this modern world. Each discipline has its own set of research

Data Science in Theory and Practice: Techniques for Big Data Analytics and Complex Data Sets, First Edition. Maria Cristina Mariani, Osei Kofi Tweneboah, and Maria Pia Beccar-Varela.

integrity norms and practices. For example, when acquiring and integrating data sources, some ethical issues include considerations of mass surveillance, data privacy, and data sovereignty. In addition, research integrity, continuous research practices, and training scientists to accomplish the following activities: selection of experimental designs, methods to reduce bias, rewards for rigorous research, and rewards and/or acknowledgement for sharing data, code, and protocols.

23.2 Data Science Ethical Principles

In this section, we will discuss five (5) data science ethics. We begin with the first ethic, enhancing value in society.

23.2.1 Enhance Value in Society

The first ethical principle of data science is to enhance value in our society. The common ethical frameworks of data science is for practitioners to produce outcomes within their work that support the improvement of public well-being. This could involve practitioners seeking to share the benefits of data science to solve societal problems in areas such as economic empowerment employment, education, environment, equality and inclusion the concept of fairness, practice examples, health and hunger, security and justice, and several others.

Next, we consider different ways to enhance the value of data science in society.

- Consider the potential impact that models have on decisions. Consider how the benefits may be distributed across society and who or what could be affected by the outcomes.
- Understand how the models will be deployed, what its impact may be, and who the target audience are.
- Understand who the relevant stakeholders are, explaining the potential biases, errors, assumptions and risks inherent in predictive modeling, inviting peer and unbiased review.
- Learn about what could mitigate any risks for unfairness. The goal is to encourage a culture which values social justice and fairness.

23.2.2 Avoiding Harm

The basic principle in moral philosophy is "nonmaleficence" that is, the ethical duty to avoid causing or inflicting harm on others. This principle also includes the responsibility to avoid and prevent existing depravity that may be likely to occur. In data science, there is the possibility to cause harm when people use data in a bad manner. Thus, ethical data guidelines provide how practitioners can avoid this by working in a manner that maintain the privacy, equality, and autonomy of individuals and groups, and speaking up about potential harm or ethical defilement. Practitioners may be subject to legal and regulatory obligations in relation to the privacy of individuals relevant to the jurisdiction in which they are working. However, they have regulatory obligations to speak up about harm or violations of legal requirements as well. This can also be applied to work relating to businesses, animals, or the environment, with consideration of commercial rights, animal welfare, and the protection of environmental resources.

An important feature of the commitment to preventing harm is closely related but does not exactly match the duty of avoidance (i.e. the action of keeping away from or not doing something). Data scientist and practitioners must use reasonable diligence when designing, creating, and implementing algorithms to avoid harm or danger. The data scientist must inform the client any real, perceived, or hidden risks when using the algorithm. After full disclosure, the client is responsible for making the decision to use or not use the algorithm. If a practitioner reasonably believes an algorithm will cause harm, he or she must take reasonable counteractive measures, including disclosure to the client, and if necessary, disclosure to the proper authorities.

23.2.3 Professional Competence

Another important ethical principle is the professional competence that expects data science practitioners to use the best practices and observe all applicable legal and regulatory rules and requirements. It entails fully comprehending data sources, error analysis, bias, noisy observations, and supporting work with vigorous algorithmic techniques that are suitable to the question being solved. Practitioners can also thoroughly assess and balance the benefits of the work corresponding to the risk. Data scientist can practice and improve on their professional competency by using the following guidelines:

- Considering and complying with all relevant professional and regulatory practices associated with the work in which they are involved.
- Being aware of any evolving legal requirements and communicating these to relevant stakeholders.
- Ensuring that the business's ethics policies, procedures, and governance are applied to data science work.
- Providing executive staff with enough information on the advantages and limitations of the work to make decisions about the use of models accordingly.
- Understanding and communicating to stakeholders and decision-makers about the ethical risks of the project like bias, uncertainty, quality issues, individual/commercial harm, methodology assumptions, disadvantages of chosen methods.

23.2.4 Increasing Trustworthiness

A data scientist can achieve the public's trust and confidence by following the ethical principle of trustworthiness. Trustworthiness refers to the ability to be relied on as honest or truthful. Practitioners can help to increase the reliability of their work by considering ethical principles throughout all stages of a project. Transparency can include but not limited to, fully explaining how algorithms are being used, why any decisions have been delegated, and being open about the risks and biases. Practitioner's should engage widely with a diverse range of stakeholders and take public perceptions and feedbacks into account both from the commencement of the project and throughout the entire projects. These activities will help to build trustworthiness with people.

23.2.5 Maintaining Accountability and Oversight

Data scientists and practitioners should maintain other principles of data ethics, that is human accountability and oversight within their work and procedure. Being accountable includes being mindful of how and when to delegate any decision-making to systems and having control in place to ensure systems deliver the intended objectives. When deciding to present any decision-making, it would be useful to fully understand and explain the potential implications of doing so, as the work could lead to introducing advanced systems which do not have adequate governance. Practitioners should note that delegating any decisions to these systems does not eliminate any of their individual responsibilities. Therefore, the data scientist can maintain human oversight of automated solutions. They can implement model authority by deciding how to monitor the model over time, such as setting review points and sign-off agreeing where responsibility lies for models in production (approval, reviews, longer-term quality assurance), agreeing a review process.

23.3 Data Science Code of Professional Conduct

We now briefly discuss the Data Science Code of Professional Conduct that are used to solve practical problems in daily life. The codes are as follows:

- *Terminology*: The terminologies used in data Science includes data, data quality, data volume, data variety, data velocity, big data, signal, noise, machine learning, deep learning, stochastic calculus, random variable, statistics, stochastic differential equation, predicted mean squared error, miss-classification rate, classifications, predictions, anomaly detection, model evaluation, filtering, and several others. Please refer to Association (2020) for a review of all the terminologies above.

- *Data Scientist–Client Relationship*: The relationship between Data Scientist and Client includes the following:
  - *Competence*: Competent data science professional services requires the knowledge, skill, preparation, and comprehensive analysis reasonably necessary for the services to be rendered.
  - *communication with clients*: The data science professional must consult with the client about means by which the client's goals are to be achieved. In addition, the data scientist must thoroughly explain the final results to the client to the extent reasonably necessary to permit the client to make informed decisions regarding the data science.
  - *Confidential information*: Confidential information is information that the data scientist creates, develops, modifies, receives, uses, or learns in the course of employment as a data scientist for a client. The data scientist has an obligation to protect all confidential information, regardless of its form or format, from the time of its creation or receipt until its authorized disposal. In fact, protecting this information is critical to a data scientist's reputation for integrity and relationship with clients, and ensures compliance with laws and regulations governing the client's organization.
  - *Conflicts of interest*: A data scientist must not provide professional data science services for a client if the services involve a concurrent conflict of interest. In this instance, a concurrent conflict of interest exists if the provision of services for one client will be directly adverse to another client; or there is a risk that providing professional data science services for one or more clients will be materially limited by the data scientist's responsibilities to another client.
  - *Duties to prospective client*: A prospective client is someone who consults with a data scientist or practitioner about the possibility of forming a client–data scientist relationship. A data scientist must not provide professional data science services to a prospective client with interests materially adverse to those of other clients in the same industry if the data scientist received information from the prospective client that could cause harm.
  - *Data science evidence, quality of data, and quality of evidence*: A data scientist must inform the client of all data science results and material facts known to the data scientist that will help the client to make informed decisions, whether or not the data science evidence are adverse. In fact disclosing all these information, such as the quality of data, and quality of evidence, helps the client to make informed decisions.
- *Integrity*: A data scientist should maintain the integrity in the data science profession. It is a misconduct if a data scientist violates or attempts to violate the Data Science Code of Professional Conduct mentioned above. It should be noted that the misuse of data science results to communicate a false reality or promote an illusion of understanding.

Please refer to Association (2020) for a comprehensive list of the data science code of professional conduct.

23.4 Application

In this section, we discuss four steps for ethics assessment of data science when applied to solve practical problems. We begin with project planning.

23.4.1 Project Planning

Project planning is the process a data scientist goes through to establish the steps required to define the project objectives, clarify the scope of what needs to be done, and develop the task list to do it. In the project planning phase,

- Relevant questions to raise are Is the data analysis in the public interest? Can data be ethically sourced? Are there any risks (privacy, harm, fairness) for individuals, groups, businesses, organizations and the environment?
- Involve the public or stakeholders to gather perceptions.
- Introduce the governance for the project, including data security and handling.
- Seek early feedback from domain experts.

23.4.2 Data Preprocessing

Data Preprocessing refers to the step in which the data get transformed to be easily interpreted by an algorithm. The steps required in data preprocessing are as follows:

- Understanding and asking questions about the origin of the data, such as how it was collected.
- Investigating the attached consents and legal uses for the data.
- Considering the privacy, dignity, and fair treatment of individuals when selecting data.
- Following good data handling practices such as data security data privacy, storing.
- Considering the impact of deriving demographics or linking with other data.
- Being transparent and providing evidence of privacy considerations to the public and/or regulators.

23.4.3 Data Management

Data management consists of acquiring, validating, storing, protecting, and processing required data to ensure the accessibility, reliability, and timeliness of the data for clients. The guidelines to be taken into consideration to achieve good data management are as follows:

- Understanding the consents and legal uses of the data.
- Practicing data cleaning, imputing, filtering the noisy observations.
- Considering the impacts of data processes to privacy, bias, and error, including linking data sources, estimating demographic, or other excluded information.

23.4.4 Analysis and Development

Analysis and development refers to the stage of modeling, deployment, and model monitoring. Steps to be taken into consideration during the analysis and development stage are as follows:

- Apply consents and permitted uses, professional and regulatory requirements.
- Monitor risks identified at planning and assess for additional risks (for example harm, bias, error, and privacy).

23.5 Problems

1 Discuss the ethical basis for undertaking a project, the project requirements for both the protection of research participants and the equitable allocation of all potential project benefits and risks.

2 Discuss four data science ethics and explain the advantages and disadvantages of each.

3 As a practitioner, how do you ensure potential ethical problems are identified in a project and what can be done to prevent them from happening?

4 Discuss how we can achieve ethical accountability in a project.

5 What are the likely misinterpretations data results and what can be done to prevent those misinterpretations?

6 Consider a potential bias introduced through the choice of datasets and variables:
   (a) Does the data include disproportionate coverage for different communities under study?
   (b) Does the data have adequate geographic coverage?
   (c) Should anyone check to address implicit biases in the data?

7 Describe the ideal data science–client relationship.

8 How do we know if we have successfully solved a problem?

9. Discuss the difference between data ethics and artificial intelligence?

10. Discuss the following with examples. To establish transparency in methods, results, and limitations,
    (a) How do we ensure that the project methods and outputs are transparent?
    (b) How do we ensure that the potential limitations of the research are clearly presented?
    (c) Should the research be used as the basis for policy action?

Bibliography

Applebaum, D. (2004). *Lévy Processes and Stochastic Calculus.* Cambridge University Press.

Association, D. S. (2020). Code of conduct. available on https://www.datascienceassn.org/code-of-conduct.html, Last accessed Aug. 2020.

Ausloos, M. and K. Ivanova (2000). Introducing false eur and false eur exchange rates. *Physica A* 286, 353–366.

Axler, S. (2002). *Applied Multivariate Analysis.* Springer.

Axler, S. (2015). *Linear Algebra Done Right*. Springer.

Ayer, M., H. Brunk, G. Ewing, W. Reid, and E. Silverman (1955). An empirical distribution function for sampling with incomplete information. *The Annals of Mathematical Statistics* 6(4), 641–647.

Balakrishnan, N. and V. Nevzorov (2004). *A primer on statistical distributions.* John Wiley & Son.

Barany, E., M. B. Varela, I. Florescu, and I. SenGupta (2011). Detecting market crashes by analyzing long memory effects using high frequency data. *Quantitative Finance* 12 (4), 623–634.

Beccar-Varela, M., H. Gonzalez-Huizar, M. Mariani, and O. Tweneboah (2016a). Use of wavelets techniques to discriminate between explosions and natural earthquakes. *Physica A: Statistical Mechanics and its Applications* 457, 42–51.

Beccar-Varela, M., H. Gonzalez-Huizar, M. Mariani, and O. K. Tweneboah (2016b). Use of wavelets techniques to discriminate between explosions and natural earthquakes. *Physica A: Statistical Mechanics and its Applications* 457, 42–51.

Beccar-Varela, M., M. Mariani, O. K. Tweneboah, and I. Florescu (2017). Analysis of the lehman brothers collapse and the flash crash event by applying wavelets methodologies. *Physica A: Statistical Mechanics and its Applications* 474, 162–171.

Bertoin, J. (1996). *Lévy Processes.* Cambridge University Press.

Billingley, P. (1986). *Probability and Measure.* Wiley Interscience.

Bingham, C., M. Godfrey, and J. W. Tukey (1967). Modern techniques of power spectrum estimation. *IEEE Trans. Audio Electroacoustic* AU-15, 55–66.

Data Science in Theory and Practice: Techniques for Big Data Analytics and Complex Data Sets, First Edition. Maria Cristina Mariani, Osei Kofi Tweneboah, and Maria Pia Beccar-Varela.

Bishop, C. M. (2006). *Pattern Recognition and Machine Learning*. Springer.

Bochner, S. (1949). Diffusion equation and stochastic processes. *Proceedings of the National Academy of Sciences, U.S.A* 35(7), 368–370.

Bollerslev, T. and H. O. Mikkelsen (1996). Modeling and pricing long memory in stock market volatility. *J. Econom.* 73, 151–184.

Breiman, L. (1996). Bagging predictors. *Machine learning* 26, 123–140.

Breiman, L. (2001). Random forest. *Machine learning* 45(1), 5–32.

Buldyrev, S., A. Goldberger, S. Havlin, R. Mantegna, M. Matsa, C. Peng, M. Simons, and H. Stanley (1995). Long-range correlation properties of coding and noncoding dna sequences: Genbank analysis. *Phys. Rev. E* 51, 5084–5091.

Casella, G. and R. L. Berger (2002). *Statistical inference*. Duxbury advanced series in statistics and decision sciences, Thomson Learning.

Chapman, C. (2019). A complete overview of the best data visualization tools. https://www.toptal.com/designers/data-visualization/data-visualization-tools.

Chartfiled, C. (2003). *The analysis of time series: An introduction*. Chapman & Hall/ CRC.

Cizeau, P., Y. Liu, M. Meyer, C. Peng, and H. Stanley (1997). Volatility distribution in the s&p500 stock index. *Physica A* 245, 441–445.

Clewlow, L. and C. Strickland (1998). *Implementing Derivatives Models*. Wiley.

Cont, R. and P. Tankov (2004). *Financial Modelling with Jumps Processes*. CRC Financial mathematics series. Chapman & Hall.

Cont, R. and E. Voltchkova (2005). Integro-differential equations for option prices in exponential lévy models. *Finance and Stochastic* 9, 299–325.

Cox, J., S. Ross, and M. Rubinstein (1979). Option pricing: A simplified approach. *Journal of Financial Econometrics* 7, 229–263.

Dajcman, S. (2013). Interdependence between some major european stock markets-a wavelet lead/lag analysis. *Prague Economic Papers*, 28–49.

Elston, R. and J. Grizzle (1962). Estimation of time-response curves and their confidence bands. *Biometrics* 8, 148–59.

Fleet, P. V. (2008). *Discrete wavelet tranformations: An Elementary approach with applications.* Wiley.

Florescu, I. (2014). *Probability and Stochastic Processes*. New York: Wiley.

Florescu, I. and C. Tudor (2014). *Handbook of Probability*. New York: Wiley.

Foufoula-Georgiou, E. and P. Kumar (1994). *Wavelet Analysis and Its applications.* Academic Press, Inc.

Fowler, J. (2005). The redundant discrete wavelet transform and additive noise. *IEEE Signal Processing Letters* 12(9), 629–632.

Glasserman, P. (2004). *Monte Carlo Methods in Financial Engineering.*, Volume 53 of Stochastic Modelling and Applied Probability. Springer New York, 2010.

Graps, A. (1995). An introduction to wavelets. *IEEE Computational Science and Engineering* 2, 50–61.

Hastie, T., Tibshirani, R., and Friedman, J. (2008, Second Edition). *The Elements of Statistical Learning: Data Mining, Inference, and Prediction*. Springer.

Hastie, T., R. TIbshirani, and J. Friedman (2008). *The Elements of Statistical Learning: Data Mining, Inference, and Prediction*. Springer.

Haugh, M. (2017). Simulating Stochastic Differential Equations. IEOR E4603: Monte-Carlo Simulation, Columbia University.

Heston, S. L. (1993). A closed-form solution for options with stochastic volatility with applications to bond and currency options. *Review of Financial Studies* 6(2), 327–43.

Huang, J., P. Shang, and X. Zhao (2012). Multifractal diffusion entropy analysis on stock volatility in financial markets. *Physica A* 391, 5739–5745.

Hull, J. C. (2008). *Options, Futures and Other Derivatives* (7 ed.). Prentice Hall.

Hurst, H. (1951). Long term storage capacity of reservoirs. *Trans. Am. Soc. Eng.* 116, 770–799.

Ivanova, K. and M. Ausloos (1999). Application of the detrended fluctuation analysis (dfa) method for describing cloud breaking. *Physica A* 274, 349–354.

James, G., D. Witten, T. Hastie, and R. Tibshirani (2013). *An Introduction to Statistical Learning with Applications in R*. Springer.

Jarrow, R. and A. Rudd (1983). *Option Pricing*. Homewood, Illinois: Dow Jones-Irwin.

Jimenez-Marquez, S. A., C. Lacroix, and J. Thibau (2002). Statistical data validation methods forlarge cheese plant database. *J. Dairy Sci.* 85, 2081–2097.

Johnson, R. and D. Wichern (2014). *Applied Multivariate Statistical Analysis*. Pearson.

Kakizawa, Y., R. Shumway, and M. Taniguchi (1998). Discrimination and clustering for multivariate time series. *Journal of the American Statistical Association* 93(441), 328–340.

Kang, H. (2013). The prevention and handling of the missing data. *Korean J Anesthesiol.* 64(5), 402–406.

Kantelhardt, J., R. Berkovits, S. Havlin, and A. Bunde (1999). Are the phases in the anderson model long-range correlated? *Physica A* 266, 461–464.

Kaufman, L. and P. Rousseeuw (1990). *Finding groups in data: An introduction to cluster analysis*. Wiley Interscience, 68–125.

Khintchine, A. Y. and P. Lévy (1936). Sur les lois stables. *C. R. Acad. Sci. Paris* 202, 374.

Kingma, D. P. and J. L. Ba (2015). Adam: a method for stochastic optimization. *International Conference on Learning Representations*, 1–13.

Koscienly-Bunde, E., A. Bunde, S. Havlin, H. Roman, Y. Goldreich, and H. J. Schellnhuber (1998). Indication of universal persistence law governing atmospheric variability. *Phys. Rev. Lett* 81, 729–732.

Koscienly-Bunde, E., H. Roman, A. Bunde, S. Havlin, and H. J. Schellnhuber (1998). Long-range power-law correlations in local daily temperature fluctuations. *Phil. Mag. B* 77, 1331–1339.

Kruskal, J. (1964). Multidimensional scaling by optimizing goodness-of-fit to a nonmetric hypothesis. *Psychometrika* 29, 1–27.

Lamberton, D. and B. Lapeyre (2007). *Introduction to stochastic calculus applied to finance*. CRC press.

Lee, B. Y. and Tarng, Y. S. (1999). Application of the discrete wavelet transform to the monitoring of tool failure in end milling using the spindle motor current. *International Journal of Advanced Manufacturing Technology* 15(4), 238–243.

Lin, J. (2016). On the dirichlet distribution. Master's thesis, Department of Mathematics and Statistics, Queens University.

Liu, Y., P. Cizeau, M. Meyer, C. Peng, and H. Stanley (1997). Quantification of correlations in economic time series. *Physica A* 245, 437–440.

Luhby, T. (2009, June 11). American's wealth drops 1.3 trillion. *CNNMoney.com*.

Mandelbrot, B. and J. Ness (1968). Fractional brownian motions, fractional noises, and applications. *SIAM Rev* 10(4), 422–437.

Mantegna, R. and H. Stanley (1994). Stochastic process with ultraslow convergence to a gaussian: The truncated lévy flight. *Phys. Rev. Lett.* 73, 2946.

Mariani, M. and I. Florescu (2020). *Quantitative Finance* (2nd ed.). John Wiley & Sons,Inc.

Mariani, M. and Y. Liu (2007). Normalized truncated lévy walks applied to the study of financial indices. *Physica A* 377, 590–598.

Mariani, M. and O. Tweneboah (2016). Stochastic differential equations applied to the study of geophysical and financial time series. *Physica A* 443, 170–178.

Mariani, M., O. Tweneboah, and M. Bhuiyan (2019). Supervised machine learning models applied to disease diagnosis and prognosis. *AIMS Public Health* 6(4), 405–423.

Mariani, M., O. Tweneboah, H. Gonzalez-Huizar, and L. Serpa (2016). Stochastic differential equation of earthquake series. *Pure and Applied Geophyisics* 173, 2357–2364.

Maruyama, G. (1955). Continuous markov processes and stochastic equations. *Rend Circ Math Palermo* 4, 48–90.

Mikosch, T. (1998). *Elementary stochastic calculus with Finance in view*. World Scientific Publishing Co. Pte. Ltd.

Milstein, G. (1974). Approximate integration of stochastic differential equations. *Teor. Veroyatnost. i Primenen.* 19(3), 583–588.

Newland, D. (1993). Harmonic wavelet analysis. *Proceedings of the Royal Society of London, Series A (Mathematical and Physical Sciences)* 443(1917), 203–225.

Øksendal, B. (2010). *Stochastic Differential Equations: An Introduction with Applications* (10 ed.). Springer.

Oksendal, B. (2010). *Stochastic Differential Equations: An Introduction with Applications*. Springer-Verlag.

Ovanesova, A. and L. Suárez (2004). Applications of wavelet transforms to damage detection in frame structures. *Engineering Structures* 26, 39–49.

Paradis, E. (2002). R for beginners. Montpellier (F): University of Montpellier.

Paul, L. (1925). *Calcul des probabilités*. Paris: Gauthier-Villars.

Peng, C., S. Buldyrev, A. Goldberger, R. Mantegna, M. Simons, and H. Stanley (1995). Statistical properties of dna sequences. *[Proc. Int'l IUPAP Conf. on Statistical Physics, Taipei], Physica A* 221, 180–192.

Peng, C., S. Buldyrev, S. Havlin, M. Simons, H. Stanley, and A. Goldberger (1994). Mosaic organization of dna nucleotides. *Phys. Rev. E* 49, 1685–1689.

Peng, C., J. Mietus, J. Hausdorff, S. Havlin, H. Stanley, and A. Goldberger (1993). Long-range anticorrelations and non-gaussian behavior of the heartbeat. *Phys. Rev. Lett.* 70, 1343–1346.

Peng, C. K., S. Havlin, H. E. Stanley, and A. L. Goldberger (1995). *Quantification of scaling exponents and crossover phenomena in nonstationary heartbeat time series.* [Proc. NATO Dynamical Disease Conference], edited by L. Glass, Chaos 5, 82–87.

Pilz, M. and S. Parolai (2012). Tapering of windowed time series.

Rencher, A. (2002). *Methods of Multivariate Analysis.* John Wiley & Sons.

Ross, S. (2007). *Introduction to probability models.* Elsevier.

Samorodnitsky, G. and M. S. Taqqu (1994). *Stable non-Gaussian random processes: Stochastic models with infinite variance.* New York: Chapman and Hall.

Sato, K. (1999). *Lévy Processes and Infinitely Divisible Distributions.*

Scafetta, N. (2011). An entropic approach to the analysis of time series. Ph.D. Thesis, University of North Texas, Denton, TX, USA, 2001.

Scafetta, N. and P. Grigolini (2002). Scaling detection in time series: Diffusion entropy analysis. *Phys. Rev. E Stat. Nonliner Soft Matter Phys* 66.

Schoutens, W. (2003). *Lévy Processes in Finance: Pricing Financial Derivatives.* Wiley Series in Probability and Stochastics.

Schoutens, W. and J. Cariboni (2009). *Lévy Processes in Credit Risk.* John Wiley & Sons.

Selesnick, I., R. Baraniuk, and N. Kingsbury (2005). The dual-tree complex wavelet transform. *IEEE Signal Processing Magazine* 22(6), 123–151.

Shao, J. (2003). *Mathematical Statistics.* Springer.

Shepard, R. (1962). The analysis of proximities: Multidimensional scaling with an unknown distance function. *Psychometrika* 27, 125–140.

Shiryaev, A. N. (2008). *Essentials of the Stochastic Finance.* World Scientific.

Shumway, R. and D. Stoffer (2010). *Time Series Analysis and Its Applications With R Examples.* Springer.

Stojanovic, V., M. Stankovic, and I. Radovanovic (1999). Application of wavelet analysis to seismic signals. *IEEE*, 13–15.

Sweigart, A. (2015). *Automate the Boring Stuff with Python: Practical Programming for Total Beginners.* No Starch Press.

Timothy, G., G. Robert, W. Nicholas, and F. C. (2016). A brief history of long memory: Hurst, mandelbrot and the road to arfima. *Entropy* 19.

Trigeorgis, L. (1991). A log-transformed binomial numerical analysis method for valuing complex multi-option investments. *Journal of Financial and Quantitative Analysis* 26(3), 309–326.

Tweneboah, O. K. (2015). Stochastic differential equation applied to high frequency data arising in geophysics and other disciplines. ETD Collection for University of Texas, El Paso, Paper AAI1600353.

Tweneboah, O. K. (2020). Applications of ornstein-uhlenbeck type stochastic differential equations. https://scholarworks.utep.edu/open_etd/3052.

Vandewalle, N., M. Ausloos, M. Houssa, P. Mertens, and M. Heyns (1999). Non-gaussian behavior and anticorrelations in ultrathin gate oxides after soft breakdown. *Appl. Phys. Lett.* 74, 1579–1581.

Vuorenmaa, T. (2005). A wavelet analysis of scaling laws and long-memory in stock market volatility. *Proc. SPIE 5848, Noise and Fluctuations in Econophysics and Finance* 27, 39–54.

Yaffee, R. and M. McGee (1999). *Introduction to Time Series Analysis and Forecasting with Applications of SAS and SPSS*. Academic Press, Inc.

Index

Data Science in Theory and Practice: Techniques for Big Data Analytics and Complex Data Sets, First Edition. Maria Cristina Mariani, Osei Kofi Tweneboah, and Maria Pia Beccar-Varela.

n